Capitalism and Conservation

Antipode Book Series

General Editor: Dr Rachel Pain, Reader in the Department of Geography, Durham University, U.K.

Like its parent journal, the Antipode Book Series reflects distinctive new developments in radical geography. It publishes books in a variety of formats – from reference books to works of broad explication to titles that develop and extend the scholarly research base – but the commitment is always the same: to contribute to the praxis of a new and more just society.

Published

Capitalism and Conservation
Edited by Dan Brockington and Rosaleen Duffy

Spaces of Environmental Justice
Edited by Ryan Holifield, Michael Porter and Gordon Walker

The Point is to Change It: Geographies of Hope and Survival in an Age of Crisis *Edited by Noel Castree, Paul Chatterton, Nik Heynen, Wendy Larner and Melissa W. Wright*

Privatization: Property and the Remaking of Nature-Society
Edited by Becky Mansfield

Practising Public Scholarship: Experiences and Possibilities Beyond the Academy
Edited by Katharyne Mitchell

Grounding Globalization: Labour in the Age of Insecurity
Edward Webster, Rob Lambert and Andries Bezuidenhout

Privatization: Property and the Remaking of Nature-Society Relations
Edited by Becky Mansfield

Decolonizing Development: Colonial Power and the Maya *Joel Wainwright*

Cities of Whiteness
Wendy S. Shaw

Neoliberalization: States, Networks, Peoples
Edited by Kim England and Kevin Ward

The Dirty Work of Neoliberalism: Cleaners in the Global Economy
Edited by Luis L. M. Aguiar and Andrew Herod

David Harvey: A Critical Reader *Edited by Noel Castree and Derek Gregory*

Working the Spaces of Neoliberalism: Activism, Professionalisation and Incorporation
Edited by Nina Laurie and Liz Bondi

Capitalism and Conservation

Edited by

Dan Brockington and Rosaleen Duffy

A John Wiley & Sons, Ltd., Publication

Library of Congress Cataloging-in-Publication Data

Capitalism and conservation / [edited by] Daniel Brockington, Rosaleen Duffy.
    p. cm. – (Antipode book series)
    Includes bibliographical references and index.
978-1-4443-3834-8 (pbk.)
1. Environmental economics. 2. Conservation of natural resources–Economic aspects. 3. Consumption (Economics). 4. Capitalism. I. Brockington, Daniel. II. Duffy, Rosaleen.
    HC79.E5.C364 2011
    333.7–dc22
                                    2011015185

9781444338348

This book is published in the following electronic formats: ePDFs (9781444391435);

Wiley Online Library (9781444391442); ePub (9781444391459)

Set in 11pt Times by Aptara

01   2011

Contents

Introduction

Capitalism and Conservation: The Production and Reproduction of Biodiversity Conservation

Dan Brockington and Rosaleen Duffy

His Royal Highness rightly says that our rainforests are worth more alive than dead. This is absolutely true. Leaving aside the immeasurable value offered by our rainforests' diversity and water conservation functions, we are facing an almost unfathomably large business opportunity, one which we can share with the Rainforest nations of the world... With an estimated 610 billion tonnes of CO_2 sequestered by our tropical rainforests, a vast $18 trillion business opportunity is before us... [I]t is increasingly clear that the solution to this problem lies not only within a free market system but also within our field of expertise. What the people of Rainforest nations need is a system that values the services locked up in their land... The rainforests are at the very centre of these countries' identities; they seek to preserve them even as they struggle against them... With capitalism at the centre of OUR identity, we must be bold enough to see the world at its widest, even as we struggle in our own way... To seize this $18 trillion business opportunity, valuing the services of our rainforest will not only require innovation in market-based mechanisms but also unprecedented global cooperation between the brightest minds of the nations of our world. Many structures and mechanisms will need to be created, but it should be our expertise that defines them, and our appetite for these markets that forces political support for them.[1]

With these words Stanley Fink, a former leading hedge fund manager, replied to Prince Charles at a banquet held at the Mansion House, in the symbolic and cultural heart of the financial district in the City of London. Prince Charles had arranged the event to communicate the work of his Rainforest Project to the financial community.[2] His purpose, simply put, is to make rainforests worth more alive than dead through fostering novel financial mechanisms that properly value the ecosystem services rainforests provide.

Fink, and Prince Charles' event, rather neatly capture the phenomenon with which this book is concerned: the current alliances between capitalism and conservation. These alliances are characterised by an aggressive faith in market solutions to environmental problems. They are actively remaking economies, landscapes, livelihoods and conservation policy and practice and they are partying in the symbolic heartlands of capitalism. Corson (this volume) provides the best summary of the shared basis of these chapters. She suggests that, by providing an avenue by which corporations and politicians can become "green", as well as through new enclosures and conservation-based enterprises, conservation fuels processes of capital accumulation. The net result is that "international biodiversity conservation is creating new symbolic and material spaces for global capital expansion" (Corson, this volume: 578). The purpose of this book is to explore the origins, manifestations and consequences of these unions.

A close relationship between capitalism and conservation is nothing new. Neoliberal conservation is but the latest stage in a long and healthy relationship between capitalism and conservation (Adams 2004; Brockington, Duffy and Igoe 2008; Neumann 1998). Rich elites have been promoting conservation of particular species for their pleasure and enjoyment long before capitalism even began; capitalist elites adopted these same privileges (Rangarajan 2001). Capitalist interests strongly advocated the first national parks in North America, and the first conservation NGOs. Indeed it is arguable that what is remarkable about the environmental NGOs and environmentalism of the 1960s and 1970s is that it took such a strong anti-capitalist line. The spirit of Edward Abbey and his Monkeywrench Gang are the exception, not the rule.[3]

The practices that link conservation and capitalism therefore are old. There is, however, something new going on. Critics and analysts of conservation are observing an increase in the intensity and variety of forms of capitalist conservation. Underlying that is a shift in the conservation movement's own *conception* of these practices. As the detailed histories that MacDonald and Corson trace in this book show, the idea that capitalism can and should help conservation save the world now occupies the mainstream of the conservation movement. There is still resistance (Walker et al 2009) but it is becoming increasingly marginal in the corridors of conservation power. This is producing changes in the imagery and rhetoric of conservation as well as its policies and practices.

The challenge for analysts has been how to observe these shifts. The task of observing conservation at work has migrated from remote rural locations where conservation policy happened to massive international meetings (the Earth Summits and World Parks Congress, World Conservation Congress, Conferences of the Parties to Conventions on Biodiversity or Climate Change), marketing seminars, elephant festivals, fundraising events in Washington, plush offices in metropoles throughout the world, film festivals, banquets and analyses of advertisements and glossy magazines. Studying conservation policy, organisations and natural resource management frequently requires multi-sited ethnographic research as well as long hours becoming familiar with the strange new languages of macroeconomics or conservation biology and the peculiar lifeways of exotic tribes of international bureaucrats and corporate sustainability officers.

Intellectually the challenge has been made easier by the thriving research into the neoliberalisation of nature (Bakker 2005, 2009; Büscher and Dressler 2007; Castree 2009; Heynen et al 2007; McCarthy 2005; McCarthy and Prudham 2004) and in particular by Noel Castree's parsing of that literature (see, for example, Castree 2008a, 2008b). This has helped set the changes we have observed in conservation into a broader economic and environmental context. As we shall see, determining exactly *how* current forms of capitalist conservation are neoliberal can be difficult. Neoliberalism itself is an unevenly applied doctrine and its practice rarely lives up to its theory (Brenner and Theodore 2002; Harvey 2005; Mansfield 2004). Nevertheless, the convergence of interest in the consequences of neoliberal policy for natural resource use have provided scholars of conservation both insight into the processes they are observing and a common language with which to communicate their work to a broader academe.

Academic interest in the changing nature of conservation has produced a flurry of activity exploring the permutations of the relationships between conservation and capitalism. Researchers working on these topics organised panels at the Society for Applied Anthropology in 2005, the Association of American Geographers in 2008 and the American Anthropological Association in 2009.[4] A group of conservation activists and analysts met in Washington in mid 2008 to compare experiences of the disciplining power of hegemonic conservation.[5] A team of researchers conducted an event ethnography of the World Conservation Congress in Barcelona in 2008.[6] Finally, a new network examining the work of celebrity and the media in conservation met in 2008 and again in 2009. The changing nature of conservation is producing new epistemic communities in academia.

The chapters published here were first presented at an international symposium in Manchester in September 2008. Twenty-two papers were discussed over 2 days; 12 were submitted to *Antipode*. Collectively they provide the strongest theoretical and empirical examinations of the convergence of capitalism and conservation that we believe has been published. They bring out the detail of the institutional and organisational alliances, some of their history and some indication of the great variety of commodity production that capitalist conservation creates.

We have sorted the chapters into two sections. The first examines the practices, networks, organisations, elites and institutions through which relationships between capitalism and conservation are cultivated. The second examines the commodities and businesses that attempt to produce conservation outcomes through capitalist means. We summarise their main arguments below.

Cultivating Conservation with Capitalism

One of the central themes of this collection is that conservation is proving instrumental to capitalism's growth and reproduction. It provides an "environmental fix" (as Harvey might put it). As Igoe and colleagues observe (this volume), where Green Marxists have predicted environmental impediments that would threaten capitalism's prosperity (O'Connor 1988), in fact these very impediments are the source of new forms of accumulation. Consumers thrive on scarcity, anxiety, fear (all help create demand), so perhaps the flourishing of capitalism in conservation, which deals in similar currency, should not be such a surprise. It is still important, however, to understand how this union is being achieved.

Tackling that question is one of the main achievements of the essay by Igoe and colleagues. Following Sklair and others they propose the existence of hegemonic "mainstream conservation" interests composed of an alliance of corporate, philanthropic and NGO interests (Sklair 2001). Mainstream conservation (one part of Sklair's "sustainable development historic bloc") proposes resolutions to environmental problems that hinge on heightened commodity production and consumption, particularly of newly commodified ecosystem services. Their views are promulgated through a mutually reinforcing collection of spectacular media productions circulated in advertisements and on the web. The power of these productions lies not in their robustness, logic or rigour, but rather because they are presented and consumed within societies dominated by spectacle (Debord 1995 [1967]). That is, these are societies where representations of, and connection to, places, people and causes have long been mediated through commodified images. In consuming these images people are given "the romantic illusion that they are adventurously saving the world" (p 502) while the deleterious ecological impacts of these very purchases, and the lifestyles they require, are neatly erased.

> By focusing consumers' attention on distant and exotic locales, the spectacular productions... conceal the complex and proximate

connections of people's daily lives to environmental problems, while suggesting that the solutions to environmental problems lay in the consumption of the kinds of commodities that helped produce them in the first place (p 504).

Spectacle is crucial to the work of capitalist conservation. It provides one of the central themes that resounds throughout this collection. The major intellectual contribution of this chapter is to demonstrate how and why it works. The data these authors use — video material on NGO websites and accounts from the world of Disney — suggest a wealth of further work. As the authors show in their conclusion, a great deal of new research would be possible into the production and reproduction of hegemony through spectacle, and its contestation in the same.

While Igoe and colleagues focus on the proliferation of capitalist conservation through society, MacDonald's chapter examines the precise mechanisms by which conservation and conservation organisations became so capitalist in the last few decades. Organisations, he observes, have specific social and historical trajectories as well as their own arenas of struggle that have to be observed if we are to understand the shifts which have taken place. MacDonald examines the development of the sustainable development agenda and ideas of ecological modernisation, contending that "biodiversity has never really driven environmental agendas" but instead has "been an instrument in larger political projects" (p 515). He suggests that during the last 30 years the institutional context of international conservation governance was forged in ways which minimised impediments to business and even advanced new rounds of enclosure and appropriation. Indeed he calls the corporate co-option and taming of environmental regulations and their decision-making bodies "institutional enclosure".

Specifically MacDonald shows how corporate interests, perceiving the first Earth Summit in 1992 to be a potential regulatory constraint on the operation of business, worked to draw the sting out of the regulatory measures that the Summit could recommend. He shows how states out-manoeuvred NGOs around the Convention on Biological Diversity to produce a convention which prioritised generating profit from genetic resources over protecting the environment. He also notes that business interests have enjoyed access to prime slots during Conferences to the Parties of the CBD. Following Sklair (2001) he argues that crucial to all these developments has been the formation of a transnational capitalist class fostering alliances and giving business interests the space and support they require.

Finally, MacDonald examines the relationship between conservation NGOs and businesses. He observes that, much as the former need the capital and legitimacy the latter provide, it has been corporate interests that have reached out to conservation groups, particularly if the latter have been perceived to be an impediment. He goes on to document how advantageous materially and symbolically these alliances can be for the corporations involved. The gains for biodiversity are less clear. Within conservation organisations the result of their embrace by commercial interests is that there has been a "near universal conflation of nature and capital [which] has established itself as a dominant view". He concludes:

> [I]t says on the WWF International website "The panda means business". But the panda didn't always mean business. At a point in the not too distant past, the easily recognizable WWF symbol meant the development of effective public engagement in the protection of wildlife habitat. The shift in the meaning of the symbol, indeed its conversion from symbol to legally protected brand, indicates a shift in the recognition of what and who are currently in a position to best contribute the support necessary for the organisation to continue its activities (p 539).

Brockington and Scholfield's chapter examines in greater detail the configuration of conservation NGOs at work in one region — sub-Saharan Africa. These authors have undertaken a unique and large-scale survey of conservation NGOs in the region (Brockington and Scholfield 2010a, 2010b), the insights from which they use to present a framework showing conservation NGO activities. Drawing on Garland's (2008) ideas of a "conservationist mode" of production, and illustrating Igoe and colleagues' work on conservation spectacle (this book), they explore how different types of NGO create symbolic and financial value from their representations of conservation need on the continent. Given that these representations often hinge on the material and/or symbolic displacement of other land uses, they call this process "appropriation for appreciation" (p 557). Interestingly the research agenda that led to this survey of NGOs resulted from the considerable attention NGOs attract from analysts and critics. However, as a result of this survey, Brockington and Scholfield conclude that these NGOs are in fact relatively minor players on much of the continent. They are important as catalysts and purveyors of neoliberal style conservation but diverse organs of the state are much more important in the day to day experience of conservation among rural Africans.

In the three chapters discussed above we have large-scale overviews of the ways in which conservation and capitalism have become intertwined in a neoliberal age. Their reach is global and continental. The remaining four chapters in this section provide more detailed case studies of particular arenas — institutionally and geographically — where neoliberal conservation has been constituted.

Corson sets out a detailed account of the work of US AID in conservation.[7] She tracks the growing funds US AID poured into conservation NGOs (at Congress' behest) and the influence this had on them and environmental politics on Capitol Hill. When the more oppositional environmental groups began to campaign against the World Bank and on domestic issues, conservation NGOs were left to set the environmental agenda. In an effort to protect the budget appropriation for biodiversity (particularly against presidents who have been hostile to environmental concerns), congressmen and women have created the bipartisan International Conservation Caucus, while NGOs formed the colourful International Conservation Caucus Foundation (ICCF). The latter enjoys substantial corporate backing and strongly promotes conservation work undertaken in partnership with business (and smaller government) in general. It matters in part because it is in the theatre of the ICCF's spectacle-producing displays that the actually existing interactions between representatives of capital and conservation are forged, celebrated and promulgated. Corson also dwells on the politics of ICCF's conservation, observing that, in reaching out to liberal and conservative congressmen and women, it has focused on foreign biodiversity, free from the hassles of domestic politics. Corson quotes one former US AID official as saying: "It is easier to do biodiversity overseas than in this country because the conflicts don't involve constituencies of Congress" (p xx).

Sachedina provides another perspective on the work of US AID in conservation, this time viewed from within one conservation NGO — the African Wildlife Foundation (AWF). AWF is the fourth largest conservation NGO working in sub-Saharan Africa. Sachedina traces its remarkable success under the leadership of its eventual CEO Patrick Bergin, who fostered a successful relationship with US AID that saw it grow dramatically in size, and re-imagine its mission from a rather parochial East African focus to a pan-African NGO with its branded "heartlands" in every part of the continent. Alarmingly, however, AWF's financial success did not equate to success on the ground in achieving its conservation goals. Sachedina shows how the organisation became orientated towards its funders such that its work in the villages where it needed to be most active was hindered. Indeed, it is probable its work may even have hindered the realisation of those objectives at the same time as it was projecting itself, and growing itself, so successfully. Sachedina holds both the funder (US AID) and recipient (AWF) responsible for the change in character of this important NGO. The result is a particularly poignant account. It was written on the basis of years spent working for and believing in the organisation whose very success rendered it less effective.[8]

Finally, the chapters by Holmes and Spierenberg and Wels focus explicitly on a persistent theme of this section — the work of conservation elites. Holmes, drawing on Sklair (2001), writes of a transantional conservation elite (elsewhere a transnational conservation class thesis; Holmes 2009) composed of bureaucrats, journalists, civil society workers, CEOs and researchers. He also describes a national conservation elite in the place of his research — the Dominican Republic — whose commitment to protected area growth (particularly that of former President Balaguer) is virtually unmatched in the world. The surprise, however, is that here national conservation elites fought with the transnational elite and resisted their intrusion. Ironically, the conservation successes of extraordinary protected area expansion owed little to the work of transnational elites and much more to home-grown, nationalistic environmentalism.[9]

Spierenberg and Wels' account focuses on the work of two of conservation's richest and most influential supporters in recent times: Anton Rupert, the South Africa billionaire tobacco and alcohol magnate, and Prince Bernhard of the Netherlands. These men were friends as well as conservation associates and constructed elite networks of corporate interests to support their work first for the WWF, and later for the Peace Parks Foundation (the fifth largest conservation NGO on the continent and the most recently established of the larger organisations). This is a spicy account, for moving in these circles involved a fair amount of skullduggery and secretive associations; their arenas mixed the intelligence networks of powerful and aggressive states with commercial and political interests. It is also valuable for the length of time it covers, extending long before neoliberal conservation, or neoliberalism itself, appeared. The chapter shows how conservation initially provided an additional means of social networking for commercial interests, while benefiting from the public (and clandestine) funds donated to strengthen parks. Later visions of conservation were much more closely integrated into commercial transformations of space and nature — although still disruptive to local resource use. In both eras commercial interests benefited from the abidingly positive gloss that associations with conservation provide. This chapter outlines rather different origins to the alliance between conservation and capitalism from that described by MacDonald. The agents of the union here were not so much business interests seeking to outflank and circumscribe potential conservation constraints. Instead the origins were men who believed in particular visions of the conservation cause and used the means and networks with which they were most familiar to advance it.

What emerges strongly from these four accounts is the contingency of the processes that have joined commercial with conservation interests. It required the absence of certain advocacy groups, and vigorous battles on Capitol Hill to produce the alliances that Corson described. It required the individual character traits of key figures (Rupert, Bernhard, Balaguer, Bergin) who made conservation causes their own, but who also, crucially, determined what forms of conservation policy would best advance their visions. At the same time, running throughout these events there is a common tendency to seek out apparently apolitical conservation fields where the complex politics and social relations can be ignored (Igoe and colleagues would call these fetishised representations "conservation spectacle"). There is also the prevailing promiscuity of the main actors involved. In the face of hostile administrations, public or economic concerns, conservation NGOs, politicians and large corporations have been repeatedly able to forge environments where they could do good business together.

Selling Conservation

The second section explores different attempts to achieve conservation goals by selling different conservation products. Carrier provides a useful theoretical perspective to start, examining the forms of commodity fetishism with which these schemes commonly deal. He argues that ethical consumption, which is bent on revealing the social and ecological context of the commodities purveyed, still entails forms of fetishism that subvert ethical consumption's central goals. He notes that ethical consumption depends upon the circulation of images that are taken to denote ethicality. Yet their validity and ability to circulate as a common currency can require a fair degree of ignorance and prejudice in the minds of the consumers in which they operate. Further he observes that ethical consumption often ignores the broader environmental and social contexts in which it operates. It achieves its ethically positive results by not counting various aspects of the production and consumption of its commodities. In sum, these constraints clearly limit the good intentions of ethical consumption to make transparent the socio-ecological origins of commodities. There is, however, a second, more insidious effect that Carrier finds at work. Underlying ethical consumption, and reinforced by every purchase, is the belief that "personal consumption decisions . . . are an appropriate and effective vehicle for correcting . . . the ill effects of a system of capitalist production" (p 683). This too has the effect of decontextualising the consumer, hiding the structures that encourage them to seek solutions in consumption in the first place.

The following three chapters then explore the construction and repercussions of ethical consumption as applied to conservation goals. West's analysis of the speciality coffee market operating from Papua

New Guinea raises a number of concerns about the propensity of the industry to effect meaningful change. It is not just that her lurid accounts of the painful assumptions and constructions of consumers and Papua New Guinea cast doubt on the enterprise. It is the very parameters under which it operates. There is a sleight of hand at work here, for speciality coffees make the poverty and otherness of the coffee producers into something exotic, while simultaneously failing to recognise the role of neoliberal capitalism in creating that poverty, and, furthermore, constraining the imagination of consumers to act against these injustices. It renders the coffee producers so exotic and alien that it stymies empathy, or even the desire to lobby for social change other than through purchasing the right sort of coffee. As she observes, "This supposed embedding of the political and social into capitalist consumption may make for a tasty cup of coffee but it makes for lukewarm political action" (p 710–711).

Neves looks at the role of whale watching in cetacean conservation. She discovers an industry that is widely portrayed by environmental NGOs (such as Greenpeace and WWF) as saving whales by providing an alternative livelihood for communities that used to depend on whaling. Yet, without any romanticism about the brutalities (to whales and whalers) that the whaling industry involved, she finds disturbing continuities between the old industry and the new one. Both result in "metabolic rifts",[10] and both fetishise the ecological costs of whaling, with whale watching doing so in the presence of whales and tourists. Whale watching plays down the ecological costs required to get to the remote islands (Azores, Canary Islands) that make whale watching possible. They also play down the ecological impacts that some whaling boats have on whales' echolocation, which produce avoidance behaviours in the whales being observed. It is just odd that such problems are omitted in the rush to proclaim whale watching as the solution to the ills of whaling.

Duffy and Moore contrast the fates of elephant back safaris in Thailand and Botswana. Along with the chapter by Katja Neves, this chapter contests the idea that tourism offers a neat solution to conservation's problems, and argues that it acts as a driver of capitalism. Since these authors identify capitalism as part of the problem, it is difficult, if not impossible, for it to offer the solution. While tourism is presented as the saviour of conservation, these chapters examine the ways that tourism extends and deepens neoliberalism. Nature-based tourism is promoted by a staggering variety of organisations, from conservation NGOs to private companies to the World Bank; they all back it as a credible and green alternative to industrial development. It is hailed as an "alternative" pathway to sustainable development. But Duffy and Moore offer a critique of this position, arguing that nature-based tourism actually allows capitalism to identify, open and colonise new spaces in nature. However, it does so in variable ways in different contexts. The chapter offers an explanation for why and how neoliberalism is shaped by its encounter with existing social and cultural dynamics. Neoliberalism, it seems, does not displace or obliterate existing ways of valuing, owing and approaching nature; instead it mixes with local dynamics to create new dynamics. The chapter presents a snapshot of neoliberalism "as it exists" on the ground.

In the final chapter in the book we change focus, scope and scale and examine some of the everyday responses that neoliberal pressures on resource use and governance produce. In a collection such as this, written by authors who are almost all characterised by their personal strong commitments to long-term research in remote locations, it is surprising that the rural responses to the pressures we have described do not feature more prominently in the writings presented. It is an indication perhaps of how broad the field of conservation studies has become. Wilshusen's rich description of Mexican ejido's responses to the dismantlings of neoliberal policies provides a welcome counterpoint. Drawing on Bourdieu (1990) he explores how neoliberal reforms of forestry that stopped short of privatisation were accommodated by local institutions and practices. He argues that these are best understood as part of decades of coping with dictates from the centre.

Revisiting Neoliberal Conservation

In the chapters included here and in other publications produced by this epistemic community, authors have frequently referred to "neoliberal conservation" (Brockington, Duffy and Igoe 2008; Büscher and Dressler 2007; Büscher and Whande 2007; Igoe and Brockington 2007; Sullivan 2006). This term derives from the literature on "neoliberalisation of nature" (Castree 2008a, 2008b; Heynen et al 2007; McCarthy and Prudham 2004) and has been one of the unifying themes behind the recent surge of scholarship.[11] There is always a risk that this sort of terminology is employed simply as a means of joining the crowd. It is important to test its worth.

The difficulty with examining the usefulness of a term like "neoliberal conservation" is that neoliberalism, in Peck and Tickell's, Harvey's and others' depictions, is a fundamentally uneven project (Brenner and Theodore 2002; Harvey 2005; Peck and Tickell 2002). It is applied with differential rigour across space, and often in direct conflict with its ideological precepts. Neoliberalism can be hard to identify as such in these circumstances.

Nevertheless some neoliberal aspects of contemporary conservation are obvious. Conservation is extensively promulgated by NGOs, and in new hybrid governance arrangements of "privatised sovereignty" that appear to be direct products of neoliberal thinking. Conservation strategies can hinge on the deregulation and reregulation of nature-based industries and environmental services. This is, plainly and simply, neoliberalism in practice. In such circumstances the environment, or as we would put it, conservation initiatives, is the actual arena wherein neoliberalism is actively constituted (McCarthy 2005).

Moreover, the fact is that we are witnessing policies and institutions that were forged at the time of neoliberalism's domination of economic and social policies globally with its concomitant trends of marketisation and commodification (Brenner and Theodore 2002). New conservation enterprises and commodities are also integral to the expansion of capitalist operations. Conservation organisations can provide endorsements that ensure an "environmental stamp of approval" (to paraphrase Corson) on new enclosures. Through mitigation, measures, partnerships, offsets and new commodity production, conservation has proven integral to capitalist growth. If anything, it would be more interesting to look for conservation strategies that are *un*touched by neoliberalism.

However, it is possible to identify vital aspects of neoliberalism that are not visible in conservation practice, or at least in the current literature on it. For instance, according to Harvey, neoliberalism is characterised by a disciplining and marginalisation of labour power. Brenner and Theodore note its "hyperexploitation of workers" (2002:342). The restrictions conservation can impose on rural livelihoods can increase the importance of wage labour. There are many instances of this occurring, and in that sense conservation can be a means of creating a proletariat. However, this aspect of conservation has rarely been presented explicitly as an aspect of labour relations (for some recent, and notable, exceptions, see Sodikoff 2007, 2009; Timms forthcoming). Nor is it easy to speak of disciplining labour within the conservation movement itself. If there is a conservation proletariat then it is a tiny group of eager volunteers sacrificing time or underpaid staff forgoing better salaries elsewhere to serve a cause.[12] These are social relations that are not well characterised by capitalist exploitation.

More importantly, according to Harvey, neoliberalism is funda-mentally about finance. It has meant "the financialisation of everything" (2005:33). It has not, however, "financialised" conservation. Some of the larger NGOs have engaged in debt for nature swaps, and there has been some interchanging of personnel, but these are peripheral developments.[13] Conservation has hardly been involved in the production of value through financialisation. Fink's bold words, with which this introduction began, were intended to shock an audience that has largely avoided the financial opportunities conservation work might provide.[14]

Thus while neoliberalism may pervade conservation, conservation remains a peripheral element of neoliberalism, certainly if measured by economic activity. But, in one final respect, conservation is integral to the neoliberal project. For Harvey, the often unacknowledged core of neoliberalism is the restoration and maintenance of class power and privilege. This was largely achieved through "exacting tribute" and was visible through heightened inequality and extraordinary concentrations of wealth among a new super-rich (2005:119). It should be abundantly clear from the work that follows that enjoying the benefits of conservation is precisely one of the privileges class power affords. Conservation is not necessarily a luxury of the rich; varieties of environmentalism include all sorts of concerns and practices that originate in poverty. However, in a neoliberal world, conservation is made into a luxury. It becomes the preserve of elites and exclusive communities. Accessing conservation's rewards, and directing its restrictions, is part and parcel of the restoration of class power that runs through the core of the neoliberal project.

The consequences of neoliberal conservation are not unremittingly negative. As we have repeatedly stated in other publications, conservation distributes both fortune and misfortune (Brockington, Duffy and Igoe 2008). Moreover the measures of neoliberal conservation have often been implemented by the most sincere and caring groups. The volunteers and employees of the conservation movement are primarily motivated by their desire to make the world a better place.

The problem is, however, that its rhetoric is relentlessly positive. It presents us only with market solutions, win–win solutions (or win–win–win and more), ethically traded commodities, saved nature, wholesome communities, integrated landscapes, sustainable development, cleansed reputations and secure conservation brands. It is because such powerful actors as Stanley Fink think only in unremittingly positive terms of the market that we must invoke a note of caution. The critical tone is a consequence of the chasms that we see opening up between the rhetoric of neoliberal conservation and its actually existing consequences. As Harvey puts it:

> It has been part of the genius of neoliberal theory to provide a benevolent mask full of wonderful-sounding words like freedom, liberty, choice and rights to hide the grim realities of the restoration or reconstitution of naked class power (2005:119).

The future challenge this epistemic community faces is to find ways of communicating its concerns more effectively with the conservation community. This collection is probably not the best way to do it, but we hope it will play some role. There are many conservationists who share our antipathy to inequality and cant. We will need to be as zealous in communicating our concerns to such audiences as we have been in sharing them here.

Acknowledgements

We thank the ESRC for support of the symposium that has produced these chapters through Dan Brockington's Fellowship "The Social Impacts of Protected Areas" (RES-000-27-0174), the Brooks World Poverty Institute, Clive Agnew, Debra Whitehead and the School of Environment and Development for their generous support. We would like to thank colleagues at Manchester University, particularly Noel Castree, Gavin Bridge, and the Society and Environment Research Group (SERG) for providing such a strong collegial atmosphere in which to work.

Endnotes

[1] Extracted from Stanley Fink's speech at the Mansion House on 10 September 2008; copy in possession of Dan Brockington. Capitals in the original.

[2] The Prince's Rainforest Project is advised by a steering group composed of Barclays, BSkyB, Climate Exchange Plc, Deutsche Bank, DLA Piper, Finsbury, Goldman Sachs, KPMG, McDonald's, Man Group plc, Morgan Stanley, Rio Tinto, Shell, Sun Media Group and Virgin Group, among others (www.rainforestsos.org/pages/steering-group viewed on 16 February 2010).

[3] Abbey's fictional book *The Monkeywrench Gang* describes the escapades of a group of environmentalists who were violently opposed to capitalist development and set about destructively opposing the construction of roads, logging camps and bridges. Dave Foreman, the founder of the radical group Earth First!, used such tactics and produced a manual on how to do so, called *A Field Guide to Monkeywrenching*.

[4] The SFAA collection resulted in two edited collections (*Conservation and Society* 2006, volume 4, issue 3 and *International Journal of Biodiversity Science and Management* 2007, volume 3, issue 2). The publication of the AAA meeting is still being deliberated.

[5] This meeting produced a number of essays in *Current Conservation* 2010, volume 3, issue 2.

[6] This produced a special issue of essays forthcoming in *Conservation and Society*. The same team may return to the Conference of the Parties to the Convention on Biological Diversity in Japan in 2010.

[7] Corson's work is based on years of doctoral research and an earlier career in Washington (Corson 2008).

[8] Sachedina's work is based on a prize-winning doctoral thesis (Sachedina 2008) which, rather unusually, was read and circulated throughout the East African conservation community where he was based. The pdf of this (100,000 word) document has probably been downloaded in excess of 600 times.

[9] In their survey of eviction for conservation, Brockington and Igoe (2006) note that active cases of eviction (in Botswana, Tanzania, Ethiopia and Thailand) are driven by strong home grown environmental debates.

[10] Metabolic rifts is defined here more broadly than Marx's original term to include ecological disruption, not just nutrient extraction.

[11] Igoe and Brockington (2007) provide an introduction to what neoliberal conservation might entail.

[12] We refer to NGO workers, not volunteers who pay to go on conservation camps and holidays with conservation NGOs for short periods of time.
[13] Mark Tereck, a managing director of Goldman Sachs, became the president and CEO of the Nature Conservancy in July 2008.
[14] The launch of investment funds like Earth Capital Partners in the City of London in December 2008, which seeks to raise $5 billion over 5 years, does not really count. These and others are seeking to make money out of new opportunities in clean technology and energy. They are looking for new profits in ecological modernisation rather than biodiversity conservation.

References

Adams W M (2004) *Against Extinction: The Story of Conservation*. London: Earthscan
Bakker K (2005) Neoliberalizing nature? Market environmentalism in water supply in England and Wales. *Annals of the Association of American Geographers* 95(3):542–565
Bakker K (2009) Neoliberal nature, ecological fixes, and the pitfalls of comparative research. *Environment and Planning A* 41(8):1781–1787
Bourdieu P (1990) *The Logic of Practice*. Stanford: Stanford University Press
Brenner N and Theodore N (2002) From the "new localism" to the spaces of neoliberalism. *Antipode* 34(3):341–347
Brockington D, Duffy R and Igoe J (2008) *Nature Unbound. Conservation, Capitalism and the Future of Protected Areas*. London: Earthscan
Brockington D and Igoe J (2006) Eviction for conservation. *Conservation and Society* 4(3):424–470
Brockington D and Scholfield K (2010a) Expenditure by conservation non-governmental organisations in Sub-Saharan Africa. *Conservation Letters* 3(2):106–113
Brockington D and Scholfield K (2010b) The work of conservation organisations in Sub-Saharan Africa. *Journal of Modern African Studies* 48(1):1–33
Büscher B and Dressler W (2007) Linking neoprotectionism and environmental governance. *Conservation and Society* 5(4):586–611
Büscher B and Whande W (2007) Whims of the winds of time? Emerging trends in biodiversity conservation and protected area management. *Conservation & Society* 5(1):22–43
Castree N (2008a) Neo-liberalising nature I: The logics of de- and re-regulation.' *Environment and Planning A* 40(1):131–152
Castree N (2008b) Neo-liberalising nature II: Processes, outcomes and effects. *Environment and Planning A* 40(1):153–173
Castree N (2009) Researching neoliberal environmental governance: A reply to Karen Bakker. *Environment and Planning A* 41(8):1788–1794
Corson C (2008) "Mapping the development machine." Unpublished PhD thesis, Environmental Science, Policy and Management, University of California, Berkeley
Debord G (1995 [1967]) *Society of the Spectacle*. New York: Zone Books
Garland E (2008) The elephant in the room. *African Studies Review* 51(3):51–74
Harvey D (2005) *A Brief History of Neoliberalism*. Oxford: Oxford University Press
Heynen N, McCarthy J, Prudham W S and Robbins P (2007) *Neoliberal Environments: False Promises and Unnatural Consequences*. London: Routledge
Holmes G (2009) "Global conservation and local resistance." Unpublished PhD thesis, Institute for Development Policy and Management, University of Manchester

Igoe J and Brockington D (2007) Neoliberal conservation: A brief introduction. *Conservation and Society* 5(4):432–449

Mansfield B (2004) Neoliberalism in the oceans. *Geoforum* 35:313–326

McCarthy J (2005) Devolution in the woods: Community forestry as hybrid neoliberalism. *Environment and Planning A* 37(6):99–1014

McCarthy J and Prudham S (2004) Neoliberal nature and the nature of neoliberalism. *Geoforum* 35:275–283

Neumann R P (1998) *Imposing Wilderness: Struggles over Livelihood and Nature Preservation in Africa*. Berkeley: University of California Press

O'Connor J (1988) Capitalism, nature, socialism: A theoretical introduction. *Capitalism, Nature, Socialism* 1:11–38

Peck J and Tickell A (2002) Neoliberalizing space. *Antipode* 34(3):380–404

Rangarajan M (2001) *India's Wildlife History*. Delhi: Permanent Black

Sachedina H (2008) Wildlife are our oil. Conservation, livelihood and NGOs in the Tarangire ecosystem, Tanzania. Unpublished D. Phil thesis, School of Geography and the Environment, University of Oxford

Sklair L (2001) *The Transnational Capitalist Class*. Oxford: Blackwell

Sodikoff G (2007) An exceptional strike: A micro-history of "People versus Park" in Madagascar. *Journal of Political Ecology* 14:10–33

Sodikoff G (2009) The low-wage conservationist: Biodiversity and perversity of value in Madagascar. *American Anthropologist* 111(4):443–455

Sullivan S (2006) The elephant in the room? Problematising "new" (neoliberal) biodiversity conservation. *NUPI Forum for Development Studies* 2006 (1): 105–135

Timms B (forthcoming) Coerced relocation from Celaque National Park, Honduras: The (mis)use of disaster as opportunity. *Antipode*

Walker S, Brower A L, Stephens R T T and Lee, W G (2009) Why bartering biodiversity Fails. *Conservation Letters* 2(4):149–157

Chapter 1

A Spectacular Eco-Tour around the Historic Bloc: Theorising the Convergence of Biodiversity Conservation and Capitalist Expansion

Jim Igoe, Katja Neves and Dan Brockington

Conservation leaders need to stop counting birds and start counting dividends that nature can pay to the people who live in it (Ger Bergkamp, Director, World Water Council[1]).

Introduction

On 8 October 2008, *New York Times* business columnist James Kantner wrote:

The World Conservation Congress is abuzz with how the conservation movement will continue to fail to achieve the objectives it has been seeking for decades unless it engages business and embraces business management techniques to further its goals.[2]

Indeed the Congress projected a strengthening consensus of synergies between growing markets and effective biodiversity conservation. IUCN leadership had recently entered a partnership with Shell Oil, and were busily forging a new one with the Rio Tinto Mining Group.[3] The entrance to the Congress was aesthetically dominated by corporate displays, while its theater featured films like "Conservation is Everybody's Business". As shocks in the world economy sent ripples of consternation through the Congress, high-profile speakers warned attendees not to view the "implosion of banks and financial institutions" as a "signal to abandon market models" (for details, see MacDonald forthcoming).[4]

High-profile conservationists, corporate leaders, and celebrities spread the same message to broader publics: capitalism is the key to our ecological future and ecological sustainability will help end our current financial crisis (Igoe forthcoming; Prudham 2009). Through online initiatives, marketing, and fundraising campaigns they not only urge people to support this vision, but also to support re-regulations (cf Castree 2007) of the environment, facilitating the commodification of nature as the solution to problems that threaten our common ecological future (Igoe forthcoming).[5] In contrast to Green Marxists' predictions that the visible and costly environmental contradictions of late capitalism would prompt alliances of environmentalists and workers to demand green socialist alternatives (esp. O'Connor 1988), such initiatives entice consumers to participate in the resolution of capitalism's environmental contradictions through advocacy, charitable giving and consumption.

Biodiversity conservation figures centrally in these transformations. The economic growth that preceded our current problems coincided with growth in protected areas, including private conservancies (Brockington, Igoe and Duffy 2008). It also accompanied dramatic growth of conservation BINGOs (Big NGOs), as they competed intensively for market shares and brand recognition and dramatically grew their budgets (Chapin 2004; Dowie 2009; MacDonald 2008; Rodriguez et al 2007; Sachedina this volume). Corporations donated millions of dollars to BINGOs and corporate representatives joined their boards (Chapin 2004; Dorsey 2005; Dowie 1995; MacDonald 2008), while conservationists sought to extend their influence inside of corporations.[6] Diverse social scientists working in areas targeted for conservation began to observe, in different locations, that partnerships between conservation and capitalism were reshaping nature and society in ways that produced new types of value for capitalist expansion and accumulation (Brockington, Duffy and Igoe 2008; Brockington and Scholfield this volume; Garland 2008; Goldman 2005; Neves this volume; West and Carrier 2004).

Such value production is not limited to specific conservation landscapes. As Garland (2008:52) has noted, conservation as a mode of production appropriates value from landscapes by "transforming it into capital with the capacity to circulate and generate further value at the global level". This includes images and narratives that circulate widely in the entertainment industry (Brockington 2009; Vivanco 2002), as well as in the marketing of new consumer goods and NGO fundraising (Brockington, Duffy and Igoe 2008; Brockington and Scholfield this volume; Garland 2008; Igoe forthcoming; Sachedina 2008). These also provide essential fodder for powerful discursive claims that markets, information technology, and expert know-how offer new possibilities of optimizing the ecological and economic functions of our planet, simultaneously allowing economic growth to continue while spreading benefits to impoverished rural communities (Luke 1997; Goldman 2008; Igoe forthcoming; McAfee 1999; Neves this volume).

These claims are not always as verifiable, or indeed as common-sensical, as their advocates appear to believe. Moreover, they are often buttressed by overly simplistic presentations of socio-environmental problems and relationships in the capitalist economy (Brockington, Duffy and Igoe 2008:ch 9). Much of the power of these ideas is derived from a currently dominant ideological context where it is believed that the attribution of economic value to nature and its submission to "free market" processes is key to successful conservation. The details of this logic are as follows. Once the value of particular ecosystem is revealed, for example an ecosystem's ability to store carbon, the ecosystem acquires economic value as a service provider or as a non-consumptive resource, as in the case of eco-tourism. The ecosystem thus putatively becomes a source of income for local communities, creating further capitalist-development opportunities (Sullivan 2009).[7] Given that within the tenets of capitalist principles the allocation of funds is directly related to associated potential returns on investment, conservationists who seek donor funds are increasingly under pressure to show the economic advantages of their conservation goals. Hence, the notion that the relationships between conservationist action and capitalist reality are necessarily beneficial becomes increasingly taken-for-granted. This idea becomes hegemonic when it is so systematically and extensively promoted that it acquires the appearance of being *the only* feasible view of how best to pursue and implement conservation goals. Alternative and critical views of this logic are consistently kept at the margins or outright silenced.

The hegemonic nature of these claims, and the ideological context from which they are derived, present major obstacles to democratic and reflexive discussions of our most pressing socio-environmental problems. Capitalism's destructive-extractive relationships with the environment are unlikely to be challenged unless we are able to understand the fundamental contradictions between capitalism's need to expand exponentially vis-à-vis the capacity of ecosystems to withstand and absorb the disturbances and stresses that this exponential growth entails. Indeed, it appears that capitalism is turning the environmental problems it creates into opportunities for further commodification and market expansion (Brockington, Duffy and Igoe 2008; Klein 2007).

The relationship between conservation and capitalism thus presents a salient site of enquiry for documenting and analyzing the kinds of relationships that facilitate capitalism's ability to reinvent itself even when its excesses appear to threaten the viability of the very ecosystems on which human economic activity depends. We thus propose a theoretical framework from which we can begin to organize our investigations and analyses of these issues. We now turn to a brief outline of this framework and its theoretical antecedents.

The Sustainable Development Historic Bloc
and the Spectacle of Nature

Our theoretical framework builds on Gramsci's concept of hegemony, which he developed in response to his discontent with monolithic and dualistic understandings of domination. Like Gramsci, we are concerned with the ways in which the ideas and agendas of particular interest groups are promoted and imposed over a world of diversity, full of conflicting values and interests. Gramsci (1971b:210–218) was especially fascinated that these processes occurred mostly without recourse to force, but rather through "the manufacture of consent" (1971a:12).

In thinking about global conservation, therefore, we are interested in understanding how such a complex and heterogeneous movement appears to be dominated by a relatively narrow set of values, ideas and institutional agendas. Here we are referring to what Brockington, Duffy and Igoe (2008:9) have labeled "mainstream conservation", which they describe as the dominant strain of global conservation. The ideas and values of mainstream conservation are most clearly visible in the operations of conservation BINGOs, which dominate conservation funding along with the discursive and spectacular representations of the values and goals of global conservation.

Fully to appreciate the power and influence of mainstream conservation, it is necessary to illuminate its larger context. This in turn refers us to another Gramscian concept; the historic bloc. The historic bloc refers to a historic period in which groups who share particular interests come together to form a distinctly dominant class. The ideas and agendas of this class thus come to permeate an entire society's understanding of the world.[8] Two aspects of the Gramscian notion of historic bloc are germane to the issues that concern us here. First, the ideologies accompanying a particular historic bloc conceal the nature of existing relations of production by presenting a naturalized view of extractive class hierarchies that comprise these relations. Second this concealment has the effect of smoothing over the contradictions, paradoxes and irreconcilable differences that exist within these relations (Gramsci 2000a).

According to Sklair (2001:8) our present historic moment is dominated by "the sustainable development historic bloc", which purports to offer easy consumption-based solutions to the environmental crises inherent in late market capitalism. They thus present a comforting alternative to the more disruptive social transformations predicted by O'Connor (1988) and other green Marxists. This historic bloc, according to Sklair, is produced and supported by a "transnational capitalist class" of corporate executives, bureaucrats and politicians, professionals, merchants and the mass media. Through their efforts, consumption and spending become indispensable to the solution of environmental problems (Sklair 2001:216).

Mainstream conservation's connections to this larger historic bloc have their roots in networks and collaborations that were forged in the creation of national parks in the American West at the end of the nineteenth century. This was also America's "Guilded Age", which was a period of rapid industrialization, extractive capitalist expansion, and the rise of iconic business tycoons, many of whom also became noted philanthropists. Ironically, these transformations simultaneously threatened the natural beauty of the American West, while producing the elites who championed the cause of protecting that natural beauty. Tsing (2005) aptly notes that, early conservationists pursued strategies that involved the enrolment of these elites in nature conservation and corporate sponsorships for the creation of protected areas (also see Spence 1999). While the contexts of these relationships has changed over time, the creation of American parks has historically entailed the intertwining and blurring of states, private enterprise and philanthropy (eg Fortwangler 2007; Mutchler 2007; Muchnick 2007).

Brockington, Duffy and Igoe (2008:9) refer to the networks and affinities that emerged from these types of processes and relationships as mainstream conservation's "collaborative legacy". These networks and affinities, which have become increasingly transnational over time, are evident in the highly visible and central position that philanthropic entities, such as the Rockefeller, Gordon and Betty Moore and the Turner Foundations, have come to occupy in mainstream conservation. Foundations like these and the conservation organizations that they fund, are frequently intertwined by tight networks of interests, values and agendas (Chapin 2004; Grandia forthcoming; Igoe and Fortwangler 2007; MacDonald 2008). They share staff, personnel and board members.[9] Moreover, the collaborative legacy forged in the American West continues to extend itself to other parts of the world and incorporate new elites.[10]

Recent expansions in the collaborative legacy of mainstream conservation have been closely tied to the growth of protected areas and conservation BINGOs briefly outlined above. During this period the five largest conservation organizations grew their collective annual budgets to billions of US dollars, thus commanding over 50% of globally available conservation funding (Chapin 2004). This growth was achieved in part by millions of dollars in corporate funding (MacDonald 2008), but also reflected an increased alignment of global conservation with bilateral donors (Corson this volume) and international financial institutions like the World Bank (Chapin 2004; Young 2002). To quote Goldman (2005:9):

In remarkable synchronicity, the sustainability crowd and the neoliberal development crowd have united to remake nature in the South, transforming vast areas of community-managed uncapitalized lands into transnationally regulated zones for commercial logging, pharmaceutical bio-prospecting, export orientated cash cropping, megafauna preservation and elite eco-tourism (cf Ferguson 2006).

Such alignments of interest are consistent with Gramsci's notion of historic bloc, but this is only part of what constitutes a hegemonic position in which particular ideologies appear as commonsensical. Gramsci (1971a:3–23) was concerned with how such common sense worldviews were produced and presented to society through the work and words of intellectuals and experts. While Gramsci defended that all men and women have intellectual capacity, he drew a distinction between "organic intellectuals", whose ideas and worldview emerged from the lived experience and interests of the masses, and "ideological functionaries",[11] ranging in stature from petty bureaucrats, to individuals publicly renowned for their intellect and expertise (see Holmes this volume), whose ideas and worldview were closely associated with the interests of ruling elites. According to Gramsci, members of both groups exhibited competencies for making statements about the world, of being "in the know", and the ability to explain the world in ways that were understandable and appealing to a broad cross-section of society. Since those in power publicly sanctioned ideologue intellectuals as the true holders of legitimate and valid knowledge, however, members of this class held a much higher position of authority, visibility, and credibility vis-a-vis the general public. Consequently, they had an enormous impact on the legitimation and propagation of the ruling class's understanding of the world. This is reminiscent of the ways in which techno-scientific knowledge is often mobilized to implement elite understandings of ecological-conservation practices, at the cost of silencing alternative types of knowledge that may be more closely associated with the interests of local communities (eg Neves 2004).[12]

Following this line of argument it is evidently crucial for ruling elites to place ideologue intellectuals in the key institutions that educate and inform the masses. Indeed, within Gramsci's paradigm ideologue intellectuals operated predominantly in the realm of civil society, an "ensemble of organisms" that he viewed as essential to the production and dissemination of hegemonic worldviews, including schools, workplaces, organized religion, trade unions. As MacDonald (forthcoming) demonstrates, we must also add professional meetings to this list. Like other Marxists of his time, Gramsci was also greatly concerned with the State's increasing reliance of mass means of communication for propagandistic purposes. What he could not have foreseen was the extent to which mass media would soon encroach on both the public and private spheres, as they were increasingly taken over and produced by capitalist interests.

Given the ascendancy of mass media and multi-media in the years since Gramsci's writing, our application of these ideas to the sustainable development historic bloc is greatly enhanced by Debord's (1995 [1967]) discussion of Spectacle, which he described as the increasingly encompassing mediation of relationships and interests by images. The following theses from Debord are apposite to the framework we are proposing:

1 Spectacle imposes a sense of unity onto situations of fragmentation and isolation (thesis 3);
2 Spectacle is an omnipresent justification of the conditions and aims of existing systems (thesis 6);
3 Spectacle conditions people to be passive while sending them a continuous message that their only viable path to action and efficacy is through consumption (thesis 46);
4 Spectacle itself is a commodity that people pay to consume, and it promotes the creation, circulation and consumption of other commodities (thesis 47);
5 In fact through Spectacle everything appears as a commodity (thesis 49);
6 This is achieved by presenting the world in terms of quantitative objects imbued with an inherent exchangeability (thesis 38);
7 Finally, and most generally, its end is its own reproduction and thus the reproduction of the conditions that produced it and the relationships that it mediates (thesis 13). In so doing, it conceals the complex and conflicted nature of those conditions and relations, presenting instead reified, and apparently problem free, generalizable forms.

Spectacular media productions, many disseminated via the internet, now play a central role in the ways that mainstream conservation portrays itself to the world. These productions represent an important form of Gramscian civil society, a realm in which celebrity intellectuals (in the terms described above) talk about the nature of pressing socio-environmental problems and the kinds of solutions they demand (cf Brockington 2009). A survey by Igoe (forthcoming) found that the spectacular productions of diverse conservation organizations are remarkably consistent, both in terms of their aesthetic and their content in several ways. They conceal the inequities and conflicts associated with particular conservation interventions, as well as the costs of global consumerism and the social and environmental contradictions it entails. They portray celebrities, corporate leaders, and high profile conservationists as a heroic vanguard in the struggle to save the planet, and invite consumers to join them. They present markets, commodification, and exchange as the most viable (if not the only) approaches to solving these problems, arguing that if nature becomes valuable enough, even poor people and their governments will want to save it. Finally, they suggest that the greening of the world economy will allow corporations to make record profits and help us out of the current global economic crisis (for details and detailed examples see Igoe forthcoming).[13]

The sense of unity these productions present facilitates NGO fundraising, as well as the marketing of new commodities and consumptive experiences. Gramsci also noted that such unified presentations of the world lend a crucial sense of coherency to the diverse, fluid and potentially fragmentary class networks that support a particular historic bloc. They not only celebrate and reproduce the dominant worldview and the action it implies, they also play an important role in concealing and managing discord and dissent. MacDonald (forthcoming) has documented the important role of Spectacle at the World Conservation Congress in managing dissent and opposition to the IUCN's private sector partnerships. He shows that previously highly visible dissent at IUCN meetings has subsided over the past several years, such that IUCN membership appears much more unified in its support of these partnerships. Thus when *New York Times* correspondent James Kantner repeatedly reported on the "singular sentiment" of Congress Participants that conservation must embrace business, he was simultaneously reporting upon and reproducing Spectacle and a particular hegemonic position through his online blog.

We have thus far confined our discussion to the realm of ideology and representation. As Gramsci emphasized, however, hegemonies and historic blocs are the product of specific relationships of production, while Debord (thesis 6) described Spectacle as the "result and product of existing modes of production". Indeed, both theorists argued that ideology and material conditions are by necessity intertwined in ways that defy dichotomous distinctions between them. We thus turn to the question of how the hegemonies and spectacles of the sustainable development historic bloc are connected to specific modes of production and new types of value and capital that circulate in the global economy.

The Currencies of Conservation Spectacle

Alliances of capitalism and conservation trade on, and are lubricated by, a number of overlapping currencies. By these we mean goods, symbols and indicators that can be shared by networks of conservation and commercial interests, and which are more effective in the production of conservation spectacle if they are shared. These currencies are protected areas, sovereignty and success. They overlap because each invokes, or requires the other, to work effectively. They are foundational the production of conservation spectacle.

Protected areas are among the most concrete, observable and comparable expression of the ways in which conservation and capitalism are remaking the world (Igoe and Brockington 2007; Brockington, Duffy and Igoe 2008). Since at least 1871, the year Yellowstone National Park was gazetted, the creation of protected areas has frequently entailed a radical separation of humans and nature (West, Igoe and Brockington 2006), which as Marx (1973 [1857–1861]) argued in the mid-nineteenth century was essential to the transformation of the natural world into objects of exchange. Specifically he argued that such transformations entail the erasure and concealment of the relationships of production from which such objects are created. He called this process fetishization. Through this process the attribution of value to a commodity is determined almost entirely by the logic of capitalist market processes. Within this logic, a commodity is valued in relation to its ability to provide returns on investment or to generate additional capital value, while other types of value based on socio-cultural factors or purely ecological criteria either fade into the background or disappear altogether.

Protected areas have been fetishized in some ways, in that their values are increasingly reduced to their ability to generate economic output and the relationships that created them are hidden from view (Carrier and Macleod 2005). This is not to imply that protected areas are only ever valued according to the logic of market processes. Though their potential market value has long motivated their creation, their creation has also been motivated by forces such as nationalist projects and the rise of ecological science. In the context of the political-economic processes unfolding since the late 1980s, however, different ways of valuing protected areas, and nature in general, have become increasingly correlated with nature's ability to generate wealth (Brockington, Duffy and Igoe 2008; Goldman 2005; Harvey 1996; McAffee 1999).

Whatever other values may come into play, however, mainstream conservation has always presented protected areas as having value that transcends all things. As Tsing (2005:97–99) argues, the worldview of early mainstream conservationists rested on the foundational belief that "nature, like god, forms the basis of universal truth, accessible through direct experience and study. To study a particular instance offers a window onto the universal." From this perspective, an imagined planetary nature was thus enfolded into the appreciation of nature in any specific protected area.

This ability of one object (a park) to stand in for the whole of that class of objects (imagined universal planetary nature) is an essential element of Marx's (1978 [1865–1870]) theory of how commodities gain exchangeability. In the process the distinct features and relationships that characterize a particular object are significantly reduced in importance. Of course, the distinctiveness of a particular park is what gives it value as a particular destination. Indeed certain parks like Yellowstone and Serengeti are popularly regarded as unique and unparalleled portals to universal nature. In our current context, however, the specific quality of parks is overshadowed by their abstract quantity, as conservationists and policy makers seek to protect designated, but growing, percentages of the earth's surface. Protected areas' great quality as a currency is that they are discrete, measured and eminently countable.

The inherent exchangeability of parks has become most salient as it is has intersected with the logic of exchangeability so pervasive in global neoliberalism. This is especially visible in mitigation policies, which assume degraded nature and environmental harm can be balanced by pristine nature and environmental protection. This allows the possibility of imagining the Earth as a virtual ledger, on which it possible to carry out a quantitative balancing of environmental goods and bads. A stark example is the creation of protected areas to mitigate ecological damage caused by large-scale extractive enterprise, such as the World Bank sponsored Chad–Cameroon oil pipline (Brockington, Duffy and Igoe 2008:3–4) and the massive World Bank sponsored Nam Theun hydro-electric project in Laos (Goldman 2005). It can also be seen in the increasingly popular idea that environmentally harmful carbon emissions can be offset through the protection of tropical rainforests, an idea now championed by Prince Charles and an expanding cadre of celebrity supporters.[14] Finally, it can be seen in the US government's recent opening of a federal office to oversee trading in ecosystem services that will be similar in function to the Security and Exchange Commission.[15] This will include overseeing the emergence of a new species credit trading scheme, in which species banks pay credit for the protection of endangered species and their habitats, which in turn can be purchased by corporations to "meet their mitigation needs" (Agius 2001; Bayon, Fox and Carrol 2007; Blundel 2006; Clark and Altman 2007; Etchart 1995; USDOI 2003).[16]

We believe that these transformations are part of the consolidation of the sustainable development historic bloc, since the concept of sustainability revolves around the possibility of trading ecological functions, services and values against ecological harm and risk (cf MacDonald this volume). Thus the paradoxes and ecological excesses of late market capitalism are recast as problems of management and the realization of market value. This recasting rests on the assumption that the quantification of nature's value, in terms of ecosystem services, is the key to financing the protection of that nature (Castree 2007; McAfee 1999; Sullivan 2009).[17]

Despite the growing importance of NGOs and companies in the sustainable development historic bloc, states remain key institutions to it. First, they are monopoly purveyors of sovereignty, which Mbembe (2001) defines as "the means of coercion that make it possible to gain advantage in struggles over resources traditionally the exclusive purview of the state". Next, by means of their sovereignty, they have the power to write laws and make policies, including those prescribed by the World Bank and other transnational financial institutions.

However, while states alone can act in the name of sovereignty, they often have to invite others to act in that capacity on their behalf in order to realize their policies. Following Mbembe (2001), we argue that networks of conservation, commerce and the state are forged in conditions of fragmented state control that exist in post-colonial contexts. They are effectively bargains to which outsiders, such as conservation NGOs, bring money, expertise and technology, on which officials from impoverished states are highly dependent. These officials in turn bring the legitimacy and power of sovereignty (Mbembe 2001:78). Ferguson (2006) has labeled such bargains "the privatization of sovereignty", and emphasizes that neoliberalism exacerbates and legitimizes this sort of fragmentation. Sovereignty is valuable to NGOs or corporate interests because it bestows legitimacy. It is only through the actions of sovereign states that protected areas can be established, tourism or hunting deregulated or people legally displaced. Exercising it is vital if the other currencies (protected areas, or success more generally) are to accrue.

Finally, success itself is an extremely valuable form of "symbolic capital" (Bourdieu 1977) that circulates far beyond the scope of specific interventions. The production of success stories is an essential marketing strategy for conservation BINGOs, whereby each seeks to distinguish itself in a highly competitive funding environment (Chapin 2004; MacDonald 2008; Sachedina this volume). We found Mosse's (2004, 2005) insights into the ways in which networks operate to make specific interventions appear coherent and successful most useful. The social reproduction of transnational, national and local institutions involved in governance, conservation and the promotion of economic growth, Mosse argues, depends heavily on the *appearance of success* according to prevailing policy paradigms. According to the hegemonic ideologies of the sustainable development historic bloc, environmental problems in late market capitalism are best repaired by capitalist solutions, and it is possible to manage our planet in ways that simultaneously maximize its economic and ecological function. The creation of protected areas is clearly a type of intervention that can deliver visible and tangible success according to the criteria of these ideologies. Indeed the establishment of protected areas by the hybrid networks of capitalist conservation has been recorded in diverse contexts (eg Bonner 1993; Dzingirai 2003; Garland 2008; Goldman 2005; McDermott-Hughes 2005; Sunseri 2005). In most recent years there has also been a proliferation of new decentralized forms of protected areas (Brockington, Duffy and Igoe 2008). Some are purchased outright, a practice established by the Nature Conservancy in the 1970s (Luke 1997). However many others are created by NGOs, private companies and states using land trusts, leases, community titles and easements. Even more so than large state-sponsored protected areas, these new forms are especially amenable to the convergence of practices by trans-institutional networks (Igoe and Croucher 2007; Diegues, pers. comm. 2008; Dowie 2009; Sachedina 2008).[18] As such interventions are occuring simultaneously all over the world, their aggregate visibility would lend significant coherence of the sustainable development historic bloc.

How do success stories, sovereignty and new (or revived) protected areas combine to produce value and profit for networks of conservation and commerce? Most obviously, protected areas transform landscapes in ways that transform them into valuable tourist destinations (West and Carrier 2004). As such, they generate economic opportunity for interests ranging in scale from the international leisure industry to local micro-enterprise. It is important to note, however, that many protected areas are created in places with limited potential, as tourist destinations still have significant potential value in other ways. As Garland (2008:52) argues, conservation and the creation of protected areas generate many kinds of "capital with potential to circulate and generate further value at the global level".

Their very existence can be rendered profitable. For example, the Green Living Project is a media and entertainment company dedicated to the documentation of conservation success stories[19] and the creation of new adventure travel markets.[20] In partnership with National Geographic Explorer, Green Living Project produces a multi-media presentation that tours REI and L.L. Bean Stores throughout North America.[21] Such productions promote conservation NGOs, market clothing and camping supplies, advertise tourist destinations, provide positive imaging for the countries in which they are set, and create possibilities for the production of nature films, coffee table books, and adventure magazines.

Other kinds of value produced by the conservation mode of production that Garland mentions include research grants, consultancy contracts, educational opportunities, travel, renown and renumerative careers (for illustration see Brockington 2009; MacDonald 2008). They also include valuable green washing services for environmentally destructive corporations (Dowie 1995) and countries with poor human rights records (Garland 2008). They are sources of valuable films, and still pictures, of landscapes and wildlife. They create new kinds of material commodities, such as Starbucks Conservation Coffee and McDonalds Endangered Species Happy Meals (Igoe forthcoming), as well as virtual commodities like carbon offsets. They produce realities in which it appears feasible to mitigate the social and environmental impacts of hydroelectric dams and oil pipelines.

From these eclectic and pervasive values, it is easy to imagine how the diverse groups that make up the transnational capitalist class would benefit from their participation in the production of the sustainable development historic bloc. As Gramsci argued almost a century ago, the maintenance and propagation of any historic bloc is directly related to its ability to render their view of the world self-evident such that it goes without questioning. The currently global production and dissemination of mass media spectacle in the realm of nature conservation is, as we explain next, a manifestation of this postulate.

Spectacular Relationships in the Global Economy of Appearances

All of the relationships and processes we have described thus far are occurring in the context of what Tsing (2005:57) calls "the global economy of appearances", in which "dramatic performance has become an essential prerequisite of economic performance". While extractive enterprise and material production remain important, economic growth in this context depends on the circulation of images and dramatic performance of institutional success. Tsing described how venture capitalists engage in "spectacular accumulation" by "conjuring profits" before they are actually realized in order to "draw an audience of potential investors". Countries, regions and towns must also dramatize their potential as places for investment (Tsing 2005:57).

Conservation NGOs, as well as the foundations, government agencies and for-profit companies that support them, also engage in spectacular performances in conjuring spaces for effective conservation interventions-cum-profitable investments. In their performances, images of dramatic landscapes and exotic people and animals are used to conjure urgent problems in desperate need of the timely solutions that the organization is uniquely qualified to offer. They present an audience of potential supporters with compelling virtual opportunities (problems that need to be solved) and the resources necessary to realize these opportunities (landscapes and animals in need of protection) if they will only make the necessary investment (a generous gift; for details see Ellison 2008; Igoe forthcoming; MacDonald 2008; Sachedina 2008).

In this context, spectacular accumulation is not geared towards direct financial returns. Rather it revolves around parlaying success as symbolic capital into other forms of capital and values that not only help grow specific conservation NGOs, but are also in the interests of a whole array of other agents and institutions (cf Brockington and Scholfield this volume). This in turn is linked to promoting the idea that the ecological ills of late market capitalism can be offset by protecting exotic nature and stimulating economic growth, as well as a pervasive and implicit message that saving the planet is ultimately best achieved by consumption, albeit of particular kinds.

The relationship of spectacular accumulation to the global economy of appearances is well illuminated by Debord's discussion of Spectacle as outlined above. Debord (thesis 1) describes Spectacle as "separation perfected", the ultimate expression of alienation: the loss of control by people over the conditions that shape their lives. Through Spectacle, he argued, the fragmented realities of life in late capitalism are given the appearance of a unified whole, which are visible everywhere.

The proliferation of new media technology over the past 20 years has rendered media spectacle less monolithic and more potentially open to contestation than under the conditions described by Debord in the late 1960s. As the work of Zygmunt Bauman reveals, however, they also have rendered spectacle more pervasive and definitive of people's lives, as many are increasingly likely to interact with a digital interface than with a real human being (2007). Bauman's (2000) prolific discussions of what he calls "liquid modernity" refine and update Debord's arguments about Spectacle. Bauman describes liquid modernity as a world of constant change and individuation, in which people must increasingly "go it alone", without support of social networks or the welfare state. It is a world of seemingly infinite possibilities and opportunities, as well as one of infinite risk (Bauman 2006). The fragmented fleetingness of these conditions are simultaneously exhilarating and terrifying, but ultimately unfulfilling. Consequently, individuals turn to mass media as a more palatable alternative to life actually lived. Here they encounter a bewildering parade of celebrities, experts and celebrity experts who provide examples and reveal secrets of how to live a successful life, as well as comforting solutions to the disturbing problems now facing humanity (cf Brockington 2009). Because they compete intensively with one another, these media celebrity experts derive their authority from their ability to "tempt and seduce" would be choosers

(Baumann 2000:64). Thus, as Adorno (1972) argued, consumption promises escape from conditions of alienation, but one so fleeting that it must be constantly renewed. It also offers the connections and safety of community, but without the inconvenient obligations that earlier forms of community demanded (Bauman 2001).

Many of the features and conditions described by Bauman, and the theoretical genealogy he invokes, are visible in media productions related to biodiversity conservation. One of the latest and most sophisticated of these is the "online community" orchestrated and mobilized by Prince Charles and his supporters to protect the world's rainforests. The website of the Prince's Rainforest Project features numerous videos of celebrities, as well as corporate and non-profit leaders. All these individuals appear on camera with a digitally animated "rainforest frog" and urge viewers to take action to save rainforests.[22] In other films, corporate leaders like Steve Easterbrook, CEO of McDonalds UK, and Sir Richard Branson, founder of the Virgin group, are cast as experts on the issue of climate change. Others feature leaders of CI (Conservation International) and Greenpeace.

Three messages are frequently repeated in these presentations. One is that tropical deforestation is a bigger cause of climate change than all the cars, trucks and airplanes in the world combined. Another is that the primary perpetrators of deforestation are poor people and their governments. Finally, the solution to this problem is to make "live trees more valuable than dead ones", by swiftly moving to an effective and efficient carbon trading mechanism (see especially the Prince's welcome video on the home page of the website, in which he outlines the entire vision of this campaign). Problems and their causes are portrayed as occuring at distant locations, while solutions revolve around new forms of commodification. Individuals are invited to join this "community" via Twitter and Facebook, submitting their own videos to the website, and "texting" world leaders.[23]

A related story in this particular world making project is McDonalds-Europe's Endangered Species Happy Meal Campaign, "designed to educate and empower children to make a difference". The boxes that the meals come in feature links to an interactive online game, which allows children to create a "virtual passport to explore the virtual world". The "virtual world" features multi-media presentations of endangered animals, as well as inviting parents to visit a virtual CI headquarters, where they may learn about McDonald's and CI's partnership to protect rainforest ecosystems, thus helping to combat climate change and make a donation.[24]

As visually compelling as these presentations are, and in spite of the appeal of the solutions they propose, their propositions about how the world works are ultimately unverifiable. To use Debord's language, they conceal the actual nature of the relationships that they mediate. Thus, for instance, there is no way of knowing the extent to which, or even if, money given to a conservation organization actually achieves its purported objectives (Brockington, Duffy and Igoe 2008; Igoe forthcoming). Concerning carbon credits, a study by Gossling et al (2007:239) indicates that for "individual customers it is currently next to impossible to judge the real value of the credits that they buy". Likewise, the presumed social and ecological relationships secured by conservation coffee remain not only unverifiable, but, as West (this volume) shows, are unlikely outcomes. Moreover, an expanding network of scholars and activists are demonstrating that such interventions frequently produce undesirable consequences, especially in terms of displacing local communities and their livelihoods (West, Igoe and Brockington 2006).

Considering the unverifiable nature of these presentations, and the negative impacts they conceal, how come they have such broad and convincing appeal? The answer, we believe, is that these spectacular productions are embedded in a much wider "society of the spectacle" that profoundly shapes the experiences of people's everyday lives (West, Igoe and Brockington 2006). One of the most successful purveyors of spectacle is the Disney Corporation. Disney's techniques for controlling spaces and the gaze of people who visit those spaces have been imitated and reproduced in settings around the world (Bryman 1999). In fact, Disney is arguably as significant to our current era of neoliberal capitalism, and its central imperative of imaginary and virtualized economies, as Ford was to the previous era of liberal capitalism and its central imperative of production and consumption of physical goods (cf Allman 2007). In the words of a Disney illusioneer (Alexander 1992:161–162):

> The environments we create are more utopian, more romanticized and more like the guest imagined they would be. The negative elements are discreetly eliminated, while positive aspects are in some cases embellished to tell the story more clearly.

This statement begins to illustrate a much wider social logic of Spectacle and liquid modernity, whereby people experience the world as a prepackaged matrix of imagined connections between things and people who do not readily present themselves as connected. So, for instance, the people in our story box (below) woke up each morning to a savannah filled with African animals, even though their hotel was in central Florida. At the end of their stay they were able to eat their way around the world.

Story box: Disney magic and cheetah conservation

The following letter is from a press release of Namibia's Cheetah Conservation Fund, http://www.cheetah.org/?html=news-press&data=news-press&key=243

Dear Disney Magic:

I want to relay the story to you of what a wonderful experience my wife, three kids and I had while visiting Disney for the first time last November. We arrived at the Animal Kingdom Lodge and our room was not ready, so we wandered around the lodge and its grounds while waiting for our room. After looking at the animals for a while we wandered into the gift shop. Our 8-year old daughter, Jillian, immediately fixated on a stuffed toy cheetah. My wife and I told our kids that this was our first day and the first shop and that they should probably wait to look at other shops before settling for a gift. Besides, we told Jillian, we could always pick the cheetah up later.

Our wandering eventually brought us out to where a Disney representative, Bianca, was stationed. She was handing out sheets with pictures of half dozen or so plants that we were supposed to find. My wife, Sharon, noticed that Bianca was from Namibia. This reminded her that Jillian, at her last birthday party, had asked all of her friends to make donations to the Cheetah Conservation Fund (which is based in Namibia) in lieu of birthday present.

Jillian heard about the Cheetah Conservation Fund (CCF) in the magazine "scholastic News" that they sent home from school. She thinks the cheetahs are beautiful, fast, interesting and wild. She was not happy to hear about the destruction of their habitat. It made her sad to hear that they are endangered. She wants wild things to be able to stay wild.

Sharon asked Bianca if she was familiar with the CCF. Bianca said that she was well aware of it and told us that not only was it based in her homeland but it was also one of the funds that Disney donates to as part of its conservation efforts. She asked us to wait while she found something for Jillian and returned with one of the buttons that you receive for making a contribution to the Disney's Wildlife Conservation Fund. Jillian and all of us were thrilled. This provided us with an opportunity to give Jillian some praise for her generosity and to teach her and our other children that when you are charitable to others, it will come back to you. Not necessarily directly but in magical ways that you would never expect.

We spent the next day at the Animal Kingdom. Jillian wore her button. Several people working there recognized her. Jillian was glowing with the recognition. This alone was gift enough.

We then moved on to the Polynesian resort and spent the next several days visiting MGM and Magic Kingdom. We had been on pretty much every ride that we had wanted to ride, at least once, so we decided to spend the final day eating our way around the world. We paced ourselves and did a good job sampling various goodies in most of the countries. We had a late dinner at the German section while listening to the fireworks booming outside.

After a fantastic evening in Epcot, we began our walk back to the monorail. Jillian then informed us that she really, really wanted the cheetah to be her final gift. She had looked at all of the other and that was the one thing she most wanted. Since she had only picked one gift to take home, we began a sharp lookout for her cheetah, unsuccessfully. We told we would find one at the Disney store near our home or perhaps order it online. To our surprise, she took it quite well.

Back in our room, the first thing we noticed was a big Disney bag taped up and sitting on Jillian's bed. Inside was the very cheetah that she wanted. At first we thought that perhaps a miracle had occurred. Inside were wonderful hand written notes from Bianca, Claire and Kim (from the Disney Wildlife Conservation Fund) and a personal email from Dr Laurie Marker, the CCF's founder and Executive Director. Jillian was beside herself with delight. It was a truly magical moment.

Jillian set to the task of naming her new cheetah. The naming of things is very important in our house. She quickly settled on Bianca as the name and Bianca she is. I want to thank the real Bianca from the bottom of my heart. Her gift goes far beyond giving a toy animal to a little girl. I have been telling this story to anyone who will sit still long enough to hear me through. I intend to see that this inspires more than just my merry little band of kids.

. . . .

The Disney Wildlife Conservation Fund has supported CCF's programs, such as the Livestock Guarding Dog, since 1999. CCF has received more than $140,000 in support from DWCF, thanks to Disney and generous guests who contribute to the DWCF. The fund has been helping ensure the survival of wildlife and wild places in all their beauty and diversity for nearly 10 years.

But the pervasiveness of these connections goes even further than giving people the impression that they are traveling around the world. They actually give people the romantic illusion that they are adventurously saving the world even as they are consuming it virtually (see story box). Guests purchase experiences that are staged as world-saving adventures, and which emulate the adventures of conservationists who are sponsored by Disney and other corporate concerns. For example, guests at Disney's Animal Kingdom Lodge are invited to "explore the riches of Africa" and "rejuvenate both body and spirit in luxurious surroundings while sharing the grandeur of the

African wilderness",[25] while throughout their visit they are continuously reminded of the links between their consumption and the conservation of biodiversity. First of all they are told that a portion of the profits go to the Disney Wildlife Conservation Fund. The Lodge is "certified green" by the state of Florida and accredited by the Association of Zoos and Aquariums. Guests are also reminded of the conservation value of the lodge through the various activities in which they participate. At Rafiki's Planet Watch, "guests can get a glimpse of conservation efforts being undertaken by Disney's Animal Program scientists to protect locally and globally threatened species".[26]

While the central argument in support of these processes is that they create value through non-consumptive experiences, they are anything but non-consumptive. Consumptive practices are normally associated with the extraction of resources from an ecosystems and/or their transformation into other goods (eg tree into wood, fish into food). Non-consumptive practices are those that at least in appearance leave resources untouched. The problem is that so-called non-consumptive practices often cause ecological and/or social disruptions which may not be immediately visible. As Neves explains in greater detail (2009, this volume) these types of disruption, along with their associated ecological costs, are rendered invisible by the consumptive experiences themselves. Carrier and Macleod (2005) further argue that eco-tourist experiences are constructed precisely to conceal their connections to the global infrastructure and relationships that made them possible in the first place. Their argument focuses especially on the links between eco-tourism and the global air travel industry.

Extending this logic even further, we have shown that the kinds of alienation and fetishization associated with the current convergence of conservation and capitalism have become so widely encompassing as to suggest that any consumptive activity can have a corresponding corrective ecological measure. The very production of this worldview creates the opportunity for continued capitalist expansion, in spite of Green Marxist predictions that these would be superseded by environmentally oriented social movements. Thus, for instance, environmental groups briefly succeeded in tarnishing McDonald's image as a friendly, family-oriented corporation, by exposing the environmental harm it was causing in Amazonia. Not only is McDonalds recovering from this tarnishing by greening its image, it is also trying to capitalize on widespread environmental concerns through the creation of new products and commodities.

The socio-ecological implications of this issue are more profound than they may at first appear. McDonalds may very well have stopped using soy-based animal feed in response to activist criticisms. It is also not to say that children learn nothing about biodiversity loss and climate change from McDonalds' Endangered Animal Happy Meals. As Schlosser (2001) so thoroughly demonstrates, however, the environmental contradictions of the fast food industry are far more pervasive and integrated than a Hotspots and Happy Meals partnership can even begin to capture. The rise of the fast food industry in North America is inextricably linked to automobile culture, expanding networks of superhighways, and the industrialization of global food production systems. By focusing consumers' attention on distant and exotic locales, the spectacular productions of McDonalds and CI conceal the complex and proximate connections of people's daily lives to environmental problems, while suggesting that the solutions to environmental problems lay in the consumption of the kinds of commodities that helped produce them in the first place.

Conclusion: Research and Resistance in the Sustainable Development Historic Bloc

So far we have said nothing about how people might resist, subvert or creatively interpret any of the phenomena we have been talking about. This is an essential area of research, since without understandings of resistance and reinterpretation historic blocs appear deterministic and immune to change. Gramsci himself held that this was not the case, and this was precisely why he was so concerned with understanding historic blocs. We strongly believe that good ethnographic research is essential for understanding both the ways in which historic blocs are reproduced and the ways in which they are transformed. We wish to conclude, therefore, with three sets of points that should be kept in mind when doing such ethnography:

1 There will always be people, things and processes that cannot be co-opted by and/or are excluded from a prevailing historic bloc. These people, things and processes are potentially counter-hegemonic. At the same time, however, historic blocs relentlessly set limits on thought, speech and action. As such, all that is potentially counter-hegemonic comes across as lacking credibility, and in most contexts is easily dismissed. The sustainable development historic bloc, especially, rests solidly on a technocratic view of the world, in which experts elected by the historic bloc are presented as the holders of fundamental truths and wisdom. Views not sanctioned by this technocracy are dismissed as ill-founded. When a potentially viable critique of the historic bloc emerges, the historic bloc is able to quickly and efficiently mobilize a seemingly endless array of experts to

counter that critique. Finally, it is essential not to forget that critiques of a prevailing historic bloc run directly counter to the economic interests of extraordinarily large and diverse groups of people.

2 Conflicts around issues of biodiversity in the context of the sustainable development historic bloc constitute what Gramsci (2000b) termed a war of position. In contrast to the types of direct frontal revolutions that occurred in Eastern Europe, Gramsci held that counter-hegemonic struggles in liberal capitalist societies would occur in the context of civil society, the political terrain of public space and media in which the dominant classes organize their hegemony and in which opposition parties and movements organize, build coalitions, and generate counter-hegemonic forms of thought, speech and action. These types of struggles, he cautioned, require a thorough knowledge of the prevailing historic bloc, careful and meticulous strategizing, and clever interventions forums in which hegemonies are produced and reproduced. In the context of biodiversity conservation these forums include meetings, workshops, congresses, summits, and the media, especially the internet. Accordingly, these are important sites at which resistance to the sustainable development historic bloc are occurring. It is important to remember, however, that increasingly sophisticated forms of Spectacle have rendered these struggles more complex than they were in Gramsci's time.

3 Spectacle continuously presents people with an aesthetic of a world that is already dead (Debord 1995 [1967]; see also Baudrillard 1993; Luke 1997).[27] This aesthetic is filled with images of life and motion, but these images themselves are dead. They cannot be changed. To the extent that consumers interact with the Spectacle it is by choosing between a set of preprogrammed consumptive experiences. But they cannot change the Spectacle through these interactions. In the context of late consumer capitalism described above, they are offered a set of prepackaged choices. Although we cannot presume to know how consumers personally conceptualize and feel about these intended metaphors, we must recognize the structures and constraints within which consumer responses will operate. However sophisticated their understanding of these choices may be (cf Carrier 2003), there is little that they can do to change the ossified spectacles of reality with which they are presented. Happily, the democratization of media technology and the internet presents new opportunities for subverting and resisting Spectacle.

We hope that this framework for understanding the sustainable development historic bloc will be useful in thinking about how future investigations of conservation and capitalism should be designed and carried out. Moreover, as intellectuals and cultural critics, it is essential that we remain mindful of our own places and spaces on the political terrain of the sustainable development historic bloc, and the ways in which we might also contribute to both its reproduction and its subversion. Hopefully, the framework we have presented in this chapter will also prove useful in doing this as well.

Endnotes

[1] From a statement at the World Conservation Congress, quoted in the New York Times, http://greeninc.blogs.nytimes.com/2008/10/08/the-failing-business-of-conservation/, accessed 4 August 2009.

[2] http://greeninc.blogs.nytimes.com/2008/10/08/user-friendly-databases-make-conservation-easier-for-business/, accessed 4 August 2009. The World Conservation Congress is the general assembly of the IUCN (World Conservation Union), http://www.iucn.org/, accessed 4 August 2009. It meets once every 4 years, and is described by the IUCN as the world's largest and most important conservation event, http://www.iucn.org/congress_08/, accessed 4 August 2009.

[3] http://www.iucn.org/about/work/programmes/business/bbp_our_work/bbp_shell/upd ate/, accessed 4 August 2009. http://cms.iucn.org/about/work/programmes/business/ bpp_news/?uNewsID=2013, accessed 4 August 2009.

[4] One of us, Jim Igoe, attended the Congress and personally observed these sorts of statements, as well as learning about them through personal communication with Ken McDonald and Saul Cohen who were conducting research on the "neoliberalization" of conservation at the congress. The statement quoted here appeared in the *New York Times* business blog in a piece called "The Failing Business of Conservation", http://greeninc.blogs.nytimes.com/2008/10/08/the-failing-business-of-conservation/, accessed 4 August 2009.

[5] http://www.youtube.com/watch?v=1CRs-7lRlPo; http://www.rainforestsos.org/; both accessed 4 August 2009.

[6] See especially IUCN President Valli Moosa's statements to the opening ceremony of World Conservation Congress in Barcelona, http://greeninc.blogs.nytimes. com/2008/10/08/the-failing-business-of-conservation/, accessed 5 August 2009, and *New York Times* columnist Tom Friedman's statement for Conservation International, http://www.youtube.com/watch?v=W1Li3O81uDs, accessed 5 August 2009.

[7] This is assuming that such income stays in the community and that it is somehow equitably dispersed.

[8] As Kate Crehan argues in *Gramsci, Culture, and Anthropology*, hegemony is far more nuanced and complex than I have briefly described it here. This book is essential reading for anyone concerned about hegemony and culture.

[9] There are many examples (see Holmes this volume, but some examples include: George Moore who was invited to join the board of Conservation International after his George and Betty Moore Foundation donated over $250 million to it; John Robinson, a Vice President of the Wildlife Conservation Society (WCS) sits on the board of the Christensen Fund, the President and Chair of which (Diane Christensen) serves as a trustee of the WCS; Yolanda Kakabadse the former President of the IUCN and Kathryn

Fuller, the former President of WWF-US have both served on the board of the Ford Foundation.
[10] AWF, for instance, has invited former heads of African States onto its board of trustees, including Benjamin Mkapa of Tanzania and Ketumile Masire of Botswana, http://www.awf.org/section/about/trustees, accessed 4 August 2009. Garland (2008) outlines the ways in which Michael Fay of the Wildlife Conservation Society forged new alliances with President Omar Bongo of Gabon, on the basis of which Bongo set aside 10% of the countries' land for protected areas. In an interview on Comedy Central, Alan Rabinowitz of the Wildlife Conservation Society describes the necessity of working closely with dictators and corrupt regimes in order to convince them of the necessity of setting aside land for conservation, http://www.colbertnation.com/the-colbert-report-videos/171137/june-10-2008/alan-rabinowitz, accessed 4 August 2009. While individuals like Bongo are unlikely to make it on the board of trustees of any conservation organizations, they are an important part of the elite networks through which conservation is achieved.
[11] For the purposes of this chapter we call them ideologue intellectuals.
[12] Neves (2004) also shows that such dichotomous distinctions hold true only as heuristic devices, since the dynamics of power relations leads to the emergence of much more fluid understandings of the world, as the articulation of different positions affects the positions themselves. The notion is indeed in accordance with Gramsci's notion of war of positions.
[13] Especially notable examples include the corporate leader videos of the Prince's Rainforest Project, http://www.rainforestsos.org/content/home/; WWF's *Let the Clean Economy Begin*, http://www.youtube.com/watch?v=Mzr4H6ZWKBE; WWF Earth hour videos, http://www.youtube.com/watch?v=C9GRh_9sQBw&feature=PlayList&p=283A669E4559A584&playnext=1&playnext_from=PL&index=2; http://www.youtube.com/watch?v=qHsJXFcSo7E, accessed 14 August 2009; http://www.youtube.com/watch?v=HxZGK_iHIXc; AWF's Starbucks video, http://www.youtube.com/watch?v=KvxdbKV4DEU; About AWF, http://www.youtube.com/watch?v=P98NDWsUf3s&feature=related; CI's video *Richard Branson on Climate Change*, http://www.conservation.org/fmg/pages/videoplayer.aspx?videoid=63; *Team Earth*, http://www.conservation.org/fmg/pages/videoplayer.aspx?videoid=47; and *McDona ld's Endangered Species Happy Meals*, http://www.conservation.org/fmg/pages/videoplayer.aspx?videoid=43; all accessed 14 August 2009.
[14] http://www.rainforestsos.org/, accessed 30 July 2009.
[15] http://ecosystemmarketplace.com/pages/article.news.php?component_id=6356&component_version_id=9499&language_id=12, accessed 10 January 2008.
[16] http://www.speciesbanking.com/, accessed 12 January 2008.
[17] For numerous examples from around the world, please visit the website of the Katoomba Group, an international network of individuals working to promote, and improve capacity related to, markets and payments for ecosystem services, http://www.katoombagroup.org, accessed 12 January 2008.
[18] Professor Antonio Diegues has been researching conservation displacement in Brazil since the late 1960s. Recently, he and his graduate students have been documenting the ways in which conservation BINGOs circumvent national laws prohibiting the sale of land to outsiders by using proxy Brazilian NGOs to purchase land, which then become mini-protected areas, where use and habitation by local people is forbidden.
[19] http://www.greenlivingproject.org, accessed 7 August 2009.
[20] http://www.prlog.org/10180118-green-living-project-and-xola-consulting-announce-strategic-partnership.html, accessed 7 August 2008.
[21] http://planetgreen.discovery.com/work-connect/green-living-project-tour.html, accessed 7 August 2009.

[22] http://www.rainforestsos.org/, accessed 30 July 2009. All of the videos sited in these paragraphs can be found at this web site.
[23] All of this extensive and rapidly expanding material can be viewed at http://www.rainforestsos.org/, accessed 10 July 2009.
[24] http://www.conservation.org/fmg/pages/videoplayer.aspx?videoid=43, accessed 10 July 2009.
[25] http://www.kingdom-travel.com/Walt_Disney_World_Resorts/Disney_Animal_King dom_Resort.shtml, accessed 9 July 2008.
[26] http://disney.go.com/disneyhand/environmentality/animals/kingdom.html, accessed 9 July 2008.
[27] We speak metaphorically here and are not making some oblique reference to the abundance of biodiversity (which is immense and exciting) or to the grave threats to it.

References

Adorno T (1972) *The Culture Industry*. London: Routledge
Agius J (2001) Biodiversity credits: Creating missing markets for biodiversity. *Environmental and Planning Law Journal* 18(5):481–503
Aldridge M and Dingwall R (2003) Teleology on television? Implicit models of evolution in broadcast wildlife and nature programmes. *European Journal of Communication* 8:114–132
Alexander A (1993) *The Culture of Nature: North American Landscapes from Disney to the Exxon Valdez*. Cambridge, MA: Blackwell Publishers
Allman T (2007) The theme-parking, megachurching, franchising, exurbing, mcmansioning of America. How Walt Disney changed everything. *National Geographic Magazine* 2007:98–115
Baudrillard J 1993. *Symbolic Exchange and Death*. London: Sage Publications
Bauman Z (2000) *Liquid Modernity*. Cambridge, UK: Polity Press
Bauman Z (2001) *Community: Seeking Safety in an Insecure World*. Cambridge: Polity Press
Bauman Z (2007) *Consuming Life*. Cambridge: Polity Press
Bayon R, Fox J and Carrol N (2007) *Conservation and Biodiversity Banking*. London: Earthscan
Blundel A (2006) Offsets could mitigate damage to biodiversity. *Nature* 442:245–246
Bonner R (1993) *At the Hand of Man. Peril and Hope for Africa's Wildlife*. London: Simon and Schuster
Bourdieu P (1977) *Outline of the Theory of Practice*. Cambridge: Cambridge University Press
Brockington D (2009) *Celebrity and the Environment: Fame, Wealth, and Power in Conservation*. London: Zed Books
Brockington D, Duffy R and Igoe J (2008) *Nature Unbound: Conservation, Capitalism, and the Future of Protected Areas*. London: Earthscan Publishers
Brockington D and Scholfield (2010) The conservationist mode of production and conservation NGOs in sub-Saharan Africa. *Antipode* 42(3):551–575
Bryman A (1999) The Disneyfication of society. In G Ritzer (ed) *The McDonaldization Reader* (pp 52–58). London: Sage Publications
Carrier J (2003) Mind, gaze, and engagement: Understanding the environment. *Journal of Material Culture* 8:5–23
Carrier J and Macleod G (2005) Bursting the bubble: The socio-cultural context of ecotourism. *Journal of the Royal Anthropological Institute* 11:315–334
Castree N (2007) Neoliberalizing nature: The logics of de- and re-regulation. *Environment and Planning A* 40:131 152

Chapin M (2004) A challenge to conservationists. *World Watch Magazine* Nov/Dec:17–31

Clark S and Altman R (2007) *A Field Guide to Conservation Finance*. Washington, DC: Island Press

Corson C (2010) Tracing the origins of neoliberal conservation through the U.S. Agency for International Development. *Antipode* 42(3):576–602

Crehan K (2002) *Gramsci, Culture, and Anthropology*. Berkeley: University of California Press

Debord G (1995 [1967]) *Society of the Spectacle*. New York: Zone Books

Dorsey M (2005) Conservation, collusion and capital. *Anthropology News* October:45–46

Dowie M (1995) *Losing Ground: American Environmentalism at the Turn of the Twenty-First Century*. Cambridge, MA: MIT Press

Dowie M (2009) *Conservation Refugees: The Hundred Year Conflict Between Global Conservation and Native People*. Cambridge, MA: MIT Press

Dzingirai V (2003) The new scramble for the African countryside. *Development and Change* 34(2):243–263

Etchart G (1995) Mitigation banks: A strategy for sustainable development. *Coastal Management* 23(3):223–227

Ellison K (2008) Business as usual. *Frontiers in Ecology and the Environment* 6(9):12

Ferguson J (2006) *Global Shadows: Africa in the Neoliberal World Order*. Berkeley: University of California Press

Fortwangler C (2007) Friends with money: Private support for national parks in the U.S. Virgin Islands. *Conservation and Society* 5(4):504–533

Garland E (2008) The elephant in the room: Confronting the colonial character of wildlife conservation in Africa. *African Studies Review* 51(3):51–74

Goldman M (2005) *Imperial Nature. The World Bank and Struggles for Social Justice in the Age of Globalisation*. Yale: Yale University Press

Gossling S, Broderick J, Upham P, Ceron J, Dubois G, Peeters P and Strasdas W (2007) Voluntary carbon offsetting schemes for aviation: efficiency, credibility, and sustainable tourism. *Journal of Sustainable Tourism* 15(3):223–248

Gramsci A (1971a) The intellectuals. In Q Hoare and G Smith (eds) *Selections from the Prison Notebooks* (pp 5–23). New York: International Publishers

Gramsci A (1971b) State and civil society. In Q Hoare and G Smith (eds) *Selections from the Prison Notebooks* (pp 210–215). New York: International Publishers

Gramsci A (2000a) Structure and superstructure. In D Forgacs (ed) *The Antonio Gramsci Reader* (p 192). New York: New York University Press

Gramsci A (2000b) War of position and war of maneuver. In D Forgacs (ed) *The Antonio Gramsci Reader* (pp 192–228). New York: New York University Press

Grandia L (forthcoming) Silent spring in the land of eternal spring: The germination of a conservation conflict. *Current Conservation*

Harvey D (1996) *Justice, Nature, and the Geography of Difference*. Oxford: Blackwell

Heynen N, McCarthy J and Robbins P and Prudham S (eds) (2007) *Neoliberal Environments: False Promises and Unnatural Consequences*. New York: Routledge

Holmes G (2010) Conservation elites and the Dominican Republic. *Antipode* 42(3):624–646

Igoe J (forthcoming) The spectacle of nature in the global economy of appearances: Anthropological engagements with the spectacular mediations of transnational biodiversity conservation. *Critique of Anthropology* 30(3)

Igoe J and Brockington D (2007) Neoliberal conservation: A short introduction. *Conservation and Society* 5(3)

Igoe J and Croucher B (2007) Conservation, commerce, and communities: The story of community-based wildlife management areas in Tanzania. *Conservation and Society* 5(3):534–561

Igoe J and Fortwangler C (2007) Whither communities and conservation? *International Journal of Biodiversity Science and Management* 3:65–76

Klein N (2007) *The Shock Doctrine: The Rise of Disaster Capitalism.* New York: Metropolitan Books

Luke T (1997) *Ecocritique: Contesting the Politics of Nature, Economy, and Culture.* Minneapolis: University of Minnesota Press

MacDonald C (2008) *Green Inc: an Environmental Insider Reveals How a Good Cause has Gone Bad.* New York, NY: Lyons Press

MacDonald K (2010) The devil is in the (bio)diversity: Private sector "engagement" and the restructuring of biodiversity conservation. *Antipode* 42(3):513–550

MacDonald K (forthcoming) Business, biodiversity and new "fields" of conservation: the World Conservation Congress and the renegotiation of organization order. *Conservation and Society*

Marx K (1973 [1857–1861) *The Grundrisse.* London, New York: Penguin

Marx K (1978 [1865–1870] *Capital,* Vol 2. New York: Vintage

Mbembe A (2001) *On the Post-Colony.* Berkeley: University of California Press

McAfee K (1999) Selling nature to save it? Biodiversity and green developmentalism. *Environment and Planning D: Society and Space* 17:133–154

McDermott-Hughes D (2005) Third nature: Making space and time in the Great Limpopo Conservation Area. *Cultural Anthropology* 20:157–184

Mosse D (2004) Is good policy unimplementable? Reflections on the ethnography of aid policy and practice. *Development and Change* 35(4):639–671

Mosse D (2005) *Cultivating Development: An Ethnography of Aid Policy and Practice.* London: Pluto Press

Muchnink B (2007) "9/10 of the law: Vigilante conservation in the American west." Unpublished paper presented to the 106th Annual Meeting of the American Anthropological Association, 26 November to 1 December, Washington DC

Mutchler (2007) "The failure of wilderness: Bureaucracy, bovines, and bullets." Unpublished paper presented to the 106th Annual Meeting of the American Anthropological Association, 26 November to 1 December, Washington DC

Neves K (2004) Revisiting the tragedy of the commons: Whale watching in the Azores and its ecological dilemmas. *Human Organisation* 63(3):289–300

Neves K (2009) The great wollemi saga: Betwixt genomic preservation and consumerist conservation. In C Casanova (ed) *Contemporary Issues in Environmental Anthropology* (pp 134–156). Lisbon: ISCSP University Press

Neves K (2010) Cashing in on cetourism: A critical ecological engagement with two whale watching business models. *Antipode* 42(3):719–741

O'Connor J (1988) Capitalism, nature, and socialism: A theoretical introduction. *Capitalism, Nature, and Socialism* 1:11–38

Prudham S (2004) Poisoning the well: Neo-liberalism and the contamination of municipal water in Walkerton, Ontario. *Geoforum* 35(3):343–359

Rodriguez J, Tabler A, Daszack P, Suckumar R, Valladares-Paudua C, Padua, S, Aguirre L, Medelin, M, Acosta, M, Aguirre A, Bonacic C, Bordino P, Bruschini J, Buchori D, Gonzales S, Matthew T, Mendez M, Mugica L, Pachecho F, Dobson A and Pearl M (2007) Globalization of conservation: The view from the south. *Science* 317:755–756

Sachedina H (2008) "Wildlife is our oil: Conservation, livelihoods, and NGOs in the Tarangire ecosystem, Tanzania." Unpublished PhD Dissertation, Oxford University

Sachedina H (2010) Conservation empire: A case study of the scaling up of the African Wildlife Foundation. *Antipode* 42(3):603–623

Schlosser E (2001) *Fast Food Nation.* New York: Houghton Mifflin

Sklair L (2001) *The Transnational Capitalist Class*. Oxford: Blackwell

Spence M (1999) *Dispossessing the Wilderness: Indian Removal and the Making of National Parks*. New York: Oxford University Press

Sullivan S (2009) An ecosystem and your service: Environmental strategists are redefining nature as a capitalist commodity. *The Land* Winter 2008/2009

Sunseri T (2005) Something else to burn: Forest squatters, conservationists, and the state in modern Tanzania. *Journal of Modern African Studies* 43:609–640

Tsing A L (2005) *Friction: An Ethnography of Global Connection*. Princeton: Princeton University Press

USDOI (2003) *Guidance for the Establishment, Use, and Operation of Conservation Banks*. Washington DC: United States Department of the Interior

Vivanco L (2002) Seeing green: Knowing and saving the environment on film. *American Anthropologist* 10(4):1195–1204

West P (2010) A wookie wouldn't drive a hummer, but would an ewok drink certified coffee? Media, consumption, and the fashioning of contemporary environmental politics. *Antipode* 42(3):690–718

West P and Carrier J G (2004) Ecotourism and authenticity: Getting away from it all? *Current Anthropology* 45:483–498

West P, Igoe J and Brockington D (2006) Parks and people: The social impacts of protected areas. *Annual Review of Anthropology* 35:251–277

Young Z (2002) *A New Green Order? The World Bank and the Politics of Global Environmentalism*. London: Pluto Press

Chapter 2
The Devil is in the (Bio)diversity: Private Sector "Engagement" and the Restructuring of Biodiversity Conservation

Kenneth Iain MacDonald

In January 2009 Friends of the Earth International (FOEI), which bills itself as "the world's largest grassroots environmental network" withdrew their membership in the International Union for the Conservation of Nature (IUCN), an organization that is made up of a wide collection of non-governmental environmental organizations and governmental environmental agencies.[1] In the letter of notification, which was addressed to the Director General of IUCN, the International Chair of FOEI clearly states that the main reason for the withdrawal was their "concern about the corporate partnership between Shell and the IUCN".[2] More specifically, the letter cited two primary concerns with the partnership. First, while the partnership would not likely have any meaningful impact on Shell's activities it seemed to be silencing IUCN's willingness to critique the negative social and environmental consequences associated with Shell's practices and thereby compromising the ability of member organizations to work in effected communities. Second, and perhaps more importantly, attempts by the membership to terminate the relationship between IUCN and Shell had been stymied by the bicameral structure of IUCN, which divides the membership according to their status as governmental or non-governmental organizations (NGOs) and effectively establishes two "houses" of membership.

The second point refers to a drama of sorts that played out during the World Conservation Congress (WCC) held in Barcelona in October 2008. The WCC is a convention of the membership of IUCN held once every 4 years and is divided into a "Forum"—with paper sessions and workshops—and a "Members Assembly" during which the entire membership votes on motions put forward by small groups of members. Motions that pass both "houses" of the assembly become resolutions that are intended to guide the activities of the IUCN Secretariat for the next 4 years. During the Barcelona WCC a number of motions put forward by members indicated a growing tension over the engagement of the Secretariat with private sector actors, and the increasing re-orientation of the organization around practices informed by ideologies of ecological modernization. The most blunt of these was Motion 107 that called for the termination of the IUCN's "agreement" with Shell. As the BBC reported, "the vote, when it came, was the most eagerly anticipated of the congress".[3] When the dust of debate had settled, the motion was defeated. Despite the fact that the motion carried over 60% of the popular vote and that 70% of NGO members had voted in support, most government delegates voted against and blocked the motion.[4] This blockage occurred because IUCN statutes mandate the calculation of votes within governmental and non-governmental "houses" and only allow a motion to pass if it secures an absolute majority in both "houses". The perception among many members was that the popular will of the membership had been blocked by an alliance between the Secretariat which, through a series of statements and other gestures,[5] had actively opposed the motion to terminate the agreement, private sector actors and State members.[6]

The subsequent withdrawal of FOEI from IUCN is but one example of a growing ideological and material divide between large international conservation organizations and smaller groups that orient themselves around "the grassroots". While visible expressions of this divide are relatively recent, in this chapter I make the case that the grounds for the shift in the ideological and practical orientation of large conservation organizations has been developing over the last two decades, and that this has happened through a structured process involving an ideological alignment of the emergent concepts of sustainable development and ecological modernization and the development of a new international institutional context for environmental governance that minimized the threat, to business, of regulatory control over access to natural capital. The first part of the chapter outlines this process of ideological alignment and the way in which it shaped the emergence of a new international institutional context. The second part of the chapter considers the implications of this shift, specifically how this new institutional context brought into being new grounds for organizational legitimacy and new sources of funding that encouraged the development of entrepreneurial practices within large conservation organizations and directed them toward intensified engagement with private sector actors.[7] In the course of making this case, I also demonstrate the importance of interpreting these actions through the lens of organizational theory and make the case that the contemporary emergence of business as a major actor in shaping contemporary biodiversity conservation is explained in part by the organizational characteristics of modernist conservation that subordinates it to larger societal and political projects such as neoliberal capitalism.

Given the significant global reach of non-governmental conservation organizations (NGCOs) and their influence in national and global environmental governance, there is a need to understand what mobilizes fundamental ideological and organizational shifts within these organizations; the processes through which such shifts in focus spread; the effect of these shifts on processes of knowledge production, policy development, and conservation programming and implementation; and how they contribute to a reorientation of biodiversity conservation targets, and outcomes, thereby creating new ecological realities through reformulated practices of biodiversity conservation. This chapter seeks to fill this gap in knowledge by exploring the conditions that have led to the restructuring of biological diversity conservation.

The Organization of Biodiversity Conservation

Since the rise of organizational environmentalism in the 1960s, business and biodiversity conservation organizations have lived in two distinct and heavily bounded worlds. The dominant view was that they embraced values, approaches and missions that were deeply incompatible. Over the past decade, however, a significant shift has occurred that has seen an increase in the forms of interaction between these two sets of actors (Secretariat of the Convention on Biological Diversity—CBD Secretariat 2004). Evidence of this is found in the restructuring of international NGCOs to accommodate various forms of relationships with private sector actors. This is revealed in programmatic initiatives that have assumed the common moniker of "Business and Biodiversity" initiatives. While the historical track of this restructuring is not well studied, such initiatives include:

1 The establishment of organizational units inside NGCOs dedicated to establishing, fostering and managing collaborative relationships between NGCOs and private sector interests;
2 Incentive programs developed at state and supra-state levels to promote the establishment of such relationships through subsidies;
3 Increase in conservation programming focused on market-based conservation incentives and grounded in a dubious equations between ecological modernization and sustainable development;
4 The establishment of NGCO/private sector networks connected through interaction among a set of common individuals.

To understand the rise in the association between private sector actors and biodiversity, it is necessary to view modernist biodiversity conservation as an organized political project.[8] There are two important dimensions to this claim. First is the recognition that the organizational dimensions of conservation exist as coordinated agreement and action among a variety of actors that take shape within radically asymmetrical power relations of ideology and practice, both of which reciprocally relate to objectives set by organizational actors. Second, the practical expression of that coordination exists as organized social groups— conservation organizations—that have emerged out of specific historical contexts. Both aspects of "organization" in this context imply the promotion of certain ideological perspectives that are worked out through processes of coordinated agreement, and implemented through the actions of conservation organizations. These are by no means exclusive processes. Indeed the actions of conservation organizations are directed through the ideological configurations brought to bear upon them by the coordinated agreement of relevant actors or what have come to be known as "stakeholders". Understanding the contemporary emergence of business as a major actor in shaping contemporary biodiversity conservation requires understanding the emergence of conservation organizations in relation to the coordinating action of the dominant ideological interests that underlie these broader political projects. Biodiversity conservation has never really driven environmental agendas. Rather, it has been an instrument in much larger political projects such as nationalism, colonialism and capitalism. This means that conservation policy and practice, whether developed within governmental or NGOs, is structured in relation to broader and longer term political goals.[9] This point is important in understanding the entrance of a seemingly new actor into the organization of biodiversity conservation—what is loosely labeled as "business"[10] in the standard international relations literature—because it signals a number of important developments: a shift in the larger political projects that drive conservation; an alteration in the ideological configurations that constitute the social ground on which conservation can be practiced; and the development of an international institutional context for biodiversity conservation that reflected those changes.

The Emergence of Sustainability

As with much of the environmental movement, biodiversity conservation underwent a significant change in the 1980s with the emergence of the concept of sustainability and, more importantly, its rapid incorporation into political rhetoric and an emerging institutional structure of global environmental governance. The concept gained credence among conservation practitioners with the emergence of the 1980s World Conservation Strategy (WCS) jointly published by the IUCN, United Nations Environment Programme (UNEP) and the World Wide Fund for Nature (WWF). The WCS had three specific goals: to maintain essential ecological processes and life support systems; to preserve genetic diversity; and to ensure the sustainable utilization of species and ecosystem, and it led conservation planners to focus on alignments between conservation, development and sustainability, most notably in the form of "sustainable use" and "incentive-based" conservation programming (Adams and Hutton 2007). The WCS, however, was more than simply a policy document. It was the initial step in an attempt to structure the establishment of coherent national conservation strategies around the world and became the basis for the rapid expansion of IUCN, WWF UNDP and, subsequently, other major conservation organizations into international project-based conservation programming. The articulation of sustainability, development and conservation expressed in the WCS and subsequent national conservation strategies successfully mobilized donor funds, most significantly from USAID and the Global Environment Facility (GEF), which drove the boom in integrated conservation and development projects and the focus on "community-based natural resource management" that assumed dominance in international conservation organizations during the 1990s.[11]

It is not difficult to read the WCS, particularly its tentative alignment with the use of "market mechanisms" to achieve conservation goals, as a document that anticipates the shift from a decade of effective state environmental regulation to a period of government resistance to such regulation that emerged under the Thatcher administration in Britain in the late 1970s and was about to emerge in the USA—one that would explicitly equate "nature" and capital and consequently subject the protection of nature to market forces (Turner 1982). In essence, the rise of sustainability in the 1980s, can be understood as a function of its position as a compromise conceptual device meant to address the environmental crisis of consumption and the apparent contradiction of capitalism—the destruction of the physical environment upon which it depends for continued growth—without alienating the governmental bodies upon which conservation had come to depend. Despite its retrospective failings, the concept of sustainable development gained popularity quickly and, partially as a function of a seductive vagueness, incorporated what had been an oppositional environmental politics into a mainstream institutional context (Brand and Görg 2008; Geisinger 1999; Hildyard 1993; O'Connor 1994; Sneddon 2000; Worster 1993). To a large extent this was achieved through a language that sought to replace protest and conflict with consensus and consent by claiming that economic and environmental goals were compatible, a message that underpinned the popularity and quick, if superficial, embrace of *Our Common Future* (WCED 1987; cf Beder 2006; Tokes 2001). Conservation organizations and other actors in the broader environmental movement quickly adopted and popularized the concept, even as they were attempting to define it. But they were not alone in this. Other actors were engaged in what was in essence an ideological struggle to gain a dominant position in the attempt to define the terms under which sustainable development would become a legitimizing instrument.

Positioning Sustainable Development as Ecological Modernization

While the concept of sustainability was under production in the 1980s a parallel perspective, ecological modernization (EM), came into being as a challenge, more than a compromise, to the popular assertions that human societies needed to seriously grapple with self-limitation (see Moland Spaargaren 2000). This response is far less a coherent theory than a loosely interrelated collection of concepts and mechanisms developed in an attempt to produce a version of capitalism that can address its own contradictions while retaining control over the regulatory tendencies of governments that might threaten the capacity for private accumulation (Ashford 2002; Hajer 1995; Keil and Desfor 2003; Tokes 2001). While the concept of EM initially developed in Europe, it has been incorporated into the antiregulatory and antigovernment ideological revolution that accompanied the ascendance of many neoliberal administrations around the world. As a response to the regulatory approaches to environmental management that accompanied the rise of popular environmentalism during the late 1960s and 1970s, ecological modernization challenged the restrictions on access to environmental resources and the impingement on capital accumulation posed by regulation. As an alternative, it framed a technocentric and interventionist variant of environmentalism that highlights the application of science, market forces and managerial ingenuity through instruments such as tradable permit schemes and markets in what have come to be known as ecosystem services— the very term being a sign of the degree to which "nature" has become an element in capitalist processes of ideological domination. EM also seeks to promote flexibility and regulatory freedom to take advantage of industry's supposed potential to engage in technological innovation; encourages more voluntarism and stakeholder participation in governance; and "promotes demand-side policies focused on mobilizing 'green' consumer behavior" (Ashford 2002:1417). The primary assumption behind EM is that "sustainable futures can be attained under conditions of a continuously growing capitalist economy" and it is in this assumption that ecological modernization asserts its unanimity with sustainable development (Keil and Desfor 2003:30).

My point here is that ecological modernization is positioned as part of an overall strategy that defines sustainable development, and provides the intellectual leverage to challenge regulatory impulses whenever and wherever they appear (cf Bernstin 2001). Its primary claim is that the historical ecological contradictions of capitalism can be resolved through new strategies of accumulation and that these should rightly be the main mode of environmental protection for the planet (cf Ervine 2007).[12] Within this frame, the survival of the natural world becomes a residual product of industrial and social processes as ecological modernization asserts the primacy of society over nature and "presupposes the hegemony of capitalist relations over other forms of social organization" (Keil and Desfor 2003:32). This is a particularly neoliberal view of the world; one that, even as it clouds the question of what kinds of nature are economically, socially and ecologically just (as opposed to efficient and marketable), speaks of win–win situations. It is also a view that has penetrated into international conservation organizations. In the words of a former IUCN employee discussing the origins of the Business and Biodiversity Unit:

> So we set up this business unit, and the idea we had was, if we were going to say "Is biodiversity a business proposition? Can people make money by saving nature? And can we get nature saved by people making money?" ... can we expand the set of instruments out there that conserve nature by using capitalist tools? This was the whole thing. And my slogan for the program was "making capitalism work for conservation." That was the slogan of the business unit.

Despite criticisms of the ways in which ecological modernization in practice seemingly contradicts the goals of conservation organizations, it has become a key element in the discursive configuration of "solutions" to the problem of biodiversity loss. It has also secured a position at the core of project activities among leading conservation organizations in a way that involves a full accommodation of the existing capitalist order. Indeed, a survey of the websites of major conservation organizations indicates that they have become important locales for the articulation of ecological modernization projects[13]. Here then we find the first hints of a new political project structuring the activities of biodiversity conservation, and it reveals a sharp move away from a high modernist phase of conservation. It also suggests a move in which some of the more marginal voices that arose in the wake of high modernism are being marginalized once again as their demands for maintaining and improving the biological diversity that underpins conditions of ecological resilience have lost primacy to concerns for efficiency, competitiveness, marketability, flexibility and development.

To understand this shift and the speed with which EM as sustainable development has penetrated conservation organizations, it is necessary to situate it within the shifting institutional and organizational context of biodiversity conservation that developed through the 1980s and 1990s. This shift had three primary components, some of which were seen as "successes" of conservation: the expansion of project-based conservation that took advantage of the extension of neoliberal practices; the appearance of new conservation organizations, like Conservation International, grounded in entrepreneurial strategies; and the emergence of a new institutional network of environmental governance that actualized external control on conservation organizations.

These were not mutually exclusive developments and all were structured by the prevalence of neoliberal forms of governance that emerged during the 1980s. These new modes of governance, in many cases imposed by multilateral financial institutions and their associated structural adjustment programs, lead to reduced state investment in long-term biodiversity protection, especially in the fiscally constrained states of the "global south" (James, Green and Paine 1999; Lapham and Livermore 2003; Mansourian and Dudley 2008; Pearce and Palmer 2001; Redclift 1995; Reed 1992). This retrenchment, however, also created the political opportunity for large conservation organizations to assume heightened responsibility for environmental management in many of those locales. Through the 1980s and 1990s conservation organizations rapidly expanded in size, and budget, and engaged in "mission shift" moving from a focus on knowledge production, and policy consultation, to fund raising and project implementation. Much of this expansion was achieved haphazardly through the establishment of country offices or regional programs that were supported by the increasing linkage between biodiversity conservation and sustainable development that had been articulated in the World Conservation Strategy. As one former IUCN staff member put it:

> The whole organization was Byzantine. Well, the organization is ... the business model of the organization is completely undeveloped. It makes no sense. But by this time in the 90s they had these so-called Regional Country Programs, so-called programs, that had developed out of a conservation and development fund-raising drive ... in the 80s. They were projects and then they became so big that IUCN, which only had about 12 people in headquarters that would service the Commissions and service the meeting every few years ... They brought it inside of the tent and it was bigger than headquarters, and you know it was all optics, there was something in

Pakistan and there was something in Kenya and it just didn't . . . there was no game plan, nothing (interview, May 2008).

Regardless of the chaotic mode of expansion, this increased global presence helped conservation organizations acquire legitimacy through their participation in the creation and implementation of international legal instruments (Flitner 1999; Jamison 1996; MacDonald 2005; Taylor and Buttel 1992). This led to the increasing use of NGCOs as vehicles to channel development funds (cf Maragia 2002), which subsequently spurred greater international growth (Reimann 2006). The combination of an international presence and institutional legitimacy reciprocally enhanced the capacity of transnational NGCOs to define global environmental problems and their solutions, and to influence national politics and decision-making. NGCOs, however, did not have the fiscal resources typical of states and needed to develop not only dependable sources of funding but also cross-sector legitimacy. Both of these needs came to be satisfied through a continued embrace of the commodification of nature under neoliberalism that had been established in the World Conservation Strategy. Organizations sought to develop modes through which biodiversity could pay for its own salvation by extending the mechanisms through which nature could be "conceived in the image of capital" into new spaces, and used this representation as the basis for the rational management of "nature as capital" (O'Conner 1994:131; Coronil 2000). Under the structuring influence of an "external" environment increasingly governed by the global institutionalization of neoliberalism, organizations that had sought to extend their spatial reach readily adjusted their operating practice and organizational structure to better align with this shifting institutional context. In need of the funds that were increasingly channeled through this institutional structure, conservation organizations not only pursued projects that readily sought to convert the use value of nature to exchange value in any number of small communities around the world, but they openly made conservation an instrument for the accumulation of capital and a vehicle through which capital interests could gain access to sites of "nature as capital". This typically occurred with the full support of relevant governments as they derived a share of revenue from what came to be known as "market-based" incentive projects. With GEF and UNDP support, these projects quickly gained ground in the 1990s as trophy hunting, bioprospecting, and ecotourism became the manifestation of a commoditized nature reoriented to serve elite and corporate interests but that would, under the rhetoric of "community-based conservation", also provide a "profit" for local communities (Hayden 2003; McAfee 1999; MacDonald 2004a, 2004b).[14]

Just as significant, this access to the potential for capital accumulation also produced a context in which "partnerships" become a vehicle through which NGCOs and private sector actors can pursue their diverse goals (Gulbrandsen and Holland 2001). With the retrenchment of state agencies, and the increasing gatekeeping role played by NGCOs, partnerships are increasingly viewed as the primary mechanism through which a negotiated form of biodiversity conservation (between various private and public actors) might be forged. Indeed, NGCOs are increasingly representing themselves as locales in which the historical opposites of private interest and environmental well-being— of profit incentive and environmental good—might be reconciled (eg Conservation International 2005; IUCN 2006). And it is this "reconciliation" that conservation organizations now represent as the "leading edge" of conservation practice. But these exaggerated claims to be on the leading edge mask the degree to which their practice is structured not only by the alignment of sustainable development and ecological modernization, but in the way this alignment was both born of and gave shape to a new institutional context of conservation in the late 1980s and through the 1990s.

Negotiating the Access of Capital: the Development of a "Global" Institutional Context for Biodiversity Conservation

A defining moment in the formalization of the new institutional context for biodiversity conservation was the development of new mechanisms of environmental governance introduced at the United Nations Conference on Environment and Development (UNCED), popularly known as the Earth Summit, held in Rio de Janeiro in 1992. The key institutional developments, relevant to NGCOs, emerging from this meeting were the CBD and the consolidation of the GEF. The development of these two institutions, however, and their implications for the work of conservation organizations was shaped by a much larger ideological struggle over the form that international environmental governance was to take. The contours of this struggle can be seen in events leading up to UNCED that positioned certain actors as central to the negotiations and others as more peripheral. Much of this struggle centered on the threat that emerging mechanisms of environmental governance posed to the private sector, particularly transnational corporations. Despite its status as an action plan based on the World Commission on Environment and Development (WCED), UNCED came to be viewed by business as a threat to the accrued benefits of neoliberalism and was, to some extent, seen as having the potential to encourage the enactment of strict regulatory control on the activities of the private sector, particularly the environmental degradation and social inequity associated with the operations of many transnational firms (Levy and Newell 2005; Rowe 2005).

In response to this perceived threat, business used its prior experience with UN activities in this area to prepare the ground for a central role in the UNCED process; particularly attempts by the UN Economic and Social Council established a Commission on Transnational Corporations (UNCTC) to monitor and provide reports on the activities of transnational corporations (TNCs) and to develop a comprehensive and legally binding UN Code of Conduct on TNCs that would enhance the capacity of developing countries to deal with serious social and environmental externalities that accompanied the tools of global capital. While these actions were curtailed by a coalition of governments, the Business and Industry Advisory Committee of the OECD and the International Chamber of Commerce (ICC), which worked through the OECD to develop an alternative, voluntary set of guidelines, this early attempt to impose environmental regulation on transnational corporations signaled the capacity of emerging international institutions to intervene and act in ways that threatened to make the accumulation of capital more complex. In response, business recognized that it needed to be much more proactive and better organized at the international level and began to organize in a coordinated way to shape the development of international environmental institutions (Bruno and Karliner 2002; Rowe 2005). This included the formation of the International Environmental Bureau within the ICC and the subsequent development of a Business Charter for Sustainable Development (Rose and Jackson 1992).

This proactive stance on the part of business meshed with a new view of global politics within the UN, manifest in the Brundtland Commission, which promoted dialogue among the world's governments and major non-governmental actors as the primary step in addressing global environmental problems: an approach that dominated planning for UNCED and screened the dialogue partners that would be involved in the UNCED process. This process based participation on the technocratic assumption that governments best expressed public interest while all other interests were, by definition, private and had an equal right to be heard by world leaders. In advance of the conference these private interests were asked to organize themselves into specific UNCED constituencies.[15] However, these constituencies were clearly not viewed as equal and the International Finance Corporation and the UNCED Secretariat promoted some private sector organizations as privileged working partners of the UNCED process (Karliner 1999). Among these was the business sector in the form of the ICC that, with UNEP support, had successfully prepared itself by convening a World Industry Conferences on Environmental Management in 1984 and 1991 (Ford 2005). In 1990 Maurice Strong, the Secretary General of UNCED, and someone with a long history in the oil and energy business, appointed Swiss businessman Stephan Schmidheiny to be his chief advisor on business and industry and to lead business participation at the UNCED. Schmidheiny quickly convened the first Business Council for Sustainable Development, composed of CEOs of major global firms, to serve as an advisory group and structure the role of business in the negotiation of preparatory conventions leading up to UNCED. The Business Council for Sustainable Development developed a business perspective on environment and development challenges and designed a business vision of sustainable development that was thoroughly grounded in ecological modernization theory.[16] As it became clear that it was no longer going to be possible to ignore environmental degradation or the environmentalism it had spawned, historical strategies of outright resistance to change gave way to strategies that sought to manage change. This move effectively sidelined the more critical stance of UNCTC, which had been working on relevant issues for over 15 years and had been slated to submit a series of recommendations to UNCED which would have imposed tough global standards on TNC activities. However, its submissions were never accepted or circulated to delegates. Rather, at the request of Strong, official "recommendations addressing transnational corporations . . . that governments might use in drafting Agenda 21" were provided by the Business Council for Sustainable Development (Bruno and Karliner 2002:36). Not surprisingly, these recommendations maneuvered to head off measures that would impose too heavy a cost on corporate activities and effectively blocked discussion of regulations as a mode of addressing the environmental impact of TNCs. Together with the ICC, and relevant national governments, the Business Council for Sustainable Development worked to ensure that Agenda 21 promoted voluntary self-regulation over other mechanisms to control the activities of TNCs (Karliner 1999; Rowe 2005).[17] This outcome is what has led many critics to claim that UNCED was in fact planned to minimize change to the status quo and to evade the central social and environmental issues posed by continued consumption associated with conventional economic expansion (Finger 2005; Hildyard 1993; Sachs 1993). This is highlighted by the text of Agenda 21 that, echoing the Brundtland Commission, advanced a view of environmental and social problems as primarily the result of insufficient capital, inadequate technology, and a lack of management expertise. Accordingly, the anticipated solutions were new modes of capital generation, technology transfer from the North to the South; and the transfer of managerial logics and expertise.

This outcome is not particularly surprising given how the UNCED Secretariat effectively positioned the best-organized and financially powerful independent sectors as privileged working partners. Notably, these included not just business and industry but also establishment-oriented NGOs, in particular IUCN, WWF and the World Resources Institute, and UN agencies like UNEP. These alliances, however, were not at all new. UNEP and IUCN, for example, had longstanding relations prior to UNCED and had cooperated in the production of the World Conservation Strategy. UNEP also had links with business through a variety of programs meant to integrate concepts of sustainable development into business practice and had worked with the ICC to produce the World Industry Conferences on Environmental Management (Trisoglio and Kate 1991). These linkages are important primarily in relation to the framework conventions that were the primary product of UNCED. While the popular attention given to UNCED was seen as a threat by business, the real threat lay in the potential of its framework conventions, particularly the CBD, to impose regulatory limits on access to "nature as capital".

Institutional Enclosure and the Containment of the CBD

In the years since UNCED, the business lobby has focused on ways to restrict the potential effects of the framework conventions on access to capital. This work was to some extent accomplished in the early drafting of the CBD which was, in its early versions, built on the platform of the World Conservation Strategy, a document already invested in an ecological modernization variant of sustainable development and used by the WCED to advocate for an international environmental convention that would designate species and genetic variability as common global heritage. This call was picked up by IUCN, which drafted legal articles that were subsequently submitted to a UNEP Ad Hoc Working Group of Experts on Biological Diversity and became the basis of draft articles for the CBD. Notably, whereas IUCN's articles dealt exclusively with biodiversity conservation and innovative mechanisms for its financing, the UNEP brief to negotiators set the tone of the final Convention text and sought to reconcile conservation with economic and technical progress by addressing:

> ...the economic dimension, including, *inter alia*, the question of adequate machinery for financial transfers from those who benefit from the exploitation of biological diversity, including through the use of genetic resources in biotechnology development, to the owners and managers of biological resources, and appropriate measures to facilitate the transfer of technical means of utilizing biological diversity for human benefit, will need to be properly considered in the negotiations of any future legal instrument for the conservation of biological diversity...(UNEP 1989)

This brief deftly recognized the crux of the debate that would emerge around the convention. While the final text of the convention—developed through a UN Intergovernmental Negotiating Committee that met five times prior to UNCED and represented most UN state members—listed the major objectives of the CBD as the conservation of biological diversity, the sustainable use of its components and the fair and equitable sharing of the benefits arising out of the use of genetic resources (Article 1), it was this last issue that has continued to be a thorn in the side of CBD negotiations. In the years since the CBD was opened, negotiations have continued to deal with the problem of "access and benefit sharing" and have failed to develop a legally binding international mechanism. The reasons for this revolve around the interests of states in protecting sovereign control over "natural resources", the interests of business in accessing genetic resources regardless of where they are located, and a realization on the part of poorer states that they could use the convention as an instrument to gain access to both technology and the bio-technology products developed through the alteration of their genetic resources (Arts 2006; Guruswamy 1999; Shime and Kohona 1992). Ironically, despite their role in the process, the CBD represents an outmaneuvering of conservation NGOs by national governments and the private sector. The loosely defined concept of sustainable development that NGCOs had used to motivate governments to deal seriously with the environmental consequences of state policy and practice was now being used to legitimize an agenda that prioritized economic growth over environmental protection and promised a new round of enclosure resulting from the imposition of new management regimes, capital flows and technology transfers.

The critiques of the CBD are well rehearsed (Brand and Görg 2008; McAfee 1999; Swanson 1999). Primary among these is that, first, it codifies a dominant perspective of nature as capital through its emphasis on sustainable use initiatives that, when translated into practice, means the use of *in situ* biodiversity to realize profit through the conversion of use value to exchange value; second, it positions biodiversity as genetic material available for exchange in a global market; and third, explicitly recognizes that states have a sovereign right to determine access to genetic resources in their territories and to allocate the benefits from the use of those resources.[18] These critiques carry over to the GEF which UNCED mandated as the financial mechanism to aid developing countries in achieving their commitments under the CBD. Donor countries, primarily the G7, fund the GEF largely as a matter of political will and replenish the fund in 4-year cycles. But these replenishments provide the opportunity for conditionality as donors articulate demands on ways in which GEF should modify structural processes, programming or funding priorities before they release new money. For example, during negotiations for the fourth replenishment of the GEF Trust Fund various donor governments led by the USA made future contributions contingent on the adoption of a performance-based resource allocation framework. They claimed that this would not only increase efficiency and equity on GEF project funding but "also guarantee project success by ensuring that GEF resources were channeled into properly developed 'enabling environments'" within recipient countries (Ervine 2007:132). Notably, "enabling environment" referred to the presence of a free market policy mix that reflected neoliberal economic practice including trade liberalization and privatization, and the presence of effective institutions to protect private property rights and promote an institutional environment conducive to business and investment.[19]

The CBD, like all international agreements, is more than simply a document; it is an institution that calls into being an active political space—an arena in which rights and interests may be negotiated and new social relations configured around those negotiations (see Strathern 2000). This arena can lead to creative opportunities for new, and previously excluded, groups to claim authority, but it also creates a context in which privileged positions and perspectives can be consolidated and codified in ways that structure policy and practice. The political space of the CBD, for example, has multiple locales. The most obvious are the biannual meetings of the Conference of the Parties (COP), convened by the Secretariat of the CBD, and the interim meetings of a variety of committees and ad hoc working groups that are open not only to CBD signatories but to a variety of civil society and private sector actors. The COP, however, is much more than simply a meeting of the parties. It is more apt to call it a stage—a space in which the range of interests that constitute a major element of environmental politics today perform and communicate their messages. This stage includes not only the continuing negotiations over the text of the convention and its programme of work, but the opportunity for a diversity of actors to make short statements advocating their positions in the presence of member states and to have these statements entered into record. More importantly, it provides a site where "stakeholders" can lobby member states and accommodates so-called "side events". These are sessions, much like those at the meetings of academic societies, which demonstrate projects or other experiences relating to implementation of the CBD and function as detailed lobby devices that seek to attract member state delegates and advocate positions in ways that might influence member state positions.[20]

Given the competition for attention at the COP meetings, programming of side events is important in influencing member state delegates, and to a remarkable degree, "the business delegation" has been able to acquire "prime-time" programming for its side events. More significantly, it has managed this with the direct assistance and resources of the CBD Secretariat (CBD Secretariat 2006a). For example, at the most recent COP-9 held in Bonn, Germany, while business-related side events ran throughout the meeting, they were concentrated during the high-level ministerial segment, a satellite meeting not open to regular attendees, in which government ministers meet together to consider some of the key political issues on the agenda of the COP. The ministerial segment is organized and hosted by the host government, which also chooses the issues for discussion. In Bonn, the German government chose to highlight its own business and biodiversity initiative at the expense of other issues of relevance to the convention. This was complemented by direct support from the CBD Secretariat which produced a calendar of business-related events at COP-9 "to help participants better plan their stay in Bonn" (CBD Secretariat 2008a). Despite the vast disparity of resources between the business delegation and other non-state actors legitimately involved in COP-9, and the organizational capacity of the business sector, business was the only delegation that received this level of support from the Secretariat. This explicit support for business is, on the surface, grounded in a decision approved at COP-8 in 2006 that encouraged "private-sector engagement" and took significant steps to incorporate business into Convention processes in ways that are not open to other participants. These included: urging national focal points of the CBD to work with national governments to encourage companies to engage in the development of national biodiversity strategies and action plans; persuading business representatives to participate in the meetings of the COP and other intergovernmental meetings; and, perhaps most significantly, encouraging:

> national focal points, where appropriate, to include private sector representatives on national delegations to meetings of the Subsidiary Body on Scientific, Technical and Technological Advice, the Conference of the Parties, and other intergovernmental meetings, and nominate them to participate in technical expert groups . . . (CBD Secretariat 2006b:260, original emphasis).[21]

The decision also committed resources to developing engagement with business by directing the Executive Secretary of the CBD Secretariat to "compile information on the business case for biodiversity and good biodiversity practice" and to "include the private sector as a target audience for its outreach materials". This directive to "engage" business, on behalf of parties, recognized that "contributions from business and industry" could be secured if work under the convention developed tools, and guidance "on biodiversity-related issues relevant to the private sector"; and "[t]ools for assessing the value of biodiversity and ecosystem services, for their integration into decision-making ... ".[22] The Executive Secretary took this decision as a green light to engage with business in an explicitly proactive way and immediately following COP-8 took the initiative to establish and staff a Business and Biodiversity initiative within the Secretariat, despite not having funding approval from the parties. Notably the head of this initiative, Nicholas Bertrand, came out of the Business and Biodiversity initiative that had earlier been established in IUCN.

This rhetoric of "engagement with business" and the degree to which the CBD is engaged in "courting" private sector actors overshadows the degree to which business has been engaged with the CBD since the UNCED and, in venues external to the realm of international environmental governance, has positioned itself as a central actor in attempts to operationalize an eco-modernist variant of sustainable development. It also obscures the foundational elements of the CBD that favour the privileged involvement of private sector actors. Primary among these was the push, from IUCN, in the initial drafts of the convention to develop new, "innovative" mechanisms for funding biodiversity conservation. This call was an implicit recognition of the constraining affect of neoliberalism on public spending for environmental protection and corresponded with IUCN's own initial ventures into market-based mechanisms to finance conservation (James, Green and Paine 1999; Lake 1997). It was also an explicit recognition that the programme of work developed by the convention and the obligations of state members, particularly those in so-called less-developed regions, would require significant infusions of capital that would not likely be generated through contributions of direct foreign aid to biodiversity conservation. In implicit and not so implicit ways, COP documents have continually recognized the capacity of the private sector to fund the work of the convention. Indeed, corporate participation in the CBD has been legitimized by member states, and the Secretariat, through an eco-modernist rhetoric of environmental management that positions corporate actors as having the will, resources and knowledge to engage in environmental repair or caretaker services to solve the environmental problems that global capitalism has itself created.

The privileged position of business within the CBD process is a stark manifestation of the way in which capital operates at an international level to shape emerging institutions that would regulate access to and use of biodiversity (Newell 2005). It is clear, however, that this influence is not simply a function of direct lobbying on the part of business, but also stems from the structural power of corporations in state economies (Clapp 2005b; Levy and Egan 1998). Given the reliance of the capitalist state's resource strength on the revenue generated through the private accumulation of capital, the state is vested in serving the international interests of its most important corporate sectors. It is not surprising then that national positions seek not only to protect these interests in international conventions like the CBD but work to extend them. In passing COP decisions like VIII/3 referred to above, member states are officially signaling explicit approval for the privileged role of business in the CBD. In fact, this is consistent with policy developments in many member states that have adopted a neoliberal-inspired approach to environmental management and facilitated the direct involvement of business in policy formulation. As a result, some state delegates see the enhanced role of business in CBD negotiations as a legitimate and direct outcome of neoliberal policies and practices implemented at the national level. It is also consistent with a national-level withdrawal from regulatory oversight that encourages flexibility and the ongoing accumulation of capital from new markets built around "ecosystem services" and biodiversity investment opportunities (eg natural cosmetics, ecotourism, etc). Within a climate of expanding financial opportunity revolving around the development of these markets, states are not only eager to protect the rights of nationally based corporations to continue to access genetic resources internationally, but to structure that access in ways that enhance the revenue-generating potential of governments. The degree of state support for this form of engagement is apparent from the involvement of a diversity of state and para-state actors in so-called business and biodiversity initiatives. For example, in the 2 years between COP-8 and COP-9, a number of events were held that reveal the alignment between business, state and NGO actors. Particular among these are a number of international conferences like the Lisbon "High Level Conference on Business and Biodiversity", sponsored by the Portuguese presidency of the EU Council as part of an official priority area on business and biodiversity engagement that brought environment ministers and their staff together with CEOs of major agribusiness and extractive industry corporations, the staff of NGCOs, and the CBD and IUCN Secretariats. Indeed, the organizing committee for the meeting brought together senior representatives from the European Commission, the Portuguese government, the CBD Secretariat, the WBCSD, WWF International and IUCN, along with a number of business and biodiveristy consultants who rotate through these organizations. The themes of these conferences rarely deviate, and centre around the incorporation of biodiversity as an element of corporate social responsibility, tools for assessing business risks and opportunities associated with "ecosystem services", markets for biodiversity goods and services, and the facilitation of partnerships between industry and NGCOs.[23] This type of event has become increasingly significant as a form of lobbying in relation to the emergence of international forms of environmental governance, for it rests upon a restricted and concentrated encounter between select representatives of business and government ministers, and is intentionally scheduled in order to allow for the lobbying of domestic governments before they send delegations off to international environmental negotiations (cf Clapp 2005a). This has long been a key strategy of business, but this kind of meeting marks a significant change in the way that governments are now allied with key business and NGO actors. It might in fact be better read as a visible expression of the way in which issues of biodiversity conservation and its governance configure particular elements of what Sklair (2000) has called a transnational capitalist class.[24] In many ways, the lobbying is complete, blocs have been formed, and events like the Lisbon "High Level" conference[25] reflect strategy and agenda setting moments in a hegemonic process that reflects the mutual capital interests of business, the state and NGCOs[26].

My point in describing the positioning of business within the CBD process is to reveal how the convention structures a political dynamic in which capitalist interests seek to secure continued access to resources by using multiple channels of influence to shape policy. This is not necessarily a new observation. Studies of similar mechanisms of international environmental governance have pointed out that many large companies, industry groups, and researchers fear the development of an international regulatory structure beyond the national scale at which they have historically exercised interest (Levy 2005). Seeing the inevitability of some form of agreement, these actors set out to find ways to structure those agreements and seek compromises that limit regulatory control on the unfettered access to biodiversity that has characterized historic patterns of capital accumulation and protect their autonomy from the threat of more extensive international regulation (Levy and Newell 2002, 2005; Lipschutz and Rowe 2005). However, while the rhetoric that surrounds business engagement with the CBD creates the impression that outreach is required to convince business to become involved in CBD processes and in the issues dealt with by the convention, it is clear from a historical analysis that a transnational capitalist class composed of corporate executives, conservation professionals, politicians and bureaucrats have actively been developing mechanisms that would provide business with a privileged position in the convention process and through which mechanisms of international environmental governance like the CBD can be shaped in ways that accommodate the interests of capital accumulation and seek to control the conditions under which regulation is imposed. The WBCSD, for example, seeks a regulatory framework that would not limit access but would facilitate the development of markets and market mechanisms that offer "new business opportunities and the chance to use ecosystems and their services to tap into previously unrealized assets" (Stigson 2008:4).[27]

This positioning amounts to a form of what we might call institutional enclosure and can be read as a direct response to the threat initially posed by the CBD. In an age of global capitalism, unhindered transnational access to raw (including genetic) materials has been key to the accumulation of wealth. A rising concern with biodiversity protection as a key element of sustainability threatened to compromise access to those raw materials. When the threat of restriction was limited to protected areas, the threat to capital accumulation was not huge. But the CBD promised to go far beyond protected areas as a mechanism for protecting biodiversity, and created a significant degree of uncertainty about both the terms of access and the continued ability to deal directly with fairly malleable national governments outside of the frame of international guidelines or regulations. While the framing of the CBD incorporated an eco-modernist stance on sustainability from the outset, the convention created political space for a variety of actors to promote policy and regulatory interventions that were potentially damaging to the interests of capital. For example, much of the NGO and indigenous and local people's participation in CBD negotiations is grounded in ideological struggle to challenge singular definitions of biodiversity and to establish a regulatory framework for access and benefit sharing that accommodates diverse understandings of material and intellectual property, recognizes multiple forms of sovereignty, and fosters widely democratic involvement in formulation of equitable "access and benefit sharing" agreements. The potential of these groups to formulate and lobby for the implementation of an equitable access and benefit-sharing regime posed a direct threat to existing patterns of capital accumulation, particularly unhindered access to and use of resources. Business was alert to this "risk" from the outset and clearly saw a need to curtail this threat. The key to subordinating those threats posed by the convention was to enclose it ideologically and materially.

The degree of support for the interests of business among the primary institutional actors responsible for organizing the COP meetings of the CBD, and particularly the explicit support of the Secretariat's Executive Secretary, suggests that business has achieved a position from which it can successfully minimize the threat that a concern with biodiversity protection poses to the continued access to resources and accumulation of capital. It has accomplished this in a number of ways including: the translation of its structural power in state economies into national support for an enhanced role in international institutions; and successfully focusing a transnational capitalist class on accumulation strategies as the solution to environmental degradation. This position also facilitates the use of the CBD and the GEF as instruments to extend practices of ideological and material domination that not only define nature as capital and assert the efficiency of market mechanisms as the most appropriate way to address environmental degradation and accordant losses of biodiversity, but that establish the conditions through which these mechanisms will be implemented. Notably, this process of institutional enclosure has precedent in the Parliamentary enclosures that accompanied the eighteenth-century annexation of the commons. A common characteristic of the spread of capitalist property relations has been the process of enclosure. Historically, enclosure amounted to a redefinition of property rights that included both the physical fencing in of lands to enable the exclusion of other potential users and the extinction of common and customary use rights. Wood (2002:108), among others, has noted that while the monarchical state in Europe initially resisted enclosure, "once the landed classes had succeeded in shaping the state to their own changing requirements ... there was no further interference, and a new kind of enclosure movement emerged in the eighteenth century, the so-called "Parliamentary enclosures". This phrase referred to the use of acts of Parliament—in a Parliament composed of landlords and lawyers—to abolish types of property rights that interfered with some landlords' powers of accumulation. We might think of institutional enclosure as akin to parliamentary enclosure—the use of the institutions of a governmental body in ways that served the interests of particular groups over others and was supported by the spread of an ideology of property which asserted the efficiency of privately enclosed land through the ability to realize economies of scale. If we view neoliberalism as a process in which a transnational capitalist class has shaped the state to its own requirements, structural control over new forms of international environmental governance that might challenge those requirements becomes a key component in reducing or eliminating obstacles to capital accumulation.

Biodiversity and the Conservation of Organizations

> You know, if you're going to greenwash for these guys, you gotta ... if you're going to be a whore, you might as well be a high-priced whore (former Earthwatch senior executive, May 2008).

The position of structural control that business has strived to achieve in the CBD process is clearly related to broader trends in the reorganization of relationships between supra-state actors and the private sector. Symbolic of this trend is the UN's Global Compact initiative launched under Kofi Annan in 1999 that makes clear UN support for voluntary self-regulation and open markets, which "offer the only realistic hope of pulling billions of people in developing countries out of abject poverty, while sustaining prosperity in the industrialized world" (Annan 2000), and the role of NGO/private sector partnerships in both defining and pursuing this elusive goal. While this rhetoric sends a signal to UN agencies and associated organizations, it masks the shifting institutional context of these organizations and the asymmetrical power relations that have emerged since UNCED to facilitate ideological and material shifts within NGCOs. To understand the contemporary practice of biodiversity conservation, then, we need to understand this shifting institutional context, the power relations it brings into being, and the structural responses they generate from conservation organizations.

The organizational environment of biodiversity conservation has become increasingly turbulent since UNCED. As mechanisms for international environmental governance have developed, new institutions and actors have become increasingly important components of the operating environment of conservation organizations. This turbulence creates substantial constraints for vulnerable organizations as the authority to set program agendas and funding become centralized and lead to the loss of a degree of autonomy in NGCOs. In the case of biodiversity conservation, the emergence of the CBD and GEF signaled a major shift in the institutional environment through their consolidation of state actors and the role of those actors in agreeing not simply on a centralized programme of work built around the convention but implicitly on ideological perspectives that guide the allocation of funds to carry out that work. In practice, then, the CBD programme of work, agreed upon by states, sets the biodiversity funding priorities and programme areas of the GEF, and diminishes the position of NGCOs in setting agendas upon which those programmes are based. In this altered institutional context, the project activities of NGCOs are structured by the environment in which the organization is embedded (Pfeffer and Salancik 2003). While it is important to recognize significant differences in the organizational histories, ideological bases and primary resource sources of conservation organizations, none of these organizations are self-contained or self-sufficient and all rely on an external environment to support their activities.[28] It is this dependence on an external environment that not only makes the control of organizational behaviour possible but almost inevitable as NGOs need to be appropriately responsive to that environment to assure continued access to the resources they need to survive. Ironically, the growth of conservation organizations facilitated by neoliberalism in the 1970s and 1980s is partially responsible for their subordination as it developed an infrastructure grounded in short-term, project-based support rather than reliable core funding. This condition of scarce capital resources was actually intensified with the emergence of the GEF and its role of coordinating resource allocation in support of the work of the CBD, as it caused a shift of donor resources away from direct core funding support to conservation organizations and into project-based funding in support of program areas aligned with the CBD. This provided the CBD Secretariat and the GEF significant power over the allocation of resources for biodiversity conservation. Just as importantly, the CBD programme of work acts as an explicit policy statement of member states, meaning that state moneys flowing to conservation organizations through other channels are more likely to come with conditionalities on their use that are tied to the CBD programme of work.

Organizational Adjustment to a New Insititutional Environment

In confronting new institutional environments and their implicit demands, there are two broadly contingent adaptive responses: the organization can change to fit environmental requirements, or the organization can attempt to alter the environment so that it fits the organization's capabilities. These, however, are not exclusive responses and in many ways the changing operational context of conservation organizations represents a diversity of responsive actions to the new, and in some ways constraining, institutional context laid by the CBD. One clear response has been to seek out alternative sources of funding that maintains a measure of autonomy for the organization (Brechin 2008). Conservation International's long list of corporate donors (rather than partners) is likely the best example of this. Largely in the guise of purchasing corporate social responsibility, and achieving significant tax benefits, business donations provide a significant portion of Conservation International's operating budget.[29] However, while Conservation International was developing alternative sources of capital, it was also engaged in the Critical Ecosystem Partnership Fund (CEPF), a GEF project, launched in 2000, that provides grants for NGOs and private sector partners to "help protect the Earth's biologically richest regions or hotspots" by creating "working alliances among diverse groups, combining unique capacities and eliminating duplication of efforts; and achieving "tangible results through an expanding network of partners working toward shared conservation goals" (GEF 2008:42).[30] The CEPF is itself a partnership between the GEF, the World Bank, Conservation International the Government of Japan, and the MacArthur Foundation. This direct engagement of Conservation International with GEF, despite its success in developing alternative funding sources, is indicative of an attempt to match an institutional context with the organizations' capabilities, but also to use the institutional environment to advance the visibility and legibility of the organization. Given the easy recognition of the term "hotspots" as a legible Conservation International trademark, it is clear that the sole conservation organization in the CEPF venture is using the partnership fund as a vehicle to extend its influence and presence across the institutional environment.[31]

This move signifies that the organizational constraints around a new institutional environment involve not only a concern with continuing access, but also the legitimacy needed to continue to secure increasingly important project-based funding. It also indicates that a significant part of gaining this legitimacy includes the willingness and capacity to develop "working alliances". In an institutional environment, shaped by neoliberalism, that increasingly accommodates and privileges the interests of business in pursuing an eco-modernist version of sustainable development, access to the resources allocated through that institutional context relies on an organization visibly and legibly aligning its activities, capacities and objectives with the ideological and material interests of the dominant actors within that institutional context (Maragia 2002). Given these structural parameters, it is not surprising that many of the so-called integrated conservation and development programmes focused on the sustainable use of biodiversity, developed during the 1990s, were oriented toward addressing the funding initiatives of GEF/UNDP and not in an empirically informed understanding of grounded problems in their implementation sites (MacDonald 2005).

This perspective on the shift in an external environment and the subordinate position of conservation NGOs within that environment help to explain why "partnerships" between conservation organizations and corporate actors and the broader involvement of NGCOs in the promotion of business and biodiversity initiatives are growing so quickly, despite a long history of warranted distance. Despite the assertions of conservation organizations that the engagement is based on their mandate to influence society in the conservation of biodiversity, signals are clearly being sent from the institutional environment, largely through the structuring of programs and the availability of funding, that "engagement" with business is a priority. Managers within conservation organizations are exposed to these signals in a variety of ways and contribute to change in the organizational environment by simultaneously focusing work on these sectors and dedicating resources to develop the institutional capacity to carry out this work. This includes hiring staff with business backgrounds, developing specific units within the organization dedicated to both the development of partnerships with corporate actors and market-based conservation mechanisms that can act as the basis of project-based partnerships; and silencing the voices of those within the organization that might be resistant to these initiatives.

If the motivation for conservation organizations to engage with business resides in organizational responses to resource scarcity and the need to demonstrate ideological alignment with a new institutional context, the motivation for business to engage with conservation organizations lies more squarely in the need to control their own external environment in order to retain access and use rights to strategic resources, and to accommodate new institutional demands as cheaply as possible (Maxwell 1997). While the direct involvement of business in biodiversity conservation is often described as a form of greenwashing, this simplistic representation overlooks the extent to which conservation organizations have historically acquired a degree of authority and a capacity to influence governments in areas of environmental management, and been heavily critical of the role of industrial development in biodiversity loss and environmental degradation. With the global expansion of conservation during the 1980s, organizations also effectively positioned themselves as gatekeepers, particularly in low-income countries through their capacity to partner with governments, influence legislation, and help to generate project-based financing. This potential of conservation organizations to structure state policy and for certain units of conservation organizations to host oppositional voices have made it important to de-center conservation organizations as a site of effective action for critics, to dislocate them from the sphere of civil society, and to engage directly in reorienting their management practices and organizational structure (see McDonald in press).

From a business perspective, managers seek to maximize control and predictability over their operations. One way to achieve this is to incorporate all relevant actors that affect those operations within an interconnected system. As concerns with biodiversity protection have threatened the security of access to resources, or the use of genetic materials, it has become important to bring conservation organizations into that system. This provides business with a better understanding of its own external environment and with ways to intervene in that environment in an effort to protect its own interests. Bringing actors into this interconnected system is not necessarily easy, but it is a process in which business has substantial experience. This is evident in a description of the way in which an oil industry lobby group—the International Petroleum Industry Environmental Conservation Association (IPIECA)[32]—initially engaged IUCN:

> [over a dinner speech] we had this little chat about looking for common ground and everything and they really loved it and so afterwards, ah, I did a few jokes and everything, and afterwards they um came up and said "well, we have this Special Issues Committee. We've got people working on this and that and oil spills and so on, but we've got this committee that looks at new and arising issues and we don't know what, we look at different things" . . . and biodiversity wasn't on the

agenda. And I said "well you ought to look at biodiversity." And they said "well . . . could we have our next meeting at your office? Could we just come down and meet you guys. So we'll have our meeting and we'll rent a room or whatever." I said, "nah, I'll give you a room. That's no problem." "And then maybe a few of your people could introduce themselves, and we could just sit and chat. We've never been to a place like IUCN." So I came back and said "well, the oil industry's going to come and have a meeting in our building." "WHAT???" HUH???" And so I said "yeah". So I got them to come and they all came. And this was Exxon, Mobil, Shell, BP. All their Heads of Health, Safety and Environment. Senior guys in the companies, you know. Alright, not the Chief Executives, but you know, you know, senior manager guys, global guys, from global headquarters. And one of the guys from one of the Australian oil companies comes and he, he kicks off . . . And the guy from Australia puts up a little PowerPoint and he shows where all the Protected Areas are according to the WCMC up in Cambridge, and he's got the World Heritage sites and the parks and everything and then he overlays the PowerPoint where all the known oil reserves are for future exploration and it's [sound of a smack] you know . . . and he says, "you know, where you guys are is where we want to be so we need to talk. We need to talk seriously" (former IUCN senior manager, May 2008).

This narrative is useful, in part, because it highlights how the engagement was initiated by business and not by the conservation organization, emphasizing that, despite the contemporary representations of outreach and exercising influence expressed by conservation organizations, business has historically sought to bring conservation organizations within their sphere of influence, and not necessarily vice versa. It also indicates that this engagement occurs when conservation exists as an obstacle blocking access to the resources that facilitate capital accumulation.

"Engagement", Influence and Emerging Interdependencies
"Engagement", however it is manifest, also helps to legitimate the presence and voice of business at international meetings like the CBD COP, where actors can profile their relationships as a way of gaining influence in the development of international regulatory mechanisms. Business, for example, readily displays representations of private sector/NGCO partnerships during COP meetings. And business delegates regularly refer to partnerships with conservation organizations in their official statements in the presence of delegates, as a way of demonstrating that they are, with jointly developed tools and the information that flows from partnerships, capable of addressing

environmental problems that their sector may have caused in the past. The intent here is to use the partnership as a vehicle through which to promote voluntary agreements as an effective means of self-correction and to minimize the potential for restrictive regulation. However, partnerships not only enhance a corporation's public profile, but they can provide a cost savings to firms by effectively using public or donated moneys to fulfill legal or ethical obligations. While publicly released partnership agreements rarely disclose financial information on costs, terms of the partnerships typically position the conservation organization as a consultant, rather than a partner; providing a service to the industry group from which the industry group benefits but at a cost to the conservation organization or the broader public that provides core funding to that organization through government donations.[33] For example, a recently signed agreement between IUCN and Holcim, a global supplier of cement, aggregates and concrete agreement, among other services, commits to the generation of "joint projects through a matching fund mechanism potentially leveraged with additional third party funds, recognizing that it will require patience and sustained effort over an extended period given the time-consuming approval processes of donors, especially in the public sector".[34] Not only are public funds involved in meeting IUCN's obligations under this agreement, but IUCN commits the labour of its employees in an effort to raise public funds, that will, in essence, compensate a resource-rich private company to (perhaps) protect biodiversity and enhance its public reputation in the process. In many ways, this sort of framework agreement with the private sector is indicative of substantive ideological change within conservation organizations. It is grounded, at least superficially, in the same rhetoric of universal values that underpin the Global Compact, but is facilitated by ideological shifts within conservation organizations that have occurred over the past 20 years (Paine 2000). In substantial ways, the engagement with capital interests, which is justified by a rhetoric of influencing their behaviour in the interest of biodiversity conservation, has in fact facilitated substantial policy and program shifts within conservation organizations in ways that situate them as vehicles for the further accumulation of capital. It is not at all clear that these shifts achieve demonstrable gains for the conservation of biological diversity (Frynas 2005).

This ideological shift has occurred gradually in many organizations. Indeed, for most organizations, even those that tout their entrepreneurial origins, it does not appear to have happened in earnest until around 2000. In many ways conservation organizations are still feeling their way and seeking out guidelines on how best to engage with the private sector (Heap 2005). But the changes inside organizations that facilitate these shifts are apparent. Not only have specific units been developed in all major conservation organizations to manage private sector relations, but new, although unequal, interdependencies have arisen between the private sector and conservation organizations. In some cases these have taken the form of secondments of employees, but more significantly, private sector actors have been invited to adopt leadership positions within conservation organizations. In most cases this amounts to advisory panels, such as IUCN's "Leaders for Nature", or WWF's Corporate Club, but in other cases it extends to more substantive positions such as Conservation International's Chairman's Council and TNC's Board of Directors.[35] This can occasionally lead to public relations problems when the increasingly close alignment between NGCO and private sector interests becomes too public and the potential for a conflict of interest too great. Perhaps the highest profile case of late has been the potential threat to IUCN's credibility when Valli Moosa, President of IUCN and former Minister of Constitutional Development and Minister of Environment in South Africa's government accepted the position as Chairperson of Eskom, South Africa's notorious state energy company while retaining his position as President of IUCN. The fact that Moosa continued to hold these positions simultaneously is a mark of how deep the interconnection between the private sector and conservation organizations like IUCN has become.

It is also a mark of the degree to which a near universal conflation of nature and capital has established itself as a dominant view within NGCOs. Other similar signs include the organizational effort with which they have engaged in the development of new forms of co-ordinated production and investment that promote programs to create markets for ecosystem services, venture capital programs designed to facilitate the growth of small biodiversity businesses, and offset programs to compensate for concentrated biodiversity impacts, among other initiatives. These shifts stem from an acceptance of eco-modernist equations of sustainable development with continued expansion in economic productivity, a recognition of the threat that environmental degradation poses to conventional modes of production, and a belief that imagining into being new markets and opportunities for accumulation are the basis for addressing these problems. As it says on the WWF International website, "The panda means business".[36] But the panda didn't always mean business. At a point in the not too distant past, the easily recognizable WWF symbol meant the development of effective public engagement in the protection of wildlife habitat. The shift in the meaning of the symbol, indeed its conversion from symbol to legally protected brand, indicates a shift in the recognition of what and who are currently in a position to best contribute the support necessary for the organization to continue its activities.[37] Increasingly, these are the individuals who anchor the business and biodiversity and private sector partnership programs that help to secure resources and confer legitimacy on the organization in relation to dominant ideological perspectives and associated resource opportunities in the institutional environment of biodiversity conservation.

Conclusion: New Contexts, New Friends and New Goals

The development of an institutional context, accordant with the demands of neoliberal capitalism, over the past 20 years has situated biodiversity conservation organizations in radically new ways. The speed with which conservation organizations have adopted both the accordant rhetoric and practice of "private sector engagement" points to the way in which "organizations are not so much concrete social entities as a process of organizing support sufficient to continue existence" (Pfeffer and Salancik 2003:24). Organizations, as coalitions, are dynamic. They alter their purposes and their spheres of authority to accommodate the interests of new dominant actors and shed parts of their structures that are overtly resistant to this change. It is this organizational characteristic of modernist conservation that subordinates it, at different points in time and space, to larger societal political projects such as imperialism, nationalism, and over the last two decades, neoliberal capitalism. The tendency to think of organizations as engaged in static, coherent, activities like "biodiversity conservation", for example, is mistaken. Once we recognize that an ideological and material project like biodiversity conservation is inseparable from larger political projects that define the constitution, and subsequent use of "biodiversity" and modes of "conservation", we can take seriously the observation that capitalist development is integrally an environmental project that operates through the restructuring of socionatural relations (Prudham 2004). But what is key here is that this restructuring is an organized practice, by which I mean simply that it occurs through organizations that are increasingly shaped by the eco-modernist imperatives of capitalist development. This makes it markedly important that we consider the role played by organizational concerns in restructuring nature–society relations.

I have described how the neoliberal restructuring of biodiversity conservation is an iterative process that is not separate from the interests of the organizations that increasingly claim responsibility for the global management and protection of biodiversity. Indeed, many of these organizations are responsible for mobilizing the conditions that led to their own subordination within an institutional environment that was not simply amendable to the interests of capital accumulation but was shaped from the outset by those interests. Much of this is related to the growth of conservation organizations, specifically during the

1980s, a growth that was itself facilitated by the neoliberal practices of multilateral financial institutions. That growth became a way to justify organized conservation and its activities, which came to be viewed as worthwhile and important, partially as a result of that growth. But it also placed conservation organizations in a vulnerable position in terms of maintaining that commitment, and increasingly tied them to an external environment dominated by neoliberal policies and practices. Removing this uncertainty was in part the motivation for pushing the development of an international convention to deal with problems of biodiversity loss and to coordinate work to reduce the impact of economic growth on biodiversity. In many ways, conservation organizations saw this as a route to extend their influence and to develop the organization as a site of greater authority, power and prestige. However, they were outmaneuvered by business as the loose rhetoric of sustainable development that they developed to legitimate that new institutional context opened up a new field of ideological struggle and created space for a particularly conservative variant of environmentalism—ecological modernization—to achieve a dominant position and to subsequently reconfigure discourses and practices of environmental management and protection according to a logic of capitalist market relations. That these discourses and practices have subsequently so successfully taken root in biodiversity conservation organizations is not simply a function of organizational dependence on dominant actors for resources, but reflects the outcome of a political class struggle that has successfully incorporated senior actors in conservation organizations within the interconnected system of a transnational capitalist class.[38] This is profoundly important because, for this class, ecological modernization has a rational and material core that does address real problems—the problems that historical modes of biodiversity conservation posed to the continued accumulation of capital. And it is through the dominant position of this class that neoliberalism and ecological modernization have not only acquired a purchase on conservation practice around the world, but have become written into the materiality of biodiversity. The organization of biodiversity conservation has been successfully restructured so that it serves capitalist expansion, just as it once served imperialist and nationalist expansion.

Endnotes

[1] IUCN, which celebrated its 60th anniversary in 2008, is a "union" composed of membership organizations; a secretariat of permanent staff and six volunteer-based Commissions. The membership has historically been composed of non-governmental organizations and State members. Private sector membership has been prohibited by statute, although there have been attempts from within the Secretariat and beyond to allow private sector organizations to join. Initially, IUCN was composed of volunteer

experts who compiled data on conditions affecting species dynamics, but through the 1980s it expanded its number of permanent staff, established regional programme offices around the world and became directly involved in project implementation (see MacDonald 2003 on the structure of IUCN). This has caused conflict with member organizations who have come to see the Secretariat as a direct competitor for the project funds that help to maintain smaller conservation organizations.

[2] http://www.foei.org/en/publications/pdfs/iucn-withdrawal (last accessed 27 February 2009). In October 2007, IUCN signed an agreement with Royal Dutch Shell with the goal of influencing Shell's biodiversity conservation performance. Similar partnerships were signed with Holcim, the leading global supplier of cement, and Total, the French oil giant. In the pipeline is an agreement with Rio Tinto, the world's largest coal extractor.

[3] http://news.bbc.co.uk/2/hi/science/nature/7654721.stm (last accessed 27 February 2009).

[4] Notably, 15 governments did vote for termination of the agreement.

[5] The IUCN Director General (head of the Secretariat and tasked with carrying out the will of the membership), for example, could be seen enthusiastically pumping her fist and mouthing the word "Yes!" when the motion was defeated.

[6] In absolute numbers, state members comprise a significant minority of IUCN membership. However, some states provide substantial framework funding to the organization and others are effective institutional partners in many parts of the world, facilitating the project work of IUCN offices. The IUCN Secretariat would have difficulty continuing to function without the support of state members, and states would have little incentive to belong to IUCN if they did not have the ability to shape organizational policy. Given this mutual reliance it is unlikely that the bicameral structure of IUCN is likely to change any time soon. At the same time, there is growing discontent among NGO members over the capacity of states to block the popular will of the membership.

[7] The observations in this chapter are grounded in fieldwork that has involved participation involving assessment of IUCN field projects, work within an IUCN Commission over the last decade, and participation in both World Conservation Congresses and CBD Conference of the Parties meetings.

[8] By modernist conservation, I mean the policies, programs and projects of large international conservation agencies, and national governments. This is not to assign any priority to this work but to distinguish it from the many small-scale conservationist practices that fall outside of this domain.

[9] Understanding the relation between conservation policy and practice and larger social and political projects is one important reason for undertaking research that tracks structural change within conservation organizations.

[10] The normalization of the term "business" in academic literature as a uniform constituency is problematic as it glosses over substantial diversity among actors that would fall within the category. When I use the term in this chapter it is in part because that is the term that has come to prominence with the biodiversity conservation community. But it is meant to signify something more; specifically lobby groups that function as "collective individuals" or those "entrusted with the activities of organizing the general system of relationships external to business itself" (Gramsci 1971:6). The International Chamber of Commerce (ICC 2008), for example, in its briefings to the Convention on Biological Diversity asserts positions on behalf of "the business delegation". Similarly, the ICC and the World Business Council on Sustainable Development (WBCSD) created the Business Action for Sustainable Development (BASD) prior to the World Summit on Sustainable Development in 2002 to, in their own words, "ensure business rallies its collective forces for the UN World Summit on Sustainable Development"; Bruno and Karliner 2003:17). The first head of the BASD was Sir Mark Moody-Stuart, a former Shell CEO.

[11] Critical analysis of many of these programs has effectively pointed out the flawed assumptions upon which they were based, including an assumed universal economic rationalism, and poorly informed constructions of the "community" concept. Notably few of these programs conducted detailed ethnographic research in their project areas to test the assumptions upon which programming was based (Brosius and Tsing 1998; Li 2007; MacDonald 2005).

[12] I use the word "rightly" here in its moral sense for the proponents of ecological modernization are making the moral claim that the market *ought* to be the primary mode of engaging in environmental protection.

[13] I use articulation in both senses of the word here. Not only are an increasing number of the publications and web pages of conservation organizations related to ecological modernization programming, but the offices of conservation organizations are sites where interactions between relevant actors occur (for example, IUCN and Shell employees are exchanged through a secondment program). See, for example, IUCN's Business and Biodiversity Programme; Conservation International's Center for Environmental Leadership in Business; WWF's International's Corporate Club; the Secretariat for the Convention on Biological Diversity's Business and Biodiversity Initiative.

[14] Not the least of the problems with these programs is their reductionist understanding of the constitution of community (Brosius and Tsing 1998).

[15] These constituencies included non-governmental organizations, inter-governmental organizations, Indigenous and local peoples, business, and education, among others.

[16] The Business Council for Sustainable Development was renamed the World Business Council on Sustainable Development and now claims over 200 CEOs as members. It was heavily involved in the "Business and Biodiversity Journey" during the 2008 World Conservation Congress and is engaged in a mutual secondment program with the IUCN Secretariat.

[17] According to Peter Hansen, former director of UNCTC, in the preparatory meetings for UNCED, "The U.S. and Japan . . . made it quite clear that they were not going to tolerate any rules or norms on the behaviour of the TNCs, and that any attempts to win such rules would have real political costs in other areas of the negotiations" (Bruno and Karliner 2002:26).

[18] While article 8j of the CBD sets out that the knowledge and practices of indigenous and local peoples are to be respected and protected, these are still subordinated to the sovereignty of the nation state.

[19] These reforms generated significant resistance from the G77 and China who saw them as the basis for excluding countries based on macro-political assessments (country scores on GEF Benefits and Performance Indexes) carried out by the World Bank. The Benefits Index claims to measure a country's potential to "generate global environmental benefits" while the Performance Index measures "capacity to successfully implement GEF programs and projects based on its current and past performance" (GEF 2005).

[20] After a side event, it is common to hear organizers ask "how many delegates attended and where were they from?"

[21] Notably several delegations, including the Netherlands and Australia, have regularly included representatives from business, "thus facilitating, *inter alia*, regular communication with the business observers to these meetings" (CBD Secretariat 2008b:11).

[22] These recommendations, however, did not come out of the blue but were based on ongoing initiatives by Brazil's Ministry of the Environment, the UK Department for Environment, Food and Rural Affairs, IUCN, the Brazilian Business Council for Sustainable Development, Insight Investment and the Executive Secretary of the CBD,

to "develop ideas, that could best be pursued through the Convention or in support of its objectives, for engaging business in biodiversity issues" (CBD Secretariat 2006b:2).

[23] Notably a framework agreement signed by IUCN and Holcim, Europe's largest cement and aggregates firm was being hailed in meeting documents as a "successful partnership", despite the fact that the agreement was signed in February 2007, just 9 months before the Lisbon meeting, before any active project work had begun, and before any independent review of the partnership.

[24] While the social relations that underpin this configuration are often hidden from view, as in the "high level ministerial" segments of the CBDCoP, or the annual economic summit at Davos—which has, for a number of years, attracted the heads of key environmental organizations—other meetings have become important as field sites that render the existence of a transnational class visible, sometimes in spectacular ways. For example, the opening plenary sessions of the 2008 World Conservation Congress were steeped in references to oligarchical relations of power that bridged a feudal past and a corporate future. The presence of European royalty, representatives of the state, and business was announced to the assembled audience who were made to wait while a procession of VIPs, distinguished by red neck straps entered the hall. Keynote speakers like Ted Turner and Mohammed Yunus served as proxy celebrities but had nothing to say about biodiversity or its conservation. However, that was not their function—they were there in an attempt to draw attention to IUCN and, by virtue of their celebrity, to confer authority on the WCC, much as Royalty represented a sort of consecration. Perhaps most importantly, they were there to facilitate the personal and organizational objectives of the senior IUCN leadership to secure access to the potentially beneficial resources of a more extensive organizational network. Even at the WCC, however, key private sector, state and organizational actors gathered in much more intimate, exclusive, and class-structured environs away from the Congress venue that included the yacht of a Saudi prince anchored in Barcelona harbor, among other sites spread around the city.

[25] Presumably a term designed to increase the attendance of those who prefer to think of themselves as "high-level" actors.

[26] For example, the closing roundtable of this meeting included the Director General of IUCN, the Executive Secretary of the CBD Secretariat, a former President of the European Investment bank, the president of Portugal's largest private employer, the Minister's of Environment of present and future EU Presidencies, and the Minister of Environment for Germany, who would hold the presidency of the CBD COP in Bonn in 2008. The topic was "the next steps for business and biodiversity in Europe".

[27] This statement by the President of the WBCSD reveals a particular Lockean view of "natural capital", suggesting that "making the business case for biodiversity" is about exposing its capacity to make a profit. Locke's nature was "waste" for one reason only—its failure to realize a profit. Accordingly, "the business case for biodiversity" is not about conserving the ecological functioning of biodiversity but of reclaiming the "waste" from its unpriced services, imagining into being new forms of economic productivity, and more specifically the conversion and application of this productivity to commercial profit. It is this application of intellectual, rather than physical, labour to nature that constitutes contemporary "improvement".

[28] As an example, Conservation International and IUCN have markedly different resource bases. Conservation International relies much more heavily on direct corporate donations to finance their work while IUCN is heavily reliant on project funds from donor governments and multi-lateral financial institutions.

[29] 56.5% of Conservation International's 2006 budget was drawn from corporations and foundations. Notably there is often an overlap between these two. Conservation International, for example, has received significant donations from the Walton Family

Foundation, much of which is derived from profits provided by investments in Wal-Mart. They also have a direct "partnership" arrangement with Wal-Mart focused on energy and waste reduction and the development of products geared toward an "environmentally conscious" consumer; in essence doing the work of cost reduction and product development for one of the world's largest corporations. TNC and WWF have also effectively tapped into corporate and private donations as a way of addressing capital constraints; cf Brechin 2008).

[30] See Litzinger (2006) for a critical appraisal of these "shared conservation goals".

[31] Notably the former Chief Conservation and Science Officer at Conservation International has become Team Leader, Natural Resources Division at GEF.

[32] IPIECA is the CSR arm of the oil and gas industry and was established in 1974 following the establishment of UNEP. IPIECA provides one of the industry's principal channels of communication with the UN (http://www.ipieca.org/).

[33] The IUCN Secretariat response to the motion to terminate the agreement with Shell, for example, warned of a direct capital loss to the Secretariat and unknown liability costs: "Core funding on the order of CHF 1,300,000 would not be available if the agreement is terminated. Further negative financial consequences may be incurred depending on the conditions of the termination of the agreement" (IUCN 2008).

[34] http://cms.iucn.org/about/work/programmes/business/bbp_our_work/bbp_holcim/ (accessed 24 June 2008).

[35] Some within the IUCN Secretariat have also openly promoted the alteration of the organization's constitution to allow private sectors members. IUCN, as a union of actors concerned with nature conservation, has historically restricted membership to state and non-governmental actors.

[36] http://www.panda.org/about_wwf/how_we_work/businesses/index.cfm

[37] Branding is not inconsequential in the sphere of conservation organizations. As organizations, under the guidance of communications consultants, become increasingly concerned with "message control", they also become increasingly focused on the value of their brand and associate trademarks. A case in point is the drawn-out legal struggle between the World Wide Fund for Nature and the former World Wrestling Federation over the right to use the acronym WWF (Davies 2002).

[38] As an indication, consult the list of VIPs present during the 2008 World Conservation Congress http://cmsdata.iucn.org/downloads/vip_list_for_congress.pdf

References

Adams W and Hutton J (2007) People, parks and poverty: Political ecology and biodiversity conservation. *Conservation and Society* 5:147–183

Annan K (2000) Secretary-General welcomes international corporate leaders to global compact meeting. Press release No UNIS/SG/2618, 27 July http://www.unis.unvienna.org/unis/pressrels/2000/sg2618.html (last accessed 21 July 2008)

Arts B (2006) Non-state actors in global environmental governance: New policy arrangements beyond the state. In M Koenig-Archibugi and M Zürn (eds) *New Modes of Governance in the Global System: Exploring Publicness, Delegation and Inclusiveness* (pp 177–201). London: Palgrave Macmillan

Ashford N A (2002) Government and environmental innovation in Europe and North America. *American Behavioral Scientist* 45:1417–1434

Beder S (2006) The changing face of conservation: Commodification, privatization and the free market. University of Wollongong, Faculty of Arts Papers No 40

Bernstin, S (2001) *The Compromise of Liberal Environmentalism.* New York: Columbia University Press

Brand U and Görg C (2008) Sustainability and neoimperial-liberal globalization: A theoretical perspective. In J Park, K Conca, M Finger and (eds) *Sustainability, Globalization, and Governance* (pp 13–33). London: Routledge

Brechin S (2008) Private sector financing of international biodiversity conservation: An exploration of the funding of large conservation NGOs. Paper presented at the Capitalism and Conservation Symposium, University of Manchester, 8–10 September

Brosius P and Tsing A (1998) Representing communities: Histories and politics of community-based natural resource management. *Society and Natural Resources* 11:157–169

Bruno K and Karliner J (2002) *Earthsummit.biz: The Corporate Takeover of Sustainable Development*. Oakland: Food First Books

CBD Secretariat (2004) *Business and Biodiversity Initiatives: A Background Document for the Business and the 2010 Biodiversity Challenge Meetings in London, 20–21 January 2005*. Montreal: Convention on Biological Diversity

CBD Secretariat (2006a) Believing in business and biodiversity. *Business 2010* 1:1, 12

CBD Secretariat (2006b) Report of the Eighth Meeting of the Parties to the Convention on Biological Diversity Decision VIII/17, Private-Sector Engagement, UNEP/CBD/COP/8/31, 15 June, http://www.cbd.int/doc/meetings/cop/cop-08/official/cop-08-31-en.pdf (last accessed 21 July 2008)

CBD Secretariat (2008a) Business at COP-9: A guide, http://www.cbd.int/cop9/business/biz-cop9-guide-en.pdf (last accessed 21 July 2008)

CBD Secretariat (2008b) Cooperation with other conventions, international organizations and initiatives and engagement of stakeholders: Addendum engagement of business. Note by the Executive Secretary, UNEP/CBD/COP/9/21/Add.1, 27 March, http://www.cbd.int/doc/meetings/cop/cop-09/official/cop-09-21-add1-en.pdf (last accessed 21 July 2008)

Clapp J (2005a) Global environmental governance for corporate responsibility and accountability. *Global Environmental Politics* 5:23–34

Clapp J (2005b) Transnational corporations and global environmental governance. In P Dauvergne (ed) *Handbook of Global Environmental Politics* (pp 284–297) Northampton, MA: Edward Elgar

Conservation International (2005) *Center for Environmental Leadership in Business: 2005 Annual Report*. Washington DC: Conservation International

Coronil F (2000) Towards a critique of globalcentrism: Speculations on capitalism's nature. *Public Culture* 12:351–374

Davies I (2002) Legal update: The panda vs. the rock. *Journal of Brand Management* 9:210–214

Ervine K (2007) The greying of green governance: Power politics and the Global Environment Facility. *Capitalism, Nature, Socialism* 18:125–142

Finger M (2005) The new water paradigm: The privatization of governance and the instrumentalization of the state. In D L Levy and P Newell (eds) *The Business of Global Environmental Governance* (pp 275–304). Boston: MIT Press

Flitner M (1999) Biodiversity: Of local commons and global commodities. In M Goldman (ed) *Privatizing Nature: Political Struggles for the Global Commons* (pp 144–166). New Brunswick: Rutgers University Press

Ford L H (2005) Challenging the global environmental governance of toxics: Social movement agency and global civil society. In D L Levy and P Newell (eds) *The Business of Global Environmental Governance* (pp 305–328). Boston: MIT Press

Frynas J G (2005) The false developmental promise of Corporate Social Responsibility: Evidence from multinational oil companies. *International Affairs* 81:581–598

GEF (2005) *Resource Allocation Framework*. Global Environmental Facility: Washington DC

GEF (2008) *Financing the Stewardship of Global Diversity*. Global Environmental Facility: Washington DC

Geisinger A (1999) Sustainable development and the domination of nature: Spreading the seed of the western ideology of nature. *Boston College Environmental Affairs Law Review* 27:43–73

Gramsci A (1971) *Selections from the Prison Notebooks of Antonio Gramsci*. New York: International Publishers

Gulbrandsen T C and Holland D C (2001) Encounters with the super-citizen: neoliberalism, environmental activism, and the American Heritage Rivers Initiative. *Anthropological Quarterly* 74:124–134

Guruswamy LD (1999) The convention on biological diversity: Exposing the flawed foundations. *Environmental Conservation* 26:79–82

Hajer M (1995) *The Politics of Environmental Discourse: Ecological Modernization and the Policy Process*. Oxford: Clarendon

Hayden C (2003) *When Nature Goes Public: The Making and Unmaking of Bioprospecting in Mexico*. Princeton: Princeton University Press

Heap J (2005) *A Survey of Guidelines for Not-For-Profit/Private Sector Interaction*. Gland: IUCN

Hildyard N (1993) Foxes in charge of the chickens. In W Sachs (ed) *Global Ecology: A New Arena of Political Conflict* (pp 22–35). London: Zed Books

ICC (2008) *Access and Benefits Sharing: General Observations and Positions Submitted for the 9th Conference of the Parties of the UN Convention on Biological Diversity*. International Chamber of Commerce, http://www.iccwbo.org/uploadedFiles/ICC/policy/intellectual_property/Statements/ABS%20submission%20COP-9%20final%2016-05-08.pdf (last accessed 18 July 2008)

IUCN (2006) *Business and Biodiversity Programme 2005 Annual Report*. Gland: The World Conservation Union

IUCN (2008) *Corrigendum, Motions: World Conservation Congress, Barcelona, Spain, 5–14 October 2008*. Gland: The World Conservation Union, http://cmsdata.iucn.org/downloads/rwg_comment_for_31mot.pdf (last accessed 15 September 2009)

James A N, Green M J B and Paine J R (1999) *A Global Review of Protected Area Budgets and Staff* (WCMC Biodiversity Series No 10). Cambridge: World Conservation Press

Jamison A (1996) The shaping of the global environmental agenda: The role of non-governmental organizations. In S Lash, B Szerszynski and B Wynne (eds) *Risk, Environment and Modernity: Towards a New Ecology* (pp 224–245). London: Sage

Karliner J (1999) *A Perilous Partnership: The United Nations Development Programme's Flirtation with Corporate Collaboration*. San Francisco: Transnational Resource and Action Centre, http://www.corpwatch.org/article.php?id=3388 (last accessed 21 July 2008)

Keil R and Desfor G (2003) Ecological modernization in Los Angeles and Toronto. *Local Environment: The International Journal of Justice and Sustainability* 8:27–44

Lake R (1997) *New and Additional?: Financial Resources for Biodiversity Conservation in Developing Countries 1987–1994*. Cambridge: Royal Society for the Protection of Birds

Lapham N P and Livermore R J (2003) *Striking a Balance: Ensuring Conservation's Place on the International Biodiversity Assistance Agenda*. Washington DC: Center for Applied Biodiversity Science, Center for Conservation and Government, Conservation International

Levy D L (2005) Business and the evolution of the climate regime: The dynamics of corporate strategies. In D L Levy and P Newell (eds) *The Business of Global Environmental Governance* (pp 73–104). Boston: MIT Press

Levy D L and Egan D (1998) Capital contests: National and transnational channels of corporate influence on the climate change negotiations. *Politics and Society* 26:337–361

Levy D L and Newell P J (2002) Business strategy and international environmental governance: Toward a neo-Gramscian synthesis. *Global Environmental Politics* 2:84–101

Levy D L and Newell P (2005) *The Business of Global Environmental Governance.* Boston: MIT Press

Li T (2007) *Will to Improve: Governmentality, Development, and the Practice of Politics.* Durham: Duke University Press

Lipschutz R D and Rowe J K (2005) *Globalization, Governmentality and Global Politics: Regulation for the Rest of Us?* London: Routledge

Litzinger R A (2006) Contested sovereignties and the Critical Ecosystem Partnership Fund. *Political and Legal Anthropology Review* 29:66–87

MacDonald K I (2003) IUCN—The World Conservation Union: A history of constraint. Paper presented to the Permanent Workshop of the Centre for Philosophy of Law Higher Institute for Philosophy of the Catholic University of Louvain (UCL), Belgium, 6 February, https://tspace.library.utoronto.ca/handle/1807/9921 (last accessed 5 February 2009)

MacDonald K I (2004a) Developing "nature": Global ecology and the politics of conservation in northern Pakistan. In J G Carrier (ed) *Confronting Environments: Local Environmental Understanding in a Globalizing World* (pp 71–96). Walnut Creek: Alta Mira Press

MacDonald K I (2004b) Conservation as cultural and political practice. *Policy Matters* 13:6–17

MacDonald K I (2005) Global hunting grounds: Power, scale and ecology in the negotiation of conservation. *Cultural Geographies* 12:259–291

MacDonald K T (in press) Business, biodiversity and new "fields" of conservation: The World Conservation Congress and the renegotiation of organizational order. *Conservation and Society*

Mansourian S and Dudley N (2008) *Public Funds to Protected Areas.* Gland: WWF International

Maragia B (2002) Almost there: Another way of conceptualizing and explaining NGOs' quest for legitimacy in global politics. *Non-State Actors and International Law* 2:301–332

Maxwell J (1997) Green schemes: Corporate environmental strategies and their implementation. *California Management Review* 39:118–133

McAfee K (1999) Selling nature to save it? Biodiversity and the rise of green developmentalism. *Environment and Planning D: Society and Space* 17:133–154

Mol A P J and Spaargaren G (2000) Ecological modernization theory in debate: A review. *Environmental Politics* 9:17–49

Newell P (2005) Business and international environmental governance. In D L Levy and P Newell (eds) *The Business of Global Environmental Governance* (pp 21–47). Boston: MIT Press

O'Connor M (1994) *Is Capitalism Sustainable? Political Economy and the Politics of Ecology.* New York: Guilford Press

Paine E (2000) *The Road to the Global Compact: Corporate Power and the Battle Over Global Public Policy at the United Nations.* New York: Global Policy Forum, http://www.globalpolicy.org/reform/papers/2000/road.htm#31 (last accessed 18 July 2008)

Pearce D and Palmer C (2001) Public and private spending for environmental protection. *Fiscal Studies* 22:403–456

Pfeffer J and Salancik G R (2003) *The External Control of Organizations: A Resource Dependence Perspective*. Stanford: Stanford University Press

Prudham S (2004) Poisoning the well: Neoliberalism and the contamination of municipal water in Walkerton, Ontario. *Geoforum* 35:343–360

Redclift M (1995) The environment and structural adjustment. *Journal of Environmental Management* 44:55–68

Reed D (1992) *Structural Adjustment and the Environment*. Boulder, CO: Westview Press

Reimann K D (2006) A view from the top: International politics, norms and the worldwide growth of NGOs. *International Studies Quarterly* 50:45–67

Rose G and Jackson S (1992) Industry's response to UNCED: Environmental management post-Rio. *Review of European Community and International Environmental Law* 1:320–324

Rowe J K (2005). Corporate social responsibility as business strategy. In R Lipschutz and J K Rowe (eds) *Globalization, Governmentality and Global Politics: Regulation for the Rest of Us?* (pp 30–46). London: Routledge

Sachs W (1993) Global ecology in the shadow of development. In W Sachs (ed) *Global Ecology: A New Arena of Political Conflict* (pp 3–21). London: Zed Books

Shine C and Kohona P T B (1992) The Convention on Biological Diversity: Bridging the gap between conservation and development. *Review of European Community and International Environmental Law* 1:278–288

Sklair L (2000) The transnational capitalist class and the discourse of globalization. *Cambridge Review of International Affairs* 14:67–85

Sneddon C S (2000) "Sustainability" in ecological economics, ecology and livelihoods: A review. *Progress in Human Geography* 24:521–549

Stigson B (2008) Business and biodiversity. *IUCN Pan-European Newsletter* 15:3–5

Strathern M (2000) Multiple perspectives on intellectual property. In K Whimp and M Busse (eds), *Protection of Intellectual, Biological and Cultural Property in Papua New Guinea* (pp 47–61). Canberra: Asia Pacific Press at the Australian National University

Swanson T (1999) Why is there a biodiversity convention? The international interest in centralized development planning. *International Affairs* 75:307–331

Taylor P J and Buttel F H (1992) How do we know we have global environmental problems? Science and the globalization of environmental discourse. *Geoforum* 23:404–416

Tokes D (2001) Ecological modernisation: A reformist review. *New Political Economy* 6: 279–291

Trisoglio A and ten Kate K (1991) *From WICEM to WICEM II: A Report to Assess the Progress in Implementation of WICEM Recommendations*. Geneva: UNEP

Turner T (1982) Scuttling environmental progress. *Business and Society Review* 42:48–52

UNEP (1989) *Preparation of an International Legal Instrument on the Biological Diversity of the Planet*. UNEP 15th Governing Council Session, 25 May, http://www.unep.org/Documents.Multilingual/Default.Print.asp?DocumentID=71&ArticleID=963&l=en (last accessed 3 August 2008)

WCED (1987) *Our Common Future*. Oxford: Oxford University Press

Wood E (2002) *The Origin of Capitalism: A Longer View*. London: Verso

Worster D (1993) The shaky ground of sustainability. In W Sachs (ed) *Global Ecology: A New Arena of Political Conflict* (pp 134–144). London: Zed Books

Chapter 3
The Conservationist Mode of Production and Conservation NGOs in sub-Saharan Africa

Dan Brockington and Katherine Scholfield

Introduction

In October 2009 the Wildlife Conservation Network held its annual "Wildlife Conservation Expo" in San Francisco. Primatologist Jane Goodall was the keynote speaker and it was billed as "the premiere wildlife conservation event in the Bay Area". Visitors had a chance to learn about conservation work locally and internationally and meet a number of prominent international conservation activists. The following day a garden party, complete with large cats, provided food, art sales, entertainment and further opportunities to meet great conservationists in a more exclusive setting, as the entrance fee was set at $1000 per person.[1] Previous events (in 2006) were celebratory dinners in which guests could, for a $1500 individual ticket or $10,000 for a table for eight, become a "global sponsor" which carried privileges of a pre-dinner VIP reception to meet some of the conservationists whose work was being celebrated.

The Wildlife Conservation Network is well connected. Its board members and advisors are wealthy philanthropists who have done well on the high-tech and software industries of the Bay area. It works by supporting individuals who are doing great deeds for charismatic wildlife. These people, labeled "conservation entrepreneurs" or "conservation heroes", each tend to have their own charitable organization which the Wildlife Conservation Network then supports. The Wildlife Conservation Network's appeal, to the public and causes selected alike, is that it offers a chance personally to meet significant conservationists at exclusive fundraising gatherings. All the sites of the conservation work sponsored are in exotic overseas locations, in South America, Africa and Asia. In addition, then, to the shopping, entertainment and variety on display that can make saving nature so enjoyable (cf Brockington and Scholfield 2009), the Wildlife Conservation Network allows supporters to connect with and support far and distant places, through the person of the celebrity conservationist (Brockington 2009).

We begin with this example because it captures clearly the ways in which conservation non-governmental organizations (NGOs) bring the wealth of the North together with the experience, knowledge and stories of exotic wild places to generate support for their causes. In this chapter, following an extensive survey of conservation NGOs operating in sub-Saharan Africa, we explore the general role of conservation NGOs supporting African conservation in transforming the region's natural capital (forests, wildlife, protected landscapes), and conservation work in the same, into symbolic capital and money. A sceptic might retort that it is obvious that NGOs raise money, and that conservation NGOs should do so from stories and images of wildlife and nature. Our argument, however, is not that simple. Following Garland's recent writings (Garland 2006, 2008) we argue that conservation NGOs are integral to a "conservationist mode" of production which intertwines wildlife and biodiversity conservation with capitalism. Conservation NGOs are not simply raising money. They are incorporating nature and wildlife into a broader capitalist system by producing images and commodities whose circulation mediates relationships between people and between people and nature (cf Igoe, Neves and Brockington this volume). They are also forging the conditions, discursively and materially, for capital to appropriate aspects or parts of wildlife and nature which had escaped being turned into commodities. In part this is achieved through legitimizing particular visions for African landscapes and wildlife, and specific types of nature production. All this highlights and demonstrates how conservation is not a domain separate and set apart from capitalism. It is produced by it, and thoroughly integrated into it, such that capital and conservation become two core complementary and mutually enforcing processes in the contemporary production of nature (Castree 2003).

Before we can proceed we must define what we mean by "NGOs" and what sort of conservation we are referring to. Neither definition is straightforward, and the term "conservation NGO" can in fact invoke a blurred area of fuzzy conceptual space rather than a precise tool for analysis. NGOs are usually understood to encompass a broad gamut of non-profit organizations which are not part of government (Bryant 2009), but the distinctions of NGOs from business and the state are becoming increasingly hard to maintain (cf Igoe, Neves and Brockington this volume).[2] NGOs may best be characterized by the high levels of expectation that surround them. Bebbington, Hickey and Mitlin's recent, and otherwise definitive, collection on the work of NGOs in development, eschewed defining what NGOs actually are. Instead they began with "the conviction ... that NGOs are only NGOs in any politically meaningful sense of the term if they are offering alternatives to dominant models, practices and ideas about development" (2008:3).[3] By "conservation" in this chapter we mean wildlife and biodiversity conservation. In African contexts it has generally encompassed state legislation and regulatory regimes that govern hunting, protected areas (national parks, wildlife sanctuaries, forest reserves and the like), wildlife trade and forestry policies. This apparently simple definition conceals considerable tension. For how can one distinguish wildlife and biodiversity conservation from a more general environmentalism and how does one deal with the overlaps between conservation and development activity? We return to these questions in the methods where we explain how we delimited and described the conservation NGO sector.

In the meantime it is important to observe that there are very good reasons to suggest that one could and should overcome the conceptual difficulties and examine the work of this rather hazy sector. For despite the problems of delimiting "conservation" and "NGOs", there does exist a large core of entities that self-indentify as "conservation NGOs". More importantly they enjoy widespread public recognition for being "conservation NGOs". While it may be difficult to argue that conservation NGOs actually exist ontologically, and are therefore quite hard to isolate epistemologically, they do exist relatively unproblematically as a social fact. Conservation NGOs matter because this very label has made them some of the most important players in wildlife and biodiversity conservation internationally. There is a space in popular, corporate and government thinking, and in policy discourse resulting, for "conservation NGOs" and those occupying it enjoy considerable influence.

In part there is their sheer size. Some of the larger international conservation NGOs are among the biggest NGOs in the world. The Nature Conservancy (TNC) is a multi-billion dollar organization, the World Wide Fund for Nature (WWF), Wildlife Conservation Society (WCS) and Conservation International annually spend hundreds of millions of dollars (Brockington, Duffy and Igoe 2008). They are significant to the conservation movement because their reach is global, and they can be particularly influential in poorer parts of the world where government expenditure on conservation issues is slight, and NGO expenditure proportionally larger. They have been favoured vehicles for spending bilateral and multilateral donor funds. They are also one of the principal means by which conservationists can channel funds from the wealthier parts of the world to the poorer parts of the world where so much biodiversity can be found.

They also matter because of their vigour and variety. Conservation NGOs in all their diversity have been at the forefront of the campaigning and image-making activities of conservation and have been since the early 1900s. They are drivers of much conservation science and policy lobbying. They are significant employers of, and provide a social habitat for, conservationists themselves. Adams, in his history of the conservation movement, found their influence profound and described them as "noisy, visionary . . . also extremely powerful, for their grip on international thinking about conservation" (Adams 2004:55).

Finally it is important to examine conservation NGOs because their influence, in common with NGOs generally, has grown in the neoliberal era. From the late 1980s NGOs enjoyed particularly privileged status within international development and conservation circles (Edwards and Hulme 1992). These years were the heyday of neoliberal policies that emphasized small government and resourced service provision from the community, or from cheaper independent suppliers (cf Corson this volume). NGOs were also thought to be essential elements of good democracies, providing independent data and reports by which governments could be held to account and active and vocal lobby groups for particular interests. Since the 1980s, NGOs have generally expanded in size and number to meet the demand for them. Conservation NGOs were no exception. Our survey of those active in sub-Saharan Africa, which we describe below, showed NGO numbers increasing dramatically from the 1980s onwards. Given the conditions of their growth it is quite likely that they will be vectors of new ideas and policies their donors were promoting at the time.

It is traditional for social scientists' attention to NGOs generally to take the form of critique, questioning the difference or viability of the alternatives that they offer, or the problems of the images and fund raising that they employ.[4] That is not, however, our purpose in this chapter. Instead we wish to provide a new framework to conceptualize their activities. We suggest that in many cases conservation NGOs are best conceived as constitutive of, and central to, the workings and spread of capitalism in sub-Saharan Africa. We argue that NGOs play a vital role in the creation of value from wildlife and nature, both in their work of protecting and reproducing wildlife and wild areas, and in creating demand for the conservation's commodities and imagery overseas. Through the activities of conservation NGOs new commodities and images are produced whose circulation comes to mediate relationships between people and between people and nature.

Within our argument there are echoes of the Gramscian critique of civil society (Forgacs 1999). Gramsci observed that an unintended effect of civil society was that it forged consent for capitalism. Even where civil society organizations fought the worst social abuses of capitalism, they ultimately sustained the legitimacy of capitalism at the same time as they decried its effects. The critics within civil society challenged specific aspects of capitalism's functioning, but they did not challenge capitalism itself. For example, tackling problems of child labour, low wages, pollution and slum housing did not in itself challenge the labour relations that lay at the basis of capitalism. But these actions did make the social reproduction of these relations more palatable and made capitalism more socially legitimate and ultimately more sustainable.

Likewise one could see environmental and conservation NGOs generally as tackling the ecological ills that capitalism produces such that ultimately capitalist economies emerge healthier but unchallenged; indeed they enjoy more legitimacy. However, our argument differs from this position in that the conservation NGOs we describe are not dealing with the contradictions internal to the operation of capitalism. Rather they are working on the frontiers of capitalism, creating the conditions conducive to its expansion. As we shall demonstrate, they provide means of turning the financial wealth produced in northern economies into new commodities, and of creating new markets for some conservation associated industries.

We have restricted our arguments to Africa partly for pragmatic reasons. It is the region we know best, a continent on which we have conducted extensive research on the conservation NGO movement. It is also the region about which our colleagues wrote on whose theories we build. Perhaps most importantly, African case materials provide some of the clearest examples of the power of audience demand in the North for marketable conservation stories and causes. We will explore the reasons for this later in the chapter.

We begin by outlining recent work exploring the general intertwining of conservation and capitalism, and more specific writings that have addressed their relationships specifically within Africa. We combine this framework with the findings of a recently completed survey of conservation NGOs in sub-Saharan Africa to explore its usefulness and consider what questions it raises for further research. Finally we link this to work on epistemic communities and transnational networks to develop a richer theoretical framework to characterize the work of conservation NGOs. Our goal is to develop a more robust theoretical and conceptual framework for understanding the actions of NGOs both in the African sites of their field activity, and their other global sites where they generate their resources, exert influence and build networks.

The Conservationist Mode of Production

In *Nature Unbound*, a general survey of conservation practice globally, Dan Brockington, Rosaleen Duffy and Jim Igoe argued that:

conservation is not merely about resisting capitalism, or about reaching necessary compromises with it. Conservation and capitalism are shaping nature and society, and often in partnership (2008:5).

To radical geographers reared on ideas of the "production of space" and "second nature", this point may seem straightforward. We know that capitalism produces nature—and vice versa (Castree 2003). The observation is required, however, because we have been slower to recognize the thorough imbrication of conservation in capitalist transformations of the world. Conservation can be seen as separate to capitalism and governed by a different logic. Brockington and colleagues, after exploring a variety of conservation practices such as mitigation, offsetting, sports hunting, certification and the creation and transformation of communities and indigenous peoples through conservation activities, found it difficult to tell whether conservation was transforming the world using capitalist instruments, or vice versa. Conservation, they said, does not so much save the world as remake and recreate it (Brockington, Duffy and Igoe 2008:6).

In the last chapter of the book, Jim Igoe elaborated the general theoretical implications of the transformative work of conservation. He observed that tourists' use of consumptive nature depends upon fetishized commodities whose origins and social relations are concealed just as much as jars of coffee on supermarket shelves. The fact that they are consumed in situ, in the landscapes and societies where they were created, does not make them one jot less alienated (cf Neves this volume). Tourists, even eco-tourists, live and move in a bubble, cut off from surrounding societies and landscapes (Brockington, Duffy and Igoe 2008:185–190; cf Carrier and Macleod 2005). Their experiences are produced for them, carefully tailored by tour companies (West and Carrier 2004). As the tourists came with expectations shaped by friends and families, or television documentaries, and as they themselves then use their travels as reference points to assess the veracity and authenticity of their experiences, the whole is brought together in a self-referential loop. Igoe further observed, following Debord (1995 [1967]), that these experiences and bubbles were not confined to the tourism circuit, but were proliferating through society in the production of fetishized images and signs as "spectacles", whose link to their referents was lost, and who were becoming the original authority against which experience of the places referred to is now measured.

> [A]n iterative process develops. The images and ideas of these landscapes become the source for the production of more virtualisms. These virtualisms in turn become the source for the production of more "wilderness" landscapes. The dialogues occurring between these landscapes and the virtualisms that informed their production become

increasingly impervious to ideas and arguments that are not derived
from the realities they describe and prescribe (Brockington, Duffy and
Igoe 2008:194–195).

Igoe went on to explore how the spectacle of nature raises money
in different contexts for different conservation actors. His writing,
however, was general. It was not tied to particular varieties of
conservation, or forms of capitalism. What we require are more detailed
analyses of particular aspects of the conservation sector, or particular
sectors of capitalism, that explore the specific processes, tensions and
contradictions at work.

Writing at almost exactly the same time, Liz Garland, after studying
the character of wildlife conservation in Africa, reached a very similar
conclusion to Brockington, Duffy and Igoe. She found that:

> [Conservation] spaces are not a bulwark against the forces of human
> activity and global capitalism, but rather products of a very particular
> mode of engagement with these forces (2008:64).

Garland's argument differs in two ways from the previous authors. First,
she was exploring the activities of a more specific arena of activities
and group of actors, namely the individuals and organizations who
lobby for wildlife conservation in Africa. Second, she outlined a more
complete theory of how a particular conservation sector transforms
African landscapes and people. Her argument is worth summarizing in
detail.

Like Igoe, Garland observed a preponderance of images of African
wildlife, whose charismatic fauna and scenery have diffused globally
such that they are "part of the natural symbolic repertoire of people all
around the world" (2008:52). Yet she also observed that Africans who
guard and live with this global heritage are relatively invisible in the
global imaginary. Saving wildlife and wildlands is a northern task in
which the work of Africans is largely hidden.

Most critiques stop there. They view conservation as a form of
the white man's burden imposed on the continent by colonial and
neocolonial powers. Garland found it more productive to argue that
wildlife conservation in Africa is:

> foremost a productive process, a means of appropriating the value
> of African nature, and of transforming it into capital with the
> capacity to circulate and generate further value at the global level
> (2008:52).

Viewed thus wildlife conservation is "a particular kind of capitalist
production" (2008:62), turning the natural capital of wildlife into
symbolic capital and, ultimately, money. Garland observes that the
"conservationist mode of exploitation" (2008:63) differs from other

use of natural resources (timber, mining, fishing, farming) because it does not depend on the physical use and appropriation of the resource. It is consumed by looking, by tourists' gazes, and by the work of photographers and filmmakers.[5]

Garland's insights came from an exploration of the hidden work of (black) Africans in the conservation sector. She observed how (white) northerners' interactions with wildlife could be turned into "Ph.D.s, research grants, jobs with international NGOs and tourism companies, academic positions, gigs on the conservation lecture circuit, popular memoirs, starring roles in *National Geographic* specials" (2008:67, and we could add newspaper articles and wildlife film documentaries). She noted that virtually identical interactions by black African conservation workers simply did not produce such profitable exchanges. Brockington, discussing the same phenomenon (2009:74–78), observes that this is a peculiarly African feature of the conservation mode of production. The racial politics simply work differently in India, Asia and South America where there is much less domination by white conservationists.

Garland goes on to argue that vital to the production of these profitable visions and transformations is the image, the expectation, that Africa *is* wild and natural (meaning free from human interference). This is the vision of Nature which West and Carrier (2004) observe at work in conservation virtualisms generally. As numerous authors have observed, and as Garland reports, the environmental histories of these "wild" places are far more anthropogenic than most visions allow (Adams and McShane 1992; Anderson and Grove 1987; Brockington 2002; Igoe 2004; Neumann 1998). Consequently investing places with conservation importance has profound social implications. Landscapes and resources are removed from one set of users, and transferred to others.

This process of appropriation for appreciation is market driven. Garland insists that the value created from the conservationist mode of production hinges on "the desires and fantasy structures" of the audiences (2008:64). Similarly Brockington has argued in *Celebrity and the Environment*, that with respect to the white celebrity conservationists with which Garland illustrates her argument, they:

> are a creation of society, a group of individuals' varyingly similar responses to market forces which demand they perform particular roles. The products they create, and the stances they adopt, are demanded of them, as much as they arise from their own breasts. They quote and mimic the dreams of their audiences (Brockington 2009:64).

Crucial then to the production of value from the natural capital of African wildlife is a hungry audience ready for images of wild Africa and a complex apparatus of people, organizations and marketing practices that reproduce and ultimately sell particular visions of what "wild" Africa looks like and how it can be experienced. The conservationist mode of production hinges on this apparatus. But what does this apparatus look like and how does it work? We wish to build on Garland's theories here and explore in more detail the functioning of conservation NGOs that we contend are an essential element of the conservationist mode of production. We believe that they are vital to the mediations through which wildlife's natural capital are converted into exchangeable and symbolic forms. To develop this idea we must examine the nature of NGOs' work in sub-Saharan Africa in more detail.

Methods

The central difficulty facing any analysis of the work of conservation NGOs in sub-Saharan Africa is that there has not, until recently, been any overview of the sector at all. We did not know how many NGOs there are, where they work, what sorts of activities they do, or how much they spend. It has therefore been difficult to make any sort of informed judgment of their role and activities.

We had to start from scratch to build up as complete a list as possible of conservation NGOs at work on the continent. Our first problem was that the uncertainties over what NGOs are, and what conservation is, makes a comprehensive list of conservation NGOs difficult to produce. There are self-identifying conservation NGOs, but there are also a great many other organizations whose identity is less certain. The task is not made easy by the general absence of published definitions. Conservation NGOs have attracted a great deal of attention in the academic literature; recent reviews include Brockington and Scholfield (2010a, 2010b) and Bryant (2009), and historical treatments include Adams (2004), Dowie (1996, 2009) and Neumann (1998). Despite this writing there have, curiously, been few, if any, attempts to define conservation NGOs. The idea of conservation NGOs has been invoked in general, but their boundaries have not been defined in particular.

It was the disputes of the nature of conservation that posed us most difficulty. First we had to distinguish the realm of conservation we were interested in from a broader set of environmentalist activity. Wildlife conservation is often distinguished from more general environmentalism. Historically, wildlife conservation was the concern of elites, whereas environmentalism was more populist, anti-establishment and activist. Environmentalists were concerned with problems of waste, pollution and nuclear power (Bryant 2009:1544). However, these distinctions have blurred considerably. The larger wildlife conservation organizations are also concerned with sustainability in its broadest forms. Pollution destroys all sorts of habitats, particularly marine and riparian ecosystems, as effectively as deliberate land conversion. The climate change agenda has seen environmental organizations make tropical forests as much their habitat as their conservation counterparts. Conservationists also like to distinguish between their cause and that of animal welfare, the latter being concerned with individuals not species and populations. However, in practical terms, the care taken over the most endangered species (counted in individuals), and the work of animal welfare activists (protecting the habitat of wild species to facilitate their release) can look remarkably similar.

Second, there was the question about what forms of poverty alleviation might also constitute conservation interventions. Some conservationists see tackling problems of poverty as vital to wildlife conservation's core agenda wherever adverse use of wildlife is driven by poverty, or where opportunities to support wildlife habitat and populations depend on their sustained, profitable use. Others see development activities as a dangerous distraction that diverts funds and attention from conservation's core business of protecting valuable nature from unwelcome change (Adams et al 2004). The debate is confusing partly because it mixes normative prescriptions of what conservationists should be interested in with more pragmatic descriptions of what actually works. It is also confusing because, as we have argued elsewhere (Brockington and Scholfield 2010b), conservation is, in all sorts of ways, simply a variety of development. It is concerned with the planned use of resources for national (and international) prosperity, and with the unplanned changes that occur as societies respond and adapt to their circumstances.

We dealt with these problems using a mixture of assertion and consultation. We examined a variety of published sources, web searches and peer review and compiled a list of over 280 organizations using a conservative definition of "conservation", which included only those that were concerned with preserving wildlife, protected areas or wild habitat. We excluded, with some exceptions, animal welfare organizations and general environmental organizations. Inclusion or exclusion initially depended on whether, in our assessment, their projects and activities contributed to specifically wildlife, habitat or protected area conservation goals. For example, we included environmental organizations if they were working specifically around the edge of protected areas in order to reduce resource use pressures upon them. We did not include development organizations with the exception of one specific wildlife management project for which we had precise data. The list of NGOs we compiled, and more details of the organizations we included and excluded, is available online (http://www.sed.manchester. ac.uk/idpm/research/africanwildlife/), and in published sources (Brockington and Scholfield 2010a, 2010b).

The pitfalls in these methods have already been discussed in these other writings. One of the main issues is whether it provides at all an accurate picture of the conservation NGO sector in the region. Our principal method for checking the validity of our list and the authority of our interpretation of it was two consultation exercises in which we sent out copies of our list and our interpretation of its implications to all the NGOs on our database. Their comments led to some organizations being removed, and more being identified. We have since also had work based on this list peer reviewed by four journals.

Despite these checks it is important to point out that the list we are using was researched remotely, and relied heavily on the internet. It poorly captures organizations that do not have websites or access to email and therefore misses a great deal of the local colour and variety on the ground. Nevertheless it has been vigorously tested and remains by far the most comprehensive of all we have come across and was most positively received by the NGOs we sent it to. It does provide a basis on which to proceed.

For each organization we examined where they were working and what sort of projects they undertook, looking at about 900 overall. We also tried to work out how much money conservation NGOs were spending, and were able to get financial data on what 87 NGOs were spending for the years 2004–2006. We have used these data to produce a new and detailed classification of the different varieties of NGOs on the continent (Brockington and Scholfield 2010b) and explore conservation finances (Brockington and Scholfield 2010a). The present chapter builds on the earlier empirical chapters by providing a richer description of how different NGOs' activities fit with our understanding of how capitalism and conservation interact. We are able to write it as a result of many months' engagement with project literature, websites and correspondence with conservationists. It is important to emphasize, however, that we have not conducted any detailed schedule of interviews with particular conservation NGOs or NGO networks. That was beyond the scope of this survey, but is the subject of ongoing research by Katherine Scholfield.

Conservation NGOs and the Conservationist Mode of Production in Sub-Saharan Africa

Our survey showed that conservation NGOs have kept pace with the global expansion of NGOs that is associated with neoliberalism. Figure 1 shows that conservation NGOs in sub-Saharan Africa expanded rapidly from the 1980s onwards, and most especially in the 1990s. Conservation NGOs have not just grown in number with time, they have also grown in size and influence. Major new organizations have come into

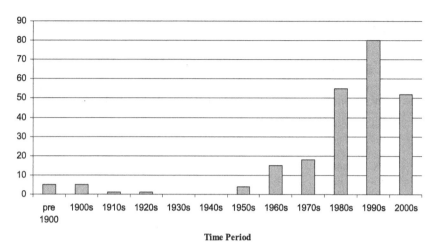

Figure 1:   The development of conservation NGOs working in sub-Saharan Africa

existence, other older players have expanded their work dramatically. Thus Conservation International (the third largest conservation NGO on the continent) was formed in 1987, the Peace Parks Foundation (the fifth largest) in 1997, the African Parks Foundation (ninth largest) in 2000. Sachedina (this volume) shows that the older African Wildlife Foundation (AWF, established in 1961) doubled its income to nearly $20 million within the last 10 years. Other, smaller organizations have also expanded dramatically in size. The Dian Fossey Gorilla Fund International (10th largest), the Rainforest Foundation and the Jane Goodall Institute (sixth largest) have all grown with the celebrity of their founders (Dian Fossey, Sting and Jane Goodall respectively).

It is also clear that these NGOs are plainly vital brokers mediating the production of value from wildlife in Africa. Half of all NGOs' headquarters are located in Europe and North America. This influence is particularly clear if we take the size of the organization into account. In the top 10 conservation organizations only one was founded and has its headquarters in the South (the Peace Parks Foundation of South Africa). Clearly mediating the relations between states, civil society and economic elites within African conservation is facilitated with a northern power base.

The sector is marked by its inequalities (Table 1). The top 10 organizations control the great majority of observed expenditure (more than 80%), and the budget and geographical activities of the largest (the WWF) is greater than the next two combined. In this respect the common concentration in the literature concerned with conservation NGOs on "BINGOs" (Big International NGOs) to the exclusion of all organizations seems merited (Chapin 2004). However, it is also plain that there are a great many smaller organizations at work whose

Table 1: The structure of the conservation NGO sector in sub-Saharan Africa

Size class	Range of expenditure inc overheads	Counted NGOs	Average expenditure inc overheads	Predicted number of NGOS for each size class	Predicted total expenditure inc overheads	Predicted structure (%)
7	Over $40 million	1	42,708,026	1	42,708,026	21
6	$10–21 million	4	15,559,663	4	62,238,652	31
5	$4.2–6.2 million	5	5,467,690	5	27,338,450	14
4	$0.8–1.9 million	10	1,351,520	18	24,327,360	12
3	$0.3–0.72 million	14	479,142	43	20,603,106	11
2	$0.1–0.3 million	26	200,090	90	18,008,100	9
1	Up to $0.1 million	27	54,927	104	5,712,408	3
Total		87		265	200,936,102	

activities are not particularly well understood. It is when we look at the varieties and hierarchies of NGOs at work on the continent that the value of Garland's framework and the idea of the conservationist mode of production become clearest, for these make plain the different strategies employed to turn natural capital into more symbolic value and financial capital.

Conservation NGOs are either fundraising organizations that then distribute funds to known causes they seek out on the ground, or they can raise funds for causes they have identified and spend money on themselves. In both instances, however, these actions are more than just fund raising. As we stated earlier, they also legitimate (particular kinds) of conservation work (in particular places) facilitating the materialization of specific views of the world and thus of the interests that are associated with them. Indeed the legitimation function is crucial to the creation of symbolic capital from conservation work, which in turn underpins the fundraising. What was interesting from our survey was the variety of means by which conservation organizations have found of turning (legitimized) interactions with wildlife into symbolic capital and then actual funds.

One particularly prominent strategy in this region is organizations' reliance on the celebrity of their founder and his/her associations with particular animals, or causes. Thus the Dian Fossey Gorilla Fund International, Iain-Douglas Hamilton's Save the Elephants, the Jane Goodall Institute, Cynthia Moss' Amboseli Trust for Elephants, the David Shepherd Foundation, the David Sheldrick Foundation and Born Free (associated with George and Joy Adamson and the actors in the *Born Free* film, Will Travers and Virginia McKenna), and Laurie Marker's Cheetah Conservation Fund all rely upon the renown and charisma of their organizations' founders. These individuals are vital mediums in the production of conservation value. One might say that they are some of the key machines at work producing money for conservation out of the ideas and images that the conservation mode of production evokes. It is books, pictures, films, magazine articles by or about them, and their lecture tours, which provide the actual products that satiate the needs of northern publics for conservation images and stories. Some of these individuals themselves benefit from the support of organizations like the Wildlife Conservation Network in California which raises funds, as we have seen, by hosting special galas and dinners that specialize in concentrating these charismatic figures into a relatively small place for one evening.

In many of the cases above, the appeal of the individuals involved is combined with the charisma of particularly well known species. Other organizations, without the support of famous people, rely on the appeal of particular taxa alone. Prominent groups include primates, elephants, rhinoceros, big cats and birds (the Bonobo Conservation Initiative, the Wild Chimpanzee Foundation, the Elephant Conservation Foundation, the Lion Conservation Fund, Rhino Ark). Other organizations (the Rainforest foundation, Savanna International, the Sahara Conservation Fund) work for particular types of habitat. Finally there are the organizations that lobby for particular places, individual national parks and protected areas. An interesting feature here is the tendency of some conservation organizations to cluster around particularly iconic parks and reserves on the continent (Kruger, Amboseli). There are few such icons, however, and the more common pattern is for NGOs to space themselves such that they are not representing the same place.

There are also isolated cases where part of the appeal of the organization is their ability to work so directly and decisively in urgent conservation situations that they can appear to usurp states' prerogative of legitimate violence. The Owens Foundation for Wildlife Conservation and the George Adamson Wildlife Preservation Trust, for example, both take pride in being on the front line defending protected areas from people, to the consternation of their critics who query the ecological rationale of this hard line stance and its impact on people (Brockington 2002; Owens and Owens 1992; Ward 1997). The African Rainforest and River Conservation Organisation (and prior to it Jean and Mathieu Laboreur; Adams and McShane 1992) tried to take a hard line against poaching in the Central African Republic, and was awarded considerable freedom over a large concession by that country's leaders (Clynes 2002; Ferguson 2006; Igoe 2002; Neumann 2004). Prince Bernhard of the Netherlands (see Wels and Spierenburg this volume) also secretly donated money to the WWF which was ring-fenced to be used in a military operation against poachers (Bonner 1993). In these instances the blurring of the lines between states and NGOs are particularly salient.

More common are organizations that act as fundraising organizations in the north to raise and channel funds for conservation causes. Perhaps the best organized is Birdlife International, a collection of independent bird conservation organizations from different countries. Birds arouse particularly strong conservation passions and have done for some time. Bird conservation organizations are among the oldest in the movement and birds are the best known of all taxa. Birdlife International capitalizes on this passion and specializes in linking wealthy bird organizations in the North with suitable partners in the South. Other examples of such organizations in the UK include Save the Rhino, perhaps best known for its marathon runners dressed as rhinoceroses; Tusk Trust, Friends of Conservation (both with royal connections). In the USA they include the Wildlife Conservation Network (see above), the Centre for Rural Empowerment and the Environment and the swaggering International Conservation Caucus Foundation, a lobby group in Washington that enrolls the support of over 100 members of congress (see Corson this volume).[6] This Foundation organizes expensive galas in Washington serving African presidents and giving prestigious awards to prominent politicians and wealthy individuals.

These NGOs are particularly important agents within the conservationist mode of production because they depend on a set of images being produced in different field projects that they then market to northern publics in order to raise money. These organizations produce many of the arenas in which nature is converted into symbolic capital, and symbolic capital is converted into financial capital. They also, with their fundraising balls, galas, fun runs, fashion parades and charity auctions, substantially influence the character of the conservationist mode of production in its northern manifestations. They make supporting conservation fun. Fundraising work involves pleasurable activities in desirable locations with much popular participation (Brockington and Scholfield 2009). This is the business of actually existing mediation. In the process these organizations also reproduce the expectations of what wildlife conservation work should look like, and thus the value of the currency of images, symbols and expectations upon which the conservationist mode of production depends.

The fundraising organizations also exemplify the common feature that conservation NGOs often function in networks. They combine because of their common geographical interests, projects, personnel and ideas as well as fundraising activities. The performance and activities of these networks could be the subject of an entirely separate chapter, and we will return to them in the conclusion. Here it must suffice to say that some are carefully planned operations. For example, the Bushmeat Crisis Task Force (BCTF) is based in the Washington DC office of the WCS but enjoys a great and varied membership from NGOs, zoos, societies and universities. Their mission is essentially to use these networks to "eliminate the illegal commercial bushmeat trade" (BCTF 2009). Others, also planned, are quite fluid, as in the case of mountain gorillas. The International Gorilla Conservation Project (IGCP) is made up of staff from AWF, WWF and Flora and Fauna International who have initiated transboundary conservation of the mountain gorilla in the Virungas region of Central and Eastern Africa. The IGCP is a formal network and a hybrid entity, funded as a distinct project supported by all participant organizations and government. In addition to these three, there are others working on mountain gorilla projects in the individual countries of Rwanda, Uganda and the Democratic Republic of Congo alongside the IGCP. For instance, there is the Dian Fossey Gorilla Fund International, the Gorilla Organisation (which split from the Dian Fossey Gorilla Fund), and the Mountain Gorilla Veterinary Project which works in partnership with the three governments of Uganda, Rwanda and the Democratic Republic of Congo. Within the mountain gorilla conservation community there is considerable fluidity of personnel. Staff have moved between NGOs, or split off to set up their own, between NGOs and donors, between NGOs and academia and between NGOs and state.

Other networks can be thrown into existence by unforeseen events. For instance, in 2005, the President of Kenya called for the downgrading of the Amboseli National Park to a national reserve, which prompted immediate outcry from a number of international conservation experts, NGO and wildlife groups. A "Save Amboseli" campaign (and website) was launched through the combined efforts of a number of conservation NGOs prominent in Kenya (although not including any of the larger and more powerful organizations).[7]

Finally we must note that the larger conservation NGOs combine all the above features. They are able to enroll the support of conservation celebrities [like Mike Fay with the WCS, as Garland (2008) shows] as well as drawing upon the popularity of familiar places and animals. They have programmes and field offices in numerous locations on the ground, as well as extensive and sophisticated fundraising departments and expensive, glittering, sponsorship events. They also act as funding bodies, forming alliances with local NGOs that will allow both funder and grantee to fulfill their objectives.

The Scope and Limits of the Conservationist Mode of Production

In the examples above, the conservationist mode of production raised its money from rich northerners, but there are other means of turning knowledge of African environments into financial capital.

The conservation NGO sector is also a community of scientists—an epistemic community (Haas 1992)—that produces and purveys authoritative knowledge about wildlife, biodiversity and their conservation to states and donors. This is a deliberate exercise. Many of the larger NGOs have their own research departments publishing in—indeed prominent contributors to the conservation content of—the more prestigious scientific journals. Viewing conservation NGOs as part of a larger epistemic community is important because it shows how the conservationist mode of production can access the large sums available from bilateral and multilateral donors and produce its greatest impacts on conservation practice.

At its most extensive, the scope of conservation NGOs' activity is visible in some of their more far-reaching impacts on states' decision makers. Their GIS models of habitat, threat and wildlife migration urge action. In Gabon, the WCS was instrumental in encouraging the country's President to set aside 11% of the land area of the country into 13 new national parks (Garland 2008). In Madagascar conservation NGOs used their influence within the donor community to facilitate the adoption of strong conservation policies by the state (Duffy 2006). The Peace Parks Foundation has maps promoting the spread of protected areas across southern, and into eastern Africa. This is mediation writ large, with the confidence, scope and ambition to transform land use and alter economies.

More generally, the larger NGOs active in sub-Saharan Africa, and indeed the rest of the world, demonstrate their potential reach through their prioritization models. These models provide mechanisms for identifying the most important places for conservationists to save, and the most urgent places demanding action (cf Sachedina this volume). There are a plethora of such measures that variously prioritize threat, wilderness and biodiversity (Redford et al 2003). Indeed through them the conservation movement collectively prioritizes 79% of the world's land surface for action (Brooks et al 2006). What is interesting here, however, is that each organization tends to have its own separate prioritizing mechanism, which acts as a fundraising brand. Conservation International has its "Hotspots", WWF its "Ecoregions", the AWF its "Heartlands", Birdlife International has "Important Bird Areas" and "Endemic Bird Areas" and the Wildlife Conservation Society "the last of the wild". The proliferation of these measures has caused some concern within the conservation community for the duplication of effort each entails (Mace et al 2000; Redford et al 2003).

In fact these measures are best understood not as devices that make it easier to decide where to work, but as fundraising tools and, as such, central to the conservationist mode of production (Smith et al 2009). This much is plain in the way that they are used, debated and devised.

First, it is difficult to tell how well these prioritizing mechanisms have actually directed expenditure, because few organizations keep those sorts of records. Where these comparisons are made they have shown that expenditure is poorly related to conservation priority mechanisms (Castro and Locker 2000; Halpern et al 2006; Mansourian and Dudley 2008). Interestingly our data on patterns of conservation expenditure are the first to show the opposite pattern, and a close correspondence between measures of threat and biodiversity at the country level (Brockington and Scholfield 2010a). It was not at all obvious, however, that we could infer cause from this correlation. We could not explain why such a close match had been achieved. When these prioritization models are defended, it is on the grounds of the money that they raised ($750 million for hotspots; Myers and Mittermeier 2003), not on the grounds of how effectively they have directed money to needy places. Finally consider how they were devised and marketed. As Sachedina shows (this volume), AWF came up with its "Heartlands" idea after advice from a consultant as to a label that would appeal to Middle American values.

While the ability of the larger conservation organizations to raise, and spend, large sums of money, both individually and as part of larger networks, should be clear, it is important that the scope of this mode of production should not be exaggerated. Significant influence is visible in the individual countries like Gabon and Madagascar, but these are unusual and isolated cases. Brockington (2006) has previously claimed that the financial power of the environmental sector in Tanzania makes it possible to speak of an "environmental–conservation complex". However subsequent work suggests that the international conservation NGOs are not dominant to the inner workings of this complex. Rather, Sachedina (2008) has observed that prominent NGOs have been effectively disciplined and tamed by the state in Tanzania, and Nelson and Agrawal (2008) have similarly noted that it is state command of wildlife revenues, not NGO payments, that are most influential in the hunting sector. An overview of the conservation NGO sector suggests that while they may be part of a dominant ideology and elite interest, within the elite itself they are relatively marginal, and much less important than the thrust of much recent writings on BINGOs suggests.

As we have argued elsewhere (Brockington and Scholfield 2010b) it is probable that conservation NGOs' influence and reach is generally constrained and limited across sub-Saharan Africa. They are particularly prominent in the North, where so many of them are based, and where they have been raising money for decades. They will therefore be prominent in the minds and perceptions of western-based academics writing about conservation issues in the South. However, their prominence here does not necessarily translate into a similar prominence on the ground in

Africa. The actual day-to-day experience of conservation by rural Africans may well have very little to do with NGOs, and there will be much more to the conservationists' mode of production than the work of NGOs.

Some simple statistics will demonstrate this. Conservation NGOs support under 15% of the protected area network, and their influence within the continent is highly uneven. In general terms it is slight in West Africa and greatest in South Africa. Within each region there are sub-metropoles of activity and larger areas where conservation NGOs are relatively absent. The work of conservation NGOs on the continent is patchy (cf Ferguson 2006). Transnational connections leave out vast swathes of African society and environments. They involve carefully chosen actors and target areas, but exclude and marginalize others. The conservationist mode of production is severely circumscribed in space and society.

The conservation NGO sector's annual turnover in sub-Saharan Africa is, in the broad scheme of things, small. We have predicted that it will be about $200 million a year (Brockington and Scholfield 2010a, 2010b). It is less than 1% of Overseas Development Aid to the region (Scholfield and Brockington 2008). It is less than the annual budget of many large individual companies. There are other varieties of the conservationist mode of production that are as significant financially. For example, the trophy hunting industry has a similar turnover ($200 million annually; Lindsey, Roulet and Romanach 2007). Hunting requires an extensive presence on the ground, for hunting operations hinge on patrolling hunting territories with (wealthy) clients. There is also the work of photographic safari tour operators who are part of the apparatus producing images of, expectations of, and experiences of particular forms of nature in the region, and turning them to profitable purpose. Both of these sectors of the conservationist mode of production are closely related to, and overlap with, the work of conservation NGOs.

It is the limits to their practical scope of action, and the contrast between that and the breadth of their vision, which underline the value of seeing conservation NGOs as part of a conservationist mode of production. The significance of conservation NGOs on the ground in sub-Saharan Africa is not so much the scale or extent of their funding and projects. Rather they are significant because of their role as brokers and introducers of new practices in Africa, as the creators of new conservation commodities, as promoters and lobbyists for more capital investment through the production of images and fundraising literature, as legitimators of conservation practice and entrepreneurship—in sum in mediating between the desires of states, economic interests and civil society nationally and internationally.

For example, conservation NGOs have been given special responsibilities in Tanzania to introduce wildlife management areas on village land. This scheme demarcates areas of village land near protected areas for wildlife use so that tourists can stay there, in return for payments to the villagers, brokered through safari companies (Igoe and Croucher 2007). However, it requires significant changes to the way in which communities and their land use are organized. Here, and in the many other cases in which NGOs have been involved in community-based wildlife management (Nelson and Agrawal 2008), NGOs are at the forefront of creating new commodities, and altering landscapes and communities as a result. Indeed some NGOs specifically see themselves as introducing market relations and market revenues. For example, the African Parks Foundation explicitly states that its role is to enable protected areas to pay their own expenses through developing strong business models. The general point is that fund raising around this work, the ideas it embodies, and peddling the new commodities (community wildlife committees, ethical ecotourism, certified hunting arrangements) raises money that can in turn fuel the commodification process on the ground and the transformation of social relationships and those between societies and nature. The actual places and societies transformed thus may be few, but the virtualisms they invoke can be quite intense, demanding substantial change (cf Brockington 2002). They could initiate new waves of similar commoditization. Indeed that is their goal.

Envoi

The findings of this chapter suggest two further enquiries. First, with respect to conservation NGOs, we have only offered an outline of the conservationist mode of production. We have offered a theory that locates conservation NGOs' work within a broader framework of capitalist endeavour, facilitating economic growth, creating new commodities, promoting and legitimizing visions that require considerable alterations of nature and society. However, to understand how these separate entities work as a sector—as a mode of production— requires understanding how they interact with each other, how they seek and acquire legitimacy, how they are joined together, and how they move around money and ideas. This would require a careful ethnography of how large and small, local and international organizations interact, and the negotiations about how money is spent and what ideas are implemented within this network.

Not all of this sort of activity is easily observable, but an ethnographic approach will clearly be informative here. Anna Tsing's (2005) study of rainforest destruction and the resistance to it in South Kalimantan,

Indonesia raises important questions about the transnational connections involved. Tsing argues that "[c]onservation inspires collaborations" and the "friction" between such collaborations shapes the conservation project (2005:6–9). That is, when certain actors come together in different ways across space and time and in different political, economic, historical and social situations, new culture and different forms of power are created and destroyed. In the case of conservation, these connections will determine how ideas are negotiated, how a conservation project is implemented, and ultimately, the outcome of such a project. Therefore, exploring individual connections, encounters, relations that take place within the conservation sector and looking in detail at the negotiations that they involve will help us to better understand the power dynamics at play within transnational conservation networks. In turn this may help us understand how particular ideas travel across distance and how they come to be implemented in a given project, or for instance, how a particular project comes to be funded over another. Exploring the communication between these partners, and in particular, how they negotiate conservation ideas and funds may help us understand how such different organizations work across transnational borders.

Second, if conservation NGOs are only working in particular places, or "enclaves", then it becomes even more important to establish the extent to which other types of conservation activities are taking place outside of these areas and what these activities might be. There are aspects of the conservationist mode of production such as the hunting and photographic tourist industries that we have barely mentioned here. Finally, and perhaps unusually for a review of the NGO sector, with respect to Africa it may actually be most important to re-examine the work of states in conservation. For these have the most extensive influence on conservation practice on the ground. If we can set these activities within the broader contexts of the work of states and other nature-based industries, and if we can build up comparative material and ethnographies from other parts of the world, we will be in a better position to understand the challenges of the conservationist mode of production that Igoe and Garland have thrown down.

Acknowledgements

Our thanks for some trenchant and incisive comments from three anonymous reviewers, to Katja Neves for her patient reading and the many people who took part in the survey of NGOs. This work was partially supported by Dan Brockington's research fellowship from the ESRC (RES-000-27-0174) which we gratefully acknowledge.

Endnotes

[1] http://wildlifeconservationnetwork.org/events/expo2009.html (accessed 7 November 2009).

[2] Numerous studies have noted the difficulties in defining boundaries between states, NGOs and donors who work alongside each other and together and who have staff who move between and within, or hold positions that straddle across, the different sectors (for instance, see Bebbington 2004; Jackson, 2005; Mitlin, Hickey and Bebbington 2007; Temudo 2005). Many NGOs develop businesses to fund their operations and market their brands to the public. There has also been slippage between NGOs and the state, with the former either taking over many state roles, or collaborating closely with them, or being sought out by donors to undertake work, and administer funds, which states are perceived not to be able to do so well by themselves. Mbembe (2001) speaks of "privatized sovereignty" and "private indirect government" in which states will lend their sovereign authority to other actors (companies or NGOs) who in turn provide the personnel, funds and equipment to give that sovereignty power on the ground. Following a Gramscian perspective (Forgacs 1999), it could be argued that NGOs have similar effects to those of state bureaucracies: the creation and implementation of policies and governance techniques that ultimately shape the world in accordance to dominant economic and political interests. Or, put differently, NGOs mediate between the interests of civil society, state power, and political-economic elites.

[3] This is a common position. Edwards (2008:58) insists that the "best of civil society [of which NGOs are part] exists to meet needs and realize rights regardless of people's ability to pay". Civil society has "distinctive roles and values" that cannot and should not be simply joined to business (2008:91). Bryant (2009:1545) notes that despite critique of particular NGOs, "there tends to be a disposition to see them as a positive and indeed an essential intervention in politics".

[4] A strong body of criticism aimed at conservation NGOs suggests that they are far too close to powerful corporate interests and governments (Chapin 2004; Dowie 1996, 2009). The broader literature on development NGOs again queries the extent to which these organisations are really part of an alternative movement (Bebbington, Hickey and Mitlin 2008; Edwards and Hulme 1995; Hulme and Edwards 1997; Igoe and Kelsall 2005). Bryant (2009:1550) observed that environmental NGOs generally are "disciplining agents in wider power structures" and are part of new architecture of "green governmentality". This can impede their autonomy and transformative potential. Others have queried the representations inherent in conservation fundraising events. All too often the places represented are simplified and decontextualized such that problematic projects can be portrayed as simply and straightforwardly good (Adams and McShane 1992; Brockington 2002; Igoe 2004; Igoe and Croucher 2007; Katz 2001).

[5] Using the case of whale watching and its "antithesis" whale hunting, Neves (this volume) demonstrates how the division between consumptive and non-consumptive is not always as clear cut as at first it may appear and, for many reasons, whale watching is a highly "consumptive", as well as a capitalist activity. Consumption here does not just refer to visions that are consumed but to actual damage caused to a whale as a result of boats used. Similar examples can arguably be found in coral reef diving, or Landrover wildlife safaris—often outwardly promoted as non-consumptive, and non-exploitative, yet with potential to cause varying levels of damage to the environments which they seek to avoid impact upon.

[6] Swaggering because it believes that "as America has exported freedom, democracy, and free enterprise, we have the ability and the interest to see that America also exports good natural resource management". http://www.iccfoundation.us/aboutus.htm (accessed 22 August 2008).

[7] The following NGOs signed an open letter to the Kenyan President: Born Free Foundation Kenya, Youth for Conservation Kenya, East African Wild Life Society, David Sheldrick Wildlife Trust, World Society for the Protection of Animals, Pwani

Environmental Resources Alliance, Animal Defenders International (UK), Animals Asia Foundation (Hong Kong), Born Free Foundation (UK), Born Free (USA), Care for the Wild International (UK), Cetacean Society International (USA), Co-Habitat (UK), David Shepherd Wildlife Foundation (UK), EIA (UK), Friends of Elephant/Vrienden van de Olifant (Netherlands), Humane Society International, Humane Society of the United States (USA), IPPL (USA), International Wildlife Coalition (Canada), Last Great Ape Organisation (Cameroon), League Against Cruel Sports (UK), One Voice (France), Pan African Sanctuary Alliance, Prowildlife (Germany), Rainforest Concern (UK), RSPCA (UK), Society for the Conservation of Marine Mammals (Germany) and African Ele-Fund. Source http://www.bornfree.org.uk/amboseli/letter2.shtml (accessed 31 July 2009).

References

Adams J S and McShane T O (1992) *The Myth of Wild Africa. Conservation without Illusion*. Berkeley: University of California Press

Adams W M (2004) *Against Extinction: The Story of Conservation*. London: Earthscan

Adams W M, Aveling R, Brockington D, Dickson B, Elliott J, Hutton J, Roe R, Vira B and Wolmer W (2004) Biodiversity conservation and the eradication of poverty. *Science* 306:1146–1149

Anderson D M and Grove R (1987) *Conservation in Africa. People, Policies and Practice*. Cambridge: Cambridge University Press

BCTF (2009) Mission, vision and goals, http://www.bushmeat.org (last accessed 21 September 2009)

Bebbington A J (2004) NGOs and uneven development: Geographies of development intervention. *Progress in Human Geography* 28(6):725–745

Bebbington A J, Hickey S and Mitlin D C (2008) *Can NGOs Make a Difference? The Challenge of Development Alternatives*. London: Zed

Bonner R (1993) *At the Hand of Man. Peril and Hope for Africa's Wildlife*. London: Simon and Schuster

Brockington D (2002) *Fortress Conservation. The Preservation of the Mkomazi Game Reserve, Tanzania*. Oxford: James Currey

Brockington D (2006) The politics and ethnography of environmentalisms in Tanzania. *African Affairs* 105(418):97–116

Brockington D (2009) *Celebrity and the Environment. Fame, Wealth and Power in Conservation*. London: Zed

Brockington D, Duffy R and Igoe J (2008) *Nature Unbound. Conservation, Capitalism and the Future of Protected Areas*. London: Earthscan

Brockington D and Scholfield K (2009) Celebrity colonialism and conservation in Africa. In R Clarke (ed) *Celebrity Colonialism: Fame, Representation and Power in (Post)Colonial Cultures* (pp 265–273). Newcastle: Cambridge Scholars Publishing

Brockington D and Scholfield K (2010a) Expenditure by conservation non-governmental organisations in sub-Saharan Africa. *Conservation Letters* 3(2):106–113

Brockington D and Scholfield K (2010) The work of conservation organisations in sub-Saharan Africa. *Journal of Modern African Studies* 48(1):1–33

Brooks T M, Mittermeier R A, da Fonseca G A B, Gerlach J, Hoffmann M, Lamoreux J F, Mittermeier C G, Pilgrim J D and Rodrigues A S L (2006) Global biodiversity conservation priorities. *Science* 313(5783):58–61

Bryant R L (2009) Born to be wild? Non-governmental organisations, politics and the environment. *Geographical Compass* 3/4:1540–1558

Carrier J G and Macleod D V L (2005) Bursting the bubble: The socio-cultural context of eco-tourism. *Journal of the Royal Anthropological Institute* 11:315–334

Castree N (2003) Commodifying what nature? *Progress in Human Geography* 27(3):273–297

Castro G and Locker I (2000) *Mapping Conservation Investments: An Assessment of Biodiversity Funding in Latin America and the Caribbean.* Washington DC: Biodiversity Support Program

Chapin M (2004) A challenge to conservationists. *World Watch Magazine* Nov/Dec:17–31

Clynes T (2002) They shoot poachers, don't they? *National Geographic Adventure Magazine* October

Corson C (2010) Shifting environmental governance in a neoliberal world: US aid for conservation. *Antipode* 42(3):576–602

Debord G (1995 [1967]) *Society of the Spectacle.* New York: Zone Books

Dowie M (1996) *Losing Ground: American Environmentalism at the Close of the 20th Century.* Cambridge: MIT Press

Dowie M (2009) *Conservation Refugees. The Hundred-Year Conflict between Global Conservation and Native Peoples.* Cambridge: MIT Press

Duffy R (2006) Global environmental governance and the politics of ecotourism in Madagascar. *Journal of Ecotourism* 5(1/2):128–144

Edwards M (2008) *Just Another Emperor. The Myths and Realities of Philanthrocapitalism.* New York: Demos

Edwards M and Hulme D (1992) *Making a Difference: NGOs and Development in a Changing World.* London: Earthscan

Edwards M and Hulme D (1995) *Non-Governmental Organisations - Performance and Accountability. Beyond the Magic Bullet.* London: Earthscan

Ferguson J (2006) *Global Shadows. Africa in the Neoliberal World Order.* Durham: Duke University Press

Forgacs D (1999) *A Gramsci Reader. Selected Writings 1916–1935.* London. Lawrence and Wishart

Garland E (2006) "State of nature: Colonial power, neoliberal capital and wildlife management in Tanzania." Unpublished PhD thesis, University of Chicago

Garland E (2008) The elephant in the room: Confronting the colonial character of wildlife conservation in Africa. *African Studies Review* 51(3):51–74

Haas P M (1992) Epistemic communities and international policy coordination. *International Organization* 46(1):1–35

Halpern B S, Pyke C R, Fox H E, Haney J C, Schlaepfer M A and Zaradic P (2006) Gaps and mismatches between global conservation priorities and spending. *Conservation Biology* 20(1):56–64

Hulme D and Edwards M (1997) *NGOs, States and Donors: Too Close for Comfort?* London: Macmillan

Igoe J (2002) Review of "Fortress Conservation". *International Journal of African Historical Studies* 35(2/3):594–596

Igoe J (2004) *Conservation and Globalisation: A Study of National Parks and Indigenous Communities from East Africa to South Dakota.* Belmont, CA: Wadsworth/Thomson Learning

Igoe J and Croucher B (2007) Conservation, commerce and communities: The story of community-based Wildlife Management Areas in Tanzania's northern tourist circuit. *Conservation and Society* 5(4):534–561

Igoe J and Kelsall T (2005) *Between a Rock and a Hard Place: African NGOs, Donors, and the State.* Durham, NC: Carolina Academic Press

Igoe J, Neves K and Brockington D (2010) A spectacular eco-tour around the historic bloc: Theorising the convergence of biodiversity conservation and capitalist expansion. *Antipode* 42(3):486–512

Jackson S (2005) The state didn't even exist: Non-governmentality. In Kivu, Eastern DR Congo. In J Igoe and T Kelsall (eds) *Between a Rock and a Hard Place: African NGOs, Donors and the State* (pp 165–196). Durham, North Carolina: Carolina Academic Press

Katz C D (2001) Whose nature, whose culture? Private production of space and the "preservation" of nature. In B Braun and N Castree (eds) *Remaking Reality. Nature at the Millenium* (pp 46–63). London: Routledge

Lindsey P A, Roulet P A and Romanach S S (2007) Economic and conservation significance of the trophy hunting industry in sub-Saharan Africa. *Biological Conservation* 134:455–469

Mace G M, Balmford A, Boitani L, Cowlishaw G, Dobson A P, Faith D P, Gaston K J, Humphries C J, Vane-Wright R I, Williams P H, Lawton J H, Margules C R, May R M, Nicholls A O, Possingham H P, Rahbek C and van Jaarsveld A S (2000) It's time to work together and stop duplicating conservation efforts. *Nature* 405(6785):393–393

Mansourian S and Dudley N (2008) *Public Funds to Protected Areas*. Gland: WWF International

Mbembe A (2001) *On the Postcolony*. Berkeley: University of California Press

Mitlin D, Hickey S and Bebbington A J (2007) Reclaiming development? NGOs and the challenge of alternatives. *World Development* 35(10):1699–1720

Myers N and Mittermeier R A (2003) Impact and acceptance of the Hotspots strategy: Response to Ovadia and to Brummitt and Lughadha. *Conservation Biology* 17(5):1449–1450

Nelson F and Agrawal A (2008) Patronage or participation? Community-based natural resource management reform in sub-Saharan Africa. *Development and Change* 39(4):557–585

Neumann R P (1998) *Imposing Wilderness. Struggles over Livelihood and Nature Preservation in Africa*. Berkeley: University of California Press

Neumann, R (2004) Moral and discursive geographies in the war for biodiversity in Africa. *Political Geography* 23:813–837

Neves K (2010) Cashing in on Cetourism: A critical ecological engagement with dominant E-NGO discourses on whaling, Cetacean conservation and whale watching. *Antipode* 42(3):719–741

Owens D and Owens M (1992) *The Eye of the Elephant*. Boston: Mariner Books

Redford K H, Coppolillo P, Sanderson E W, Da Fonseca G A B, Dinerstein E, Groves C, Mace G, Maginnis S, Mittermeier R A, Noss R, Olson D, Robinson J G, Vedder A and Wright M (2003) Mapping the conservation landscape. *Conservation Biology* 17(1):116–131

Sachedina H (2008) Wildlife are our oil. Conservation, livelihoods and NGOs in the Tarangire ecosystem, Tanzania. Unpublished D. Phil thesis, School of Geography and Environment, University of Oxford

Sachedina H T (2010) Disconnected nature: The scaling up of African Wildlife Foundation and its impacts on biodiversity conservation and local livelihoods. *Antipode* 42(3):603–623

Smith R J, Verissimo D, Leader-Williams N, Cowling R M and Knight A T (2009) Let the locals lead. *Nature* 462:280–281

Temudo M P (2005) Western beliefs and local myths: A case study on the interface between farmers, NGOs and the state in Guinea-Bissau rural development interventions. In J Igoe and T Kelsall (eds) *Between a Rock and a Hard Place: African Ngos, Donors and the State* (pp 253–278). Durham, NC: Carolina Academic Press

Tsing A L (2005) *Friction: An Ethnography of Global Connection*. Princeton: Princeton
   University Press
Ward S (1997) Boy Tarzan vs Rambo of the Bush. *The South African Trumpet* June
Wels H and Spierenburg M (2010) Conservative philanthropists, royalty and business
   elites in nature conservation in southern Africa. *Antipode* 42(3):647–670
West P and Carrier J G (2004) Ecotourism and authenticity. Getting away from it all?
   Current Anthropology 45(4):483–498

Chater 4
Shifting Environmental Governance in a Neoliberal World: US AID for Conservation

Catherine Corson

Changing Public–Private Non-profit Power Relations under Neoliberalism

The reduction of the state under neoliberalism,[1] and the resulting reconfiguration of state, market, and civil society relations, has shifted the landscape of twenty-first century environmental governance, in particular opening up room for private actors to influence state policy. This chapter explores how the rise of neoliberalism in the 1980s and its institutionalization in the 1990s underpinned the formation of a dynamic alliance among members of the US Congress, the US Agency for International Development (USAID), an evolving group of environmental non-governmental organizations (NGOs)[2] and the corporate sector around biodiversity conservation funding. By focusing strictly on *international* biodiversity conservation this alliance—driven to a great extent by non-elected agents who are perceived to represent civil society despite their corporate partnerships—has been able to shape public foreign aid policy and in the process create new spaces for capital expansion.

The arguments presented here forge new ground in academic conversations about conservation and neoliberalism by illuminating the concrete practices within US foreign aid through which new forms of environmental governance under neoliberalism are produced. Specifically, they draw on the work of intellectuals who document the opportunities for civil society groups provided by the downsizing of the neoliberal state (eg Castree 2008; Peck and Tickell 2002) to address a lacuna in three interrelated bodies of literature. Together, these works examine the neoliberalism of nature (eg Castree 2008; Heynen et al 2007), the growth of the big international conservation NGOs (BINGOs)[3] and their increasing corporate linkages (eg Brockington,

Duffy and Igoe 2008; Büscher and Whande 2007), and the contemporary move in conservation away from engaging local actors (eg Brosius and Russell 2003; Dressler and Buscher 2008).

While these scholars unveil critical transformations in human–environment relations taking place in the name of conservation under neoliberalism, they have often elided the intricacies of the shifting and uneven power dynamics among state, market and civil society organizations through which such changes have emerged. By focusing on the inter-organizational relations entailed in US environmental foreign aid policy-making, this chapter helps to launch critical engagement with policy issues related to nature's neoliberalization, as called for by Castree (2007). At the same time, it responds to appeals for analysis of the micro-politics of foreign aid donors (Cooper and Packard 1997; Watts 2001), and particularly the sponsors of international conservation (King 2009), to advance an emerging scholarship that applies ethnographic methods to elucidate the internal workings of conservation and development funding institutions (eg Crewe and Harrison 1998; Lewis and Mosse 2006). In doing so, it illustrates how collaboration among the public and non-profit sectors have both reflected and contributed to a move within global environmentalism from an anti-capitalist stance in the 1960s and 1970s to its twenty-first century embrace of the market.

Since the 1970s, environmental NGOs have successfully lobbied the US Congress to support US foreign assistance for environmental issues. In particular, a group of environmental advocacy organizations catalyzed and shaped USAID's initial environment program. However, two interrelated transitions in the relations among USAID, the US Congress, an evolving group of environmental NGOs and the private sector—which have entailed both reactions to and the embracing of neoliberal ideology and reforms — underpin the agency's contemporary emphasis on biodiversity conservation. The first comprised congressional and Democratic administration efforts to direct USAID funding to NGOs—moves that both resulted from and reacted to state privatization in the 1980s and 1990s. The second encompassed NGO-mobilized efforts to protest against neoliberal reforms and protect the environment, the most recent of which, ironically, has invoked neoliberal rhetoric toward this aim.

To summarize briefly, in the context of the burgeoning interest in biodiversity in the 1980s, the Democratic Congress directed USAID to fund biodiversity conservation.[4] At the same time, in an effort to counter Reagan's privatization of state functions and associated turn to private contractors, the Congress mandated the agency to support NGOs. As a result, USAID funded conservation NGOs to implement its emergent biodiversity portfolio. Concurrently, many of the environmental advocacy groups that had launched USAID's environmental portfolio in the 1970s shifted their advocacy efforts to fighting for domestic environmental issues and to protesting World Bank projects. This move eventually left the growing conservation NGOs—now with a special interest in preserving USAID's biodiversity funding—to take up the endeavor to promote environmental foreign aid. The Clinton Administration's embrace of the global environmental agenda, combined with continued privatization of government services and the privileging of NGOs, then reinforced opportunities for the conservation NGOs to benefit from USAID funding. In reaction to internal USAID budget pressures that threatened biodiversity funding in the late 1990s, these NGOs launched a campaign to protect the funding. They consolidated this campaign during the second Bush Administration when concurrent disregard for environmental issues and massive foreign aid reforms again endangered biodiversity funding. In the twenty-first century, the NGOs have attracted powerful corporate and bipartisan political support behind USAID's biodiversity program.

Based on the analysis presented in this chapter, I make three broad claims that offer important insights into the nature of modern neoliberal conservation. First, throughout these transitions, conservation NGOs have capitalized on idealized visions of themselves as representatives of a civil society operating to counter the force of private interests thought to be behind environmental degradation. This vision has sustained their access to policy-makers and influence on public policy despite the multinational corporate partnerships that characterize the BINGOs' twenty-first century operations.

Second, the strict focus on *international* biodiversity has been fundamental to the development of an alliance among the BINGOs, USAID, corporate leaders and members of the US Congress behind US environmental foreign aid. By defining "the environment" as foreign biodiversity, to be protected in parks away from competing economic and political interests and in foreign countries, the BINGOs and allied partners have enticed US politicians and corporate leaders to support environmental foreign aid. They have created an avenue through which they can become "environmentally friendly" without confronting the environmental degradation caused by excessive resource consumption in the USA or the foreign and domestic investments of US corporations.

These successful political strategies, aimed at mobilizing funding for foreign environmental issues, have contributed to the process by which environmentalism has become enrolled in the promotion of capitalist expansion. In fact, I contend that the international biodiversity conservation agenda has created new symbolic and material spaces for global capital expansion. First, it supplies a critical stamp of environmental stewardship for corporate and political leaders. Second, not only does it carve out new physical territories for capitalist accumulation through both the physical demarcation and enclosure of common lands as protected areas, but also through the growing capitalist enterprise that is forming around the concept of biodiversity conservation.

New Forms of Environmental Governance

With its ideological and material antipathy toward state regulation and influence, neoliberalism has become manifest not only in deregulation, but also in re-regulation designed to create new commodities and new governing structures that sustain neoliberalism. As states have faced cuts to fiscal and administrative resources and functions under neoliberal reforms, there has been an associated move toward public–private partnerships, which bring increasing influence by the private and non-profit sectors on what was once state policy. This transition has diffused environmental governance among states, individuals, NGOs, private companies, transnational institutions and local communities. In particular, as the boundaries among the state, private sector and non-profit worlds have become more porous under neoliberalism, certain NGOs have stepped into the vacuum of state social provision (Büscher and Dressler 2007; Castree 2008; Ferguson and Gupta 2002; Igoe and Brockington 2007; Jepson 2005; McCarthy and Prudham 2004; Peck and Tickell 2002). They have become, as Harvey (2005:177) writes, "the Trojan horses of global neoliberalism".

Despite the recognition of these shifts, relatively little empirical work exists on how this dispersed governance in the international development agenda has created new avenues for the intertwining of capitalism and conservation. Through analysis of the everyday politics among NGOs, branches of the state, and the private sector, this chapter uses the lens of USAID to concurrently address this gap and respond to calls for ethnographic information about conservation and development donors. In doing so, it links literature on the neoliberalization of nature with critiques of conservation policy and practice, and it situates itself in three related debates on neoliberalism, conservation and development.

The first of these examines the neoliberalism of nature—seen in measures such as privatization and regulatory rollback, commodification of nature, and new enclosures (for overviews see Castree 2008; Heynen et al 2007). For example, it analyzes the manifestation of neoliberal ideology in environmentalism, evident in tradable emission permits, transferable fishing quotas, user fees for public goods and utility privatization, as well as through corporations' use of environmental discourses (eg McAfee 1999; McCarthy and Prudham 2004). In biodiversity conservation specifically, hegemonic practice now values nature based on its potential market price. The enclosure, commoditization, and privatization of nature has resulted in an emphasis not just on ecotourism, but also on mechanisms like direct payments and public–private partnerships to promote conservation, and management of parks by private entities (Hutton, Adams and Murombedzi 2005; Igoe and Brockington 2007; West and Brockington 2006). Adding a critical new dimension to this literature, I explore how public and private entities, in their endeavor to mobilize funding for environmental foreign aid, have embraced neoliberal ideology, and in turn, contributed to its further expansion.

The second body of literature explores the entwined growth of the BINGOs and their expanding corporate linkages. As conduits of conservation's commoditization, networks of states, foreign aid donors, philanthropists, corporations and conservation organizations have attracted escalating financial support for international conservation since the 1980s. In particular, the BINGOs have turned to corporations and private sponsors to finance conservation (Bailey 2006; Büscher and Whande 2007; Chapin 2004; Dowie 2005). In the process, they have created what I term a "conservation enterprise", in which funds are shifted among public, private and non-profit entities in the name of conservation, without ever being used "on-the-ground". The analysis presented below reveals how transformations in US foreign aid politics and policies have contributed to the emergence of this conservation enterprise.

Accompanying the rise of big conservation is a move within conservation away from engaging local actors. As protected area networks spread across the globe, a third group of researchers continue to document new enclosures of common lands under the guise of biodiversity conservation and the associated displacement of local and indigenous peoples (Brockington, Duffy and Igoe 2008; Brockington, Igoe and Schmidt-Soltau 2006; Chapin 2004). A subset of the aforementioned body of literature looks at the relationship between neoliberal discourse, community conservation, and the privatization and commercialization of conservation (Dressler and Buscher 2008; King 2009; McCarthy 2005). As Igoe and Brockington (2007:446) aptly summarize:

> neoliberalism's emphasis on competition, along with its rolling back of state protection and the social contract, creates spaces in which local people are not often able to compete effectively in the face of much more powerful transnational interests.

At its extreme, the turn away from community conservation is evidenced by the call for a return to exclusionary parks, or what critiques call "fortress conservation", in which local people are excluded, by force if necessary, from utilizing resources within park boundaries (for an analysis, see Adams and Hutton 2007; Brechin et al 2002; Wilshusen et al 2002). Nevertheless, critiques have also contended that ecoregional and transboundary approaches, which aim to extend conservation beyond parks to landscape scales, have simultaneously furthered the influence of state agencies, international and national NGOs and private companies in conservation and reduced investment in local communities (Brosius and Russell 2003; Gezon 2000; Wolmer 2003). Ultimately, this scholarship illuminates how the foreign aid donor politics about which I write manifest on-the-ground and in people's daily lives. However, while these authors show the enclosures happening as a result of biodiversity conservation, they have elided how such endeavors are embedded in and productive of power-laden relationships among financing organizations—a gap that I aim to fill.

Finally, as King (2009) points out, there have been relatively few detailed empirical studies that uncover the internal debates and politics of the organizations behind international conservation. I would add to his critique that few studies explore the *inter-organizational* relationships that comprise the international biodiversity conservation agenda. In attending to this omission, I join a nascent group of researchers who use ethnographic analysis to investigate the micro-politics of foreign aid donors (eg Bebbington and Kothari 2006; Crewe and Harrison 1998; Goldman 2005; Lewis et al 2003; Lewis and Mosse 2006; Mosse 2005).

The findings presented here draw on a decade of experience working in Washington, DC politics, particularly with USAID and the US Congress, as well as on specific research, carried out between 2005 and 2008. In this regard, long-term participant observation informed my appreciation of the political dynamics and bureaucratic cultures analyzed here and fundamentally shaped the original research design, research process and final analysis. The focused research then entailed the analysis of 30 years of USAID policy and program documents related to environment and natural resources issues, as well as USAID congressional presentations, congressional appropriations and authorization bills, hearing records, and NGO lobbying material, for example.[5] It was then supplemented with 70 key-informant interviews with current and former staff from USAID, the US Congress, NGOs, other lobby organizations, consultant groups, research institutions and universities. Interviewees were chosen through a combination of snowball and targeted sampling, in which individuals were chosen for their ability to provide critical information and/or to represent a range of perspectives and organizations. In particular, I selected individuals who were personally pivotal in shaping the environment program— such as those who directed lobbying efforts, wrote particular pieces of legislation, or oversaw USAID's environment program—as well as individuals who could offer extensive historical perspectives. Many of the interviewees were current or former senior officials in these organizations, and as is typical in Washington, many of the interviewees had worked in more than one relevant position and/or organization. Interview data were analyzed and triangulated using content analysis software. However, in order to protect confidentiality, all information is reported anonymously in that sources are identified only by general position and interview dates are not disclosed.

State-Led Environmental Foreign Aid

USAID's official environmental program began in the 1970s, in the context of the Keynesian emphasis on the role of the state in regulating private activities and protecting human welfare. At that time, state-coordinated development ideals still pervaded foreign aid policy, and the rising domestic environmental movement hinged on a belief in state regulation. As McCarthy and Prudham (2004:278) remind us, "Increasing environmental protection was one of the major achievements of the Keynesian state." Moreover, 1970s environmentalism, as manifested in the 1972 United Nations Conference on the Human Environment in Stockholm and 1973 Club of Rome's *Limits to Growth* study, for example, was organized around redressing the negative environmental effects of capital expansion and belief in natural resource limits to economic growth. In this context, the US Congress passed a number of domestic environmental regulatory reforms in the 1970s, including the National Environmental Policy Act (NEPA).

A group of environmental advocacy NGOs catalyzed the creation of USAID's environmental program by first pushing the agency to ameliorate projects' negative environmental impacts and then calling on it to fund environmental projects. USAID's initial program began as the result of a lawsuit brought by the Environmental Defense Fund (EDF) and the Natural Resources Defense Council (NRDC) (EDF v. AID 6 ELR 20121 (D.D.C. 1975)),[6] for the agency's failure to comply with NEPA requirements for environmental impact statements (22 CFR part 216; 41 Fed Reg. 26913; Burpee, Harrigan and Remington 2000; Sirica 1975). Capitalizing on expanding American awareness of international environmental issues, the NGOs then convinced the US Congress to enact an amendment to the Foreign Assistance Act (FAA), which authorized USAID to provide assistance for the protection and management of "environmental and natural resources...upon which depend economic growth and human well being, especially that of the poor" (US Congress 1977). Together with later amendments (US Congress 1981, 1983), these efforts launched the agency's environmental agenda with a focus on state intervention to manage natural resource supplies for the poor.[7]

Despite the belief in state-led development and environmental regulation, the roots of the neoliberal revolution were already forming in the 1970s. These were manifested in, for example, the 1973 "Chicago Boys'" restructuring of the Chilean economy according to neoliberal economic ideals; the Organization of the Petroleum Exporting Countries oil embargo that followed on the heels of the 1971 abandonment of fixed exchange rates; and the resulting recycling of petrodollars through New York investment banks to third world governments (Gore 2000; Gowan 1999; Hart 2006; Harvey 2005).

Evolving Environmentalism under Neoliberal Reforms

With this groundwork laid, neoliberalism rose to prominence in mainstream economic policy in the 1980s, particularly under Margaret Thatcher in the United Kingdom and Ronald Reagan in the United States. While Paul Volcker had begun the initial move toward neoliberal monetary policy during the Carter administration, Reagan furthered deregulated various industries, reduced corporate taxes dramatically and promoted the reduction of big government and the expansion of the private sector. Neoliberal champions blamed foreign aid "failures" on Keynesian state-coordinated development, and USAID began to promote both greater US private sector involvement in foreign aid and private enterprise development in aid recipient countries (Berríos 2000).

Concurrently, the international environmental movement embraced the concept of "sustainable development", defined in the now well-known language of the Brundtland Report (UN 1987, and compare MacDonald this volume). The concept provided an avenue by which environmentalists could engage with and influence an international development agenda heretofore dominated by development economists (Adams 1995; Redclift 1987). However, with its endorsement of economic growth, sustainable development directly contradicted the 1970s environmental movement's stress on "limits to growth". Historical divides between international development and environmental agendas crumbled as the idea of sustainable development offered sufficient flexibility to justify their historically different causes (Raco 2005).

In this context, articulating the economic importance of environmental issues became a critical means of integrating environment issues into USAID's development agenda. Reports and letters that informed the 1970s congressional amendments emphasized the need to protect the natural resources upon which poor people in developing countries relied (Blake et al 1980; Scherr 1978). However, as the program enlarged in the ensuing decades, the agency's environmental advocates found that they needed to highlight environmental projects' contributions to a country's overall economic growth in order to mobilize political support. As one former senior USAID official recounted:

> Articulating the rationale for environment as an economic issue was an important part of advancing the environmental agenda. We had to devise a rationale that was consistent with the agency's mission, and AID's primary mission was economic development.

In an article in which she described the emerging USAID environmental portfolio, influential USAID environmental advisor Molly Kux highlights six "investment rationales" for foreign aid donors to invest in environmental conservation, including maintaining ecosystem services; addressing the rural populations' economic aspirations; increasing nature-based tourism; protecting endangered species; investigating natural economic products; building on indigenous conservation; and promoting sustained yield harvesting (Kux 1991:298–299).

Her priority list reveals historical roots of the contemporary faith in market-based conservation. These roots reflect strategic policy designed to appeal to the constituent groups that support environmental foreign aid: politicians and bureaucrats committed to economic development and, as I will show, congressional advocacy organizations, who were primarily interested in species conservation. Nevertheless, the 1980s tactic of embracing economic growth and sustainable development in order to access the development agenda laid the groundwork for the later rise in market-based conservation approaches and ultimately the process through which conservation became a conduit for capitalist expansion.

The concept of sustainable development provided a platform for NGOs wanting to push back against Reagan's neoliberal efforts to reduce regulation and privatize government services. In 1981, a group of US NGOs mobilized around the idea of sustainable development to form the Global Tomorrow Coalition (GTC). Led by Thomas Stoel of the NRDC and Russell Peterson, President of the National Audubon Society, it aimed to build public and congressional support for global environmental issues and to stem a feared rollback of 1970s environmental programs by the Reagan administration (GTC 1981). The GTC helped to push forward additional amendments to the FAA in the mid 1980s.[8]

It is critical to appreciate that many of the environmental NGOs active on both domestic and international environmental issues at the time, such as the National Wildlife Federation, World Wildlife Fund (WWF-US), and National Audubon Society, drew their mandates, as well as their operating budgets, from the general public's interest in saving animals. Thus, when growing data on forest loss and species extinction, particularly in the tropics, began catalyzing public interest in "biological diversity" in the early-to-mid 1980s (Wilson 1988), these species-focused groups latched on to the biodiversity campaign and helped to push through the 1980s FAA amendments, which emphasized endangered species, forests and biological diversity (Shaffer et al 1987; US Congress 1983, 1986b; USAID 1988).[9] In turn, this legislation laid the foundation for the Appropriations Committee's fiscal year (FY)[10] 1987 mandate that USAID spend US$4 million annually on biodiversity programs (US Congress 1986a).[11]

As the NGOs mobilized around the concept of biodiversity, they profited from a concurrent trend to prioritize participatory development and involve civil society in international development policy formulation and implementation. Foremost in this modified agenda were idealized visions of civil society (Mohan and Stokke 2000). Its proponents saw NGOs as embodiments of civil society and as alternatives to failing states, more efficient and cost effective than governments, and a means to downsize the state by privatizing foreign aid. In the 1980s and 1990s, multilateral and bilateral donors turned more and more to NGOs to implement their agendas (cf Brockington and Scholfield this volume). Edwards and Hulme (1996: 962) cite Organization for Economic Cooperation and Development figures that bilateral foreign aid directed through NGOs rose from 0.7% in 1975 to 5% in 1993–94 (US$2.3 billion in absolute terms). Riding this wave, environmental NGOs across the globe became conduits for new state-sponsored sustainable development programs (Edwards and Hulme 1996; Smillie 1997; Zaidi 1999).

Reflecting this trend, in its own effort to counter Reagan's turn to private contractors, the Democratic Congress began emphasizing the need to implement federal programs through the not-for-profit sector. For example, the 1986 biodiversity amendments also included language that stated:

> Whenever feasible, the president shall accomplish the objectives of this section through projects managed by private and voluntary organizations [PVOs] or international, regional, or national nongovernmental organizations that are active in the region or country where the project is located (USAID 1988; US Congress 1986b).

A former congressional appropriations aide recalled the rationale for the congressional language as a response to environmental NGO complaints about the ineffectiveness of contractors in implementing environmental programs:

> [T]he contractors didn't have a clue how to go outside of their own little boxes. And if they had a clue, they had no desire [so] ... We specifically encouraged [USAID] to work with NGOs.

As a result of the congressional pressure to fund NGOs, USAID set up a PVO office specifically to train NGOs on how to develop proposals for USAID grants.[12]

Under concurrent congressional mandates to expand the biodiversity conservation program and to use NGOs, USAID turned to the wildlife conservation NGOs to implement its biodiversity program. It established an informal biological diversity consultative group with NGOs and began providing small grants of around US$300,000 to the conservation organizations, particularly WWF-US and The Nature Conservancy (TNC) (Shaffer et al 1987; Vincent 1991).[13] It was, as a former USAID environmental officer admitted, "A marriage of convenience for both USAID and the NGOs."

However, as the environmental campaign against Reagan grew, activists who had previously engaged with international environmental issues turned homeward to save domestic environmental programs from the Reagan administration cuts. An environmental NGO representative recalled:

> We got to a point by the early 80s [in which] our ability to impact [USAID] from the outside had really diminished. When the Reagan Administration basically came in and declared war on the environment... To the extent that there had been a focus in the environmental movement internationally all of the sudden it refocused domestically, to try and save the environment at home.

Furthermore, by the end of the 1980s, many of the Washington-based environmental advocacy groups, such as NRDC, the staff of which had previously pushed to institutionalize USAID's environmental programs and which did not accept USAID funds for the most part, began redirecting their attention to the World Bank, building the campaigns against the Indian Narmada Dam and Brazilian Polonoroeste highway projects (Goldman 2005; Rich 1994; Wade 1997). A former congressional aide remembered:

> A lot of the international people [from environmental NGOs] splintered off and formed the Bank Information Center... because they wanted to focus really heavily on the banks. USAID was secondary to them.

At this critical turning point, the environmental consequences of the World Bank's programs and the Reagan revolution drew previous advocates' attention away from USAID's environmental agenda. As a result, the composition of the organizations tracking USAID's environmental programs began to change. The task of mobilizing members of the US Congress to support environmental foreign aid was gradually taken up, over the course of the next two decades, by the growing wildlife conservation NGO. A former USAID official reflected, "WWF-[US] and TNC are still there pushing very hard. But in terms of energy, climate change, pollution issues, that lobby has gone away". As grantees of USAID biodiversity funding, the conservation NGOs developed a rising vested interest in protecting the agency's biodiversity program. The advocacy alliance behind USAID environmental programs was transformed from the initial group of environmental activists who protested against neoliberal policies to a collection of USAID grantee organizations with a specific programmatic interest, and who would ultimately endorse neoliberalism.

"Civil Society"-Led Environmentalism

Building on the 1980s momentum around neoliberalism and sustainable development, the 1990s marked a turning point in which environmentalism began promoting the commodification of nature. Global environmental discourses associated with the rise of international environmental institutions in the 1990s recast environmental degradation as the result of policy failures that could be corrected through market solutions (McAfee 1999). Reflecting this turn, the Clinton Administration embraced both the global environmental agenda and market-based incentives.

However, enthusiastic support for the environmental program was confronted with substantial reductions in financial and personnel resources under combined Republican congressional and Democratic Clinton administration neoliberal government reforms. As part of the 1994 Republican congressional revolution and move to shrink the federal deficit, the Congress dramatically reduced discretionary programs—including USAID's. US foreign aid fell from over US$12 billion in 1993 to US$9 billion in 1996 (Lancaster 2007). Simultaneously, the Clinton administration continued previous Republican administrations' efforts to reduce government and contract out state programs. According to the General Accounting Office (GAO), between FY 1992 and FY 2002, 37% of the agency's direct hire employees either left and were not replaced or were laid off in the reduction-in-forces (GAO 2003; Zeller 2004).

As the agency eliminated permanent slots, it began hiring temporary contractors overseas and in Washington DC to do the work that permanent employees had previously done (Zeller 2004). In this manner, USAID changed from "an agency of US direct-hires that largely provided direct, hands-on implementation of development projects to one that manages and oversees the activities of contractors and grantees" (GAO 2003:6). As the agency reduced the ratio of staff to funds, grant sizes increased, and staff became less involved with the technical aspects of projects, turning it over to contractors and grantees instead.[14]

The government privatization reforms converged with the Clinton administration's view of NGOs as key constituents and belief in the important role of civil society (Lancaster 1999). Vice President Al Gore spearheaded the National Performance Review (NPR), which aimed to streamline government, making it more efficient, effective and businesslike; to instill in federal agencies private sector accountability standards; and to achieve measurable results (Gore 1993). The final NPR USAID report encouraged the funding of US and foreign NGOs (Atwood 1993; Corneille and Shiffman 2004; Gore 1993). Then, in 1995, Gore introduced USAID's New Partnership Initiative as a "new model of combating poverty", which underscored engaging civil society and the private sector in USAID projects and specifically required the agency to channel 40% of aid funds through NGOs (Esman 2003; USAID 1995). By FY 2000, according to one GAO study, USAID directed about $4 billion of its US$7.2 billion program to NGOs (GAO 2002:6).[15]

As the agency was downsizing under Clinton–Gore neoliberal reforms designed to reduce the state, the congressional appropriators tried to protect biodiversity and continued to mandate that the agency finance NGOs to conduct this work. By FY 1995, the appropriators were requiring that USAID spend US$25 million specifically on biodiversity conservation, and the FY 1996 Senate report that accompanied the USAID appropriations bill stated:

> As USAID makes efforts to downsize, it should remain active in regions that are significant for global biodiversity . . . [and] NGOs are often the most cost-effective channels for delivering development assistance (US Congress 1995).

This belief in NGOs as more cost-effective and less biased than contractors persists. For example, as one interviewee, a former senior USAID official and current NGO senior staff member, reflected on the congressional preference for obtaining information about environmental issues from NGOs, "If you're a not-for-profit you have the moral high ground." Thus, while "businesses" are often thought to have greater relative access to Congress in general, NGOs are seen as the protectors of moral issues, environment being one of them.

Nonetheless, in the second half of the 1990s, the Appropriations Committee stopped mandating that the agency spend specific amounts on biodiversity; it simply reiterated annually the desire that the agency continue to support biodiversity funding. Toward the end of the decade, key USAID and NGO staff became concerned about internal agency decisions to prioritize funding for other aid programs over biodiversity conservation, and the conservation NGOs launched a campaign to convince the US Congress to restore USAID's biodiversity expenditures.[16] Their campaign met with success, and the Appropriations Committee responded by stating that "[t]he managers direct AID to restore overall biodiversity funding as well as funding to the Office of Environment and Natural Resources to levels that reflect the proportion of funding of development assistance provided in fiscal year 1995" (US Congress 1999).

In the meantime, in the late 1990s, the BINGOs experienced tremendous growth. Chapin (2004:22) claims that investments by WWF-US, TNC, and Conservation International (CI) in conservation in the developing world grew from roughly US$240 million in 1998 to close to US$490 million in 2002. Rodriguez et al (2007) argue that CI's "Hot Spots" strategy accompanied an increase in overall annual expenditures from US$27.8 million in 1998 to US$89.3 million by 2004, and WWF[-US]'s "Ecoregions" program accompanied a rise in expenditures from US$80 million to US$121.7 million between 1997 and 2005. USAID was a contributor to this growth. In the 1990s, according to Chapin:

> USAID provided a total of roughly US$270 million to NGOs, universities, and private institutions for conservation activities. The lion's share of this amount destined for NGOs was harvested by WWF[-US], which received approximately 45 percent of the available money (Chapin 2004:24).

Citing Chapin's figures, Dowie (2005:4) adds that "The five largest conservation organizations, CI, TNC, and WWF[-US] among them, absorbed over 70 percent of that expenditure."

Thus, by the turn of the century, conservation organizations were expanding, while USAID had shrunk. In the context of reforms that downsized the government and privatized government services, USAID became more of a grants management organization, turning over much of the project management to contractors and grantees. The convergence of the downsizing with reiterated congressional backing of biodiversity meant that the agency was reducing its staff just as Congress was requiring it to spend increasing amounts on biodiversity.[17] At the same time, Clinton Administration and congressional policies promoted the funding of NGOs to carry out the agency's environmental agenda. With a vested interest in USAID's biodiversity program, the conservation NGOs stepped up their congressional advocacy when its funding became threatened in the late 1990s. In the context of the second Bush administration anti-environmentalism, they further consolidated their collaboration.

Redefining Environmentalism after 9/11

The election of George W. Bush in 2000 marked the ascension to power of a neo-conservative administration, which, in the wake of the destruction of the World Trade Center in September 2001, explicitly blurred the lines between foreign aid and military policy. In a major overhaul of foreign aid, the Bush administration created a new foreign aid agency, the Millennium Challenge Account (MCA), designed to assist countries that met US-determined standards for governance and economic reform, the latter of which were solidly neoliberal in orientation (Mawdsley 2007; MCC 2007). Ironically, while a key goal of the MCA was to separate foreign aid from foreign policy (Nowels 2005), the administration tied USAID assistance more closely to foreign policy and military interests, specifically through the 2002 National Security Strategy, which emphasized development as one of three strategic areas of national security, and the 2004 white paper, which aspired to promote anti-terrorism and make foreign aid more effective (The White House 2002; USAID 2004). The administration's vision of foreign aid as a tool of national security was closely connected to its belief in using foreign aid to promote US business abroad (The White House 2002). Initiatives such as the Global Development Alliance (GDA) aimed to leverage private sector funding for development, citing as justification the fact that:

> In the 1970s, 70 percent of resource flows from the United States to the developing world were from official development assistance and 30 percent were private. Today, 80 percent of resource flows from the United States to the developing world are private and 20 percent are public (USAID 2003:3).

Finally, in 2006, Secretary of State Condeleezza Rice and the new USAID Administrator, Randall Tobias initiated a major foreign aid reform, which moved USAID's policy office to the State Department and classified all USAID programs into one of five categories: peace and security; governing justly and democratically; investing in people; economic growth; and humanitarian assistance.

While funds for foreign aid rose "at one of the fastest rates in the history of US aid-giving, expanding by roughly 40 percent between 2001 and 2005" (Lancaster 2007:91), the environment was not a Bush priority. The MCA basically ignored environmental issues until NGOs pushed Congress to mandate that it add an environmental indicator as one of the economic and governance standards that recipient countries had to meet. The 2006 reform placed environmental programs as a subcategory under economic growth, with specific prioritization on biodiversity conservation, natural resources and reducing pollution (US Department of State 2007). The Bush administration's relegation of environment to the economic growth portfolio meant that USAID environment officials had to articulate environmental programs in terms of their contribution to economic growth. Moreover, through the GDA, USAID became a linchpin in the growing NGO–corporate conservation partnerships, which included, for example the Sustainable Forest Products Global Alliance, a US$23 million initiative among the Home Depot, Metafore, the US Forest Service and WWF-US. In this context, articulating the economic benefits of conservation and reaching out to the private sector were necessary political moves to secure USAID funds.

As in the Reagan revolution, the demotion of environmental issues catalyzed a revived congressional–NGO partnership, which aimed to protect the environmental gains of the previous two decades, but which focused this time primarily on biodiversity conservation. The Appropriations Committee—with Senator Patrick Leahy and his foreign operations subcommittee appropriations aide, Tim Rieser, as the key champions—began including biodiversity conservation mandates in the appropriations bill itself.[18] For FY 2002, the committee set aside US$275 million for:

> programs and activities which directly protect tropical forests, biodiversity and endangered species, promote the sustainable use of natural resources, and promote a wide range of clean energy and energy conservation activities (US Congress 2002).

By the next year, the appropriators had included US$145 million just for biodiversity conservation (US Congress 2003) and by FY 2008, US$195 million for biodiversity (US Congress 2007)—a substantial increase over the US$4 million mandate the appropriators had started with in the 1980s.

Thus, while the revived campaign was a reaction to Bush reforms, linkages between conservation and neoliberalism solidified. Under concurrent administration reforms and congressional biodiversity mandates, USAID's environmental program was protected through three avenues: its contribution to economic growth, private sector partnerships through mechanisms like the GDA, and congressionally mandated biodiversity funds. Embracing the private sector and emphasizing the economic value of conservation became necessary political survival strategies. Nonetheless, as the next section describes, new NGO activities on Capitol Hill catapulted these associations to a new level.

Creating the New Conservation Enterprise

The beginning of the twenty-first century also witnessed a mounting collaboration among the four large conservation NGOs: WWF-US, TNC, CI and Wildlife Conservation Society (WCS). In 2003, building on the mobilization started at the end of the Clinton administration to protect biodiversity funds, the four joined forces under an entity called the International Conservation Partnership (ICP). The ICP aimed to build widespread congressional support for conservation through activities such as congressional briefings and lunches, jointly endorsed letters, and overseas congressional trips to priority biodiversity sites.[19] One of the ICP's primary activities was the annual publication of an International Conservation Budget (ICB), which recommended appropriations levels for the major US government-funded international biodiversity conservation programs, including USAID's. Its successful circulation to members of congress speaks for itself in that the amount legislated in the appropriations bills each year generally reflected those promoted in the ICB. For example, for FY 2008, it recommended US$195 million for USAID, which was the amount that the Appropriations Committee included for USAID later that year (US Congress 2007).

The ICP also inspired the 2003 creation of a bipartisan House International Conservation Caucus (ICC),[20] which, with an eclectic membership of 150 ranging from the far left to the far right, had become one of the largest bipartisan caucuses in the House by the end of 2007. Representatives Hal Rogers, a Republican from Kentucky; John Tanner, a Democrat from Tennessee; Ed Royce, a Republican from California; and Tom Udall, a Democrat from New Mexico co-chaired the caucus (DePhillis 2007; ICC 2008). In 2005, a parallel caucus was created in the Senate, and as of 2009, it was chaired by Senators Sam Brownback, a Republican from Kansas; Dick Durbin, a Democrat from Illinois; Olympia Snowe, a Republican from Maine; and Sheldon Whitehouse, a Democrat from Rhode Island.

The caucuses' strict focus on foreign environmental issues has underpinned their ability to bring together a bipartisan coalition that includes a broad spectrum of political perspectives.[21] First, organizing around international biodiversity has enabled the coalition to continue to draw on reliable US public concern about, and therefore congressional interest in, saving charismatic megafauna in other countries. It has also allowed many congressional members to embrace environmentalism without confronting domestic constituents. As one former USAID official said:

> It is easier to do biodiversity overseas than in this country because the conflicts don't involve constituencies of Congress. When there are problems with local communities [overseas], they don't call up their congressman.

As such, the caucuses have attracted individuals who might consider themselves anti-environmentalist on domestic issues by providing a way, as one congressional aide told me, "to be proactive when it comes to the environment without being labeled a traditional environmentalist". What has brought these diverse individuals together is, as an NGO congressional liaison summarized, "They [the members] all like wildlife, and they have all at one time or another visited international park sites abroad." Here, NGO-organized trips for congressional members and staff to biodiversity sites overseas have been important mechanisms to mobilize congressional interest in funding international biodiversity conservation.[22] Most congressional staff I interviewed had been on overseas jaunts with one or more of the four ICP partners.

In July 2006, the ICP formed the International Conservation Caucus Foundation (ICCF), a separate 501C(3) organization, with the mission to support the ICC, and specifically to provide "an educational forum on Capitol Hill, where we keep Members of Congress and their staff constantly updated with information we synthesize from our base of NGO supporters on the most pressing and timely issues in international conservation" (ICCF 2007b). To this end, the ICCF has provided congressional briefings on topics such as WalMart's commitment to sustainability, the USAID-funded Living in a Finite Environment program in Nambia, and the ecosystem payments program in Costa Rica.

Initially funded by the BINGOs, the ICCF has since attracted a number of corporate advisors and sponsors. The members of its advisory "conservation council" have included corporate giants such as Exxon Mobil, International Paper, and Unilever. In putting together this sponsorship, the ICCF has drawn on the corporate linkages of some of its founding NGOs. Bailey (2006) reports that TNC's corporate associates and major contributors at various times have included 3M, Shell Oil, General Motors, Ford Motor Company, BP Exploration, MCI Telecommunications Company, MBNA America Bank, Enron Corporation, Georgia-Pacific, Johnson and Johnson, Weyerhaeuser Company, Waste Management Inc. Monsanto Company and Dow Chemical. Similarly, Chapin (2004:24) writes that "some 1,900 corporate sponsors" donated a total of US$225 million to TNC in 2002, and that "CI's website lists over 250 corporations, which donated approximately US$9 million to its operations in 2003". In 2008, these corporations included, among others, Anglo-American, Chevron and Rio Tinto (CI 2008). Likewise, TNC has chapters at the state and country level, many of which have powerful political and corporate ties.

The ICCF's *Partners in Conservation* brochure showcases a number of public–private partnerships undertaken by the organization's sponsors (ICCF 2007a). For example, it cites the Goldman Sachs and WCS partnership to protect 680,000 acres on the island of Tierra del Fuego, Chile and the American Forest & Paper Association, Indonesia Ministry of Forestry, and CI partnership, entitled the Alliance to Combat Illegal Logging, which uses remote sensing to monitor illegal logging. Other partnerships include the WalMart and the National Fish and Wildlife Foundation Acres for America program, which conserves 1 acre of critical wildlife habitat for every acre of land developed for an existing WalMart facility or new one created in the United States, and ExxonMobil's support for the Save The Tiger Fund, which, it boasts, "represents the largest single corporate commitment to saving a species" (ICCF 2007a:20).

Perhaps most striking is the promotional material put out by the ICCF, including the widely circulated invitations to the ICCF's annual galas. These galas provide vehicles for colossal shifts of funds among *US-based* state, private and non-profit sectors in the name of *foreign* conservation, and as such, contribute to a growing biodiversity conservation enterprise. As colorful collages of corporate and conservation NGO logos, their invitations provide striking symbols of the merging of conservation and capitalism. It is hard to identify where conservation ends and capitalism begins. Attendance at such fundraising events costs, for example in 2006, between US$1000 and 50,000. These galas have honored various celebrities, including former UK Prime Minister Tony Blair, actor Harrison Ford and Chad Holliday, Chairman and CEO of DuPont, for their contributions to international conservation. The 2006 and 2007 invitations boasted meals prepared by "Texas Cowboy Chef Tom Perini", who was "the Caterer to the President of the United States".

Importantly, despite the organization's efforts to increase government expenditures on biodiversity conservation, the ICCF invokes anti-governmental rhetoric to attract conservative and corporate members. For example, ICCF president David Barron underscored the bipartisan nature of the foundation and its neoliberal tenets at the ICCF's September 2006 inaugural gala. In a published letter to the gala attendees, he stated (emphasis in original):

> We are *not* advocating more government. Quite the contrary, we are advocating private sector solutions ... *We are pro-development and pro-business. We are pro-people, pro-wildlife and pro-wilderness.*

The ICCF's outreach to conservative and corporate leaders reflects, like the sustainable development agenda and the framing of environmental issues in economic terms, a successful strategy designed to raise funds and awareness for environmental conservation. Similar to sustainable development, international biodiversity conservation has become a nucleus around which public and private organizations can find common interests. By defining "the environment" as a foreign concern, the ICCF's high-profile effort has provided an avenue for organizations and individuals who have been heretofore considered "anti-environmentalists" to appear environmental. By providing this stamp of environmental approval, however, the biodiversity conservation movement has enabled global capital expansion.

Reconfiguring Environmental Governance around Biodiversity Conservation

In this chapter, I have illustrated how dynamic power relations among environmental NGOs, USAID, the US Congress and private corporations since the 1970s have both reflected and contributed to the contemporary rise of neoliberal conservation. I argue that the ascent and hegemony of neoliberal economic orthodoxy in the 1980s and 1990s established the conditions for two critical changes in relations among these entities, in which the neoliberal state ceded the field of environmental governance to NGOs, and NGOs in turn took up the cause. The resulting reconfiguration of interests and power both led to the rise of biodiversity conservation within the USAID environmental portfolio and has been reinforced by it.

These transformations have entailed intertwined responses to and the embracing of neoliberal ideology and reforms. While NGOs benefited from state privatization, the state's turn to NGOs also aimed to counter the privileged position of the private sector in the privatization process. Similarly, while the environmental advocacy organizations' protest against the World Bank and Reagan administration policies reflected a movement against neoliberal expansion, the more recent conservation NGO-driven endeavor to protect biodiversity funding has invoked neoliberal rhetoric to attract bipartisan and corporate support. In contrast to their 1970s predecessors, NGOs today have built their arguments and legitimacy upon a neoliberal conception of governance.

The BINGOs have attracted corporate and bipartisan support not just through neoliberal rhetoric, but also by focusing strictly on *international* conservation. This emphasis has enabled politicians and corporate leaders alike to become environmentalists without engaging in controversial environmental issues or confronting anti-environmental constituents, and it holds together the alliance among the NGOs, USAID, corporate leaders and members of the US Congress behind environmental foreign aid.

As such, this agenda has both contributed to and been reproduced through new forms of environmental governance, in which non-elected agents—both not-for-profit and private sector—have been able to shape public policy. Here, idealized visions of NGOs as representing civil society has sustained their influence on policy formulation and implementation. Despite the transformed composition of the environmental foreign aid lobby from a loose alliance of environmental advocacy organizations into the contemporary coalition of large conservation NGOs with bipartisan congressional ties and corporate support, the perception of environmental NGOs as a countering force to anti-environmental corporations continues. While advocating against state interference, the alliance has been able to direct public funding. In this process, as the state has turned to private organizations to implement its work, it has in turn become dependent on these entities not just to design and implement programs, but also to mobilize political support for their existence.

Finally, the intertwining of conservation and neoliberalism in Washington DC politics, through public/private/non-profit "partnerships", has facilitated capital accumulation in the United States, as well as created new spaces for capitalist expansion overseas. The biodiversity conservation movement, with its expanding corporate partnerships, has enabled capitalist expansion by not only supporting the enclosure of common lands and exclusion of former resource users, but also by labeling otherwise exploitative corporations as environmental stewards and by building a capitalist enterprise centered on biodiversity conservation. In this enterprise, funds are shifted among government, private and non-profit sectors; conservation organizations have grown into corporate-like entities; and annual galas provide venues for exorbitant expenditures of wealth in the name of conservation.

In this manner, neoliberal conservation has reinforced the separation of environmental concerns from their broader political economic drivers. It has allowed for the conceptualization of environmental goals without changes in existing political institutions, or distributions of economic power or resources flows (Adams 1995; McAfee 1999; Redclift 1987), and in fact, it has reinforced these institutions and resource flows. While there remains a hopeful nod to environmentalism as a potential political movement in opposition to neoliberalism (McAfee 1999; McCarthy and Prudham 2004), it is clear that, in the twenty-first century, an environmental movement, once organized in opposition to economic growth, has instead become its conduit.

Acknowledgements

The research and writing presented here would not have been possible without the generous support of the Mellon Foundation/American Council of Learned Societies, the National Science Foundation, the Rural Sociological Society, the Woodrow Wilson International Center for Scholars, and the University of California at Berkeley. I would also like to thank the numerous people who were willing to be interviewed for this research project, as well as research assistants Caitlin Hachmyer, Sylvia Ewald and Kim Howell. Finally, I owe a great deal to Nancy Peluso, Louise Fortmann, Gillian Hart, Bill

Adams, Jennifer Casolo, Daniel Graham, Elizabeth Shapiro and Desmond Fitz-Gibbon and two anonymous reviewers, all of whom provided insightful commentary on earlier versions of this chapter.

Endnotes

[1] I draw on Castree's (2008:142) "loose consensus" that the "generic elements of neoliberal thought and practice" include privatization, marketization, deregulation, reregulation, market proxies in the residual public sector, and the construction of flanking mechanisms in civil society (state-led encouragement of civil society groups to provide services that state previously did).

[2] Environmental NGOs refer to the broad group of NGOs that undertake environmental activities. Within this category, I distinguish between environmental advocacy groups as those that use the court system to push for environmental change (ie Sierra Club, Natural Resources Defense Council, Environmental Defense Fund), and conservation NGOs as those primarily concerned with wildlife, habitat, and biodiversity (in particular Conservation International, Wildlife Conservation Society, The Nature Conservancy and World Wildlife Fund).

[3] The BINGOs typically refer to Washington-DC based World Wildlife Fund (although sometimes includes its international and country-specific counterparts), Conservation International and The Nature Conservancy. I also discuss the activities of the Wildlife Conservation Society.

[4] I explore these themes in more detail in other articles currently under review.

[5] The term lobby refers to the broad range of educational, outreach and advocacy work that non-profits can undertake under sections 501(c)(3) and 501(c)(4) of the Internal Revenue Code (26 U.S.C. § 501(c)(3)).

[6] Other parties to the suit were the National Audubon Society and the Sierra Club.

[7] This emphasis undoubtedly reflected the passage in 1973 of the New Directions legislation, which, reflecting the 1970s 'basic needs' era, amended the FAA to prioritize issues of relevance to the poor.

[8] Interview with a former environmental NGO representative.

[9] Interviews with environmental NGO representatives.

[10] The federal government fiscal year begins on 1 October of the previous calendar year and ends on 30 September of the year of its title: ie fiscal year 2008 runs from 1 October 2007 to 30 September 2008.

[11] Where previous policy directives by the authorizing committees had ordered the agency to undertake biodiversity projects and authorized spending levels, the Appropriations Committee mandated the *amount* that USAID had to spend on biodiversity conservation.

[12] Interviews with USAID officials.

[13] Interviews with current and former USAID officials.

[14] Interviews with USAID officials.

[15] This report included private voluntary organizations, consulting firms and universities as NGOs.

[16] Interviews with former USAID staff and NGO senior staff.

[17] Interview with a former USAID official.

[18] Interviews with congressional aides, environmental NGO representatives and USAID staff; note that, in contrast to congressional directives in the report language that accompanies appropriations bills, bill language is legally binding and generally reserved for major issues.

[19] Interviews with conservation NGO congressional liaisons and congressional aides.

[20] A Congressional caucus is a group of congressional members who convene to pursue common legislative objectives.

[21] Interviews with conservation NGO congressional liaisons and senior staff.
[22] Interviews with a conservation NGO congressional liaison and a former congressional aide.

References

Adams W M (1995) Green development theory? Environmentalism and sustainable development. In J Crush (ed) *The Power of Development* (pp 87–99). London: Routledge

Adams W M and Hutton J (2007) People, parks and poverty: Political ecology and biodiversity conservation. *Conservation and Society* 5(2):147–183

Atwood J B (1993) *Statement of Principles on Participatory Development*. US Agency for International Development, Washington DC, 16 November

Bailey J (2006) "The limits of largess: International environmental NGOs, philanthropy and conservation." Unpublished PhD thesis, University of California, Berkeley,

Bebbington A and Kothari U (2006) Transnational development networks. *Environment and Planning A* 38:849–866

Berríos R (2000) *Contracting for Development: The Role of For-Profit Contractors in US Foreign Development Assistance*. Westport, CT: Praeger Publishers

Blake R O, Lausche B J, Scherr S J, Thomas B, Stoel J and Thomas G A (1980) *Aiding the Environment: A Study of the Environmental Policies, Procedures, and Performance of the US Agency for International Development*. Washington DC: Natural Resources Defense Council, February

Brechin S R, Wilshusen P R, Fortwangler C L and West P C (2002) Beyond the squarewheel: Toward a more comprehensive understanding of biodiversity conservation as social and political process. *Society and Natural Resources* 15:41–64

Brockington D, Duffy R and Igoe J (2008) *Nature Unbound: The Past, Present and Future of Protected Areas*. London: Earthscan

Brockington D, Igoe J and Schmidt-Soltau K (2006) Conservation, human rights, and poverty reduction. *Conservation Biology* Volume 20(1):250–252

Brosius J P and Russell D (2003) Conservation from above: An anthropological perspective on transboundary protected areas and ecoregional planning. *Journal of Sustainable Forestry* 17(1/2):39–65

Burpee G, Harrigan P and Remington T (2000) *A Cooperating Sponsor's Field Guide to USAID Environmental Compliance Procedures* (2nd ed). Baltimore, MD: Program Quality and Support Department, Catholic Relief Services

Büscher B and Dressler W (2007) Linking neoprotectionism and environmental governance: On the rapidly increasing tensions between actors in the environment-development nexus. *Conservation and Society* 5(4):586–611

Büscher B and Whande W (2007) Whims of the winds of time? Emerging trends in biodiversity conservation and protected area management. *Conservation and Society* 5(1):22–43

Castree N (2007) Neoliberal ecologies. In N Heynen, J McCarthy, S Prudham and P Robbins (eds) *Neoliberal Environments: False Promises and Unnatural Consequences* (pp 281–286). London: Routledge

Castree N (2008) Neoliberalising nature: The logics of deregulation and reregulation. *Environment and Planning A* 40:131–152

Chapin M (2004) A challenge to conservationists. *World Watch Magazine* November/December:17–31

CI (2008) Corporate partners. http://dev2.conservation.org/discover/partnership/corporate/Pages/default.aspx (last accessed 4 April 2008)

Cooper F and Packard R (1997) Introduction. In F Cooper and R Packard (eds) *International Development and the Social Sciences: Essays on the History and Politics of Knowledge* (pp 1–41). Berkeley, CA: University of California Press

Corneille F and Shiffman J (2004) Scaling-up participation at USAID. *Public Administration and Development* 24:255–262

Crewe E and Harrison E (1998) *Whose Development? An Ethnography of Aid.* London: Zed Books

DePhillis L (2007) Foundation shuns "polarizing" issues, enlists strange bedfellows. *Greenwire* 12 July

Dowie M (2005) Conservation refugees. *Orion Magazine* (November/December)

Dressler W and Buscher B (2008) Market triumphalism and the CBNRM "crises" at the South African section of the great limpopo transfrontier park. *Geoforum* 39:452–465

Edwards M and Hulme D (1996) Too close for comfort? The impact of official aid on nongovernmental organizations. *World Development* 24(6):961–973

Esman M (2003) *Carrots, Sticks, and Ethnic Conflict: Rethinking Development Assistance.* Ann Arbor, MI: University of Michigan Press

Ferguson J and Gupta A (2002) Spatializing states: Toward an ethnography of neoliberal governmentality. *American Ethnologist* 29(4):981–1002

GAO (2002) *USAID Relies Heavily on Nongovernmental Organizations, but Better Data Needed to Evaluate Approaches.* Report to the Chairman, Subcommittee on National Security, Veterans Affairs, and International Relations, Committee on Government Reform, House of Representatives, Report No. GAO-02-471. April

GAO (2003) *Strategic Workforce Planning Can Help USAID Address Current and Future Challenges.* Report No. GAO-03-946. Foreign Assistance: Report to Congressional Requesters. August

Gezon L (2000) The changing face of NGOs: Structure and communitas in conservation and development in Madagascar. *Urban Anthropology* 29(2):181–215

Goldman M (2005) *Imperial Nature: The World Bank and Struggles for Social Justice in the Age of Globalization.* New Haven, CT: Yale University Press

Gore A (1993) *From Red Tape to Results: Creating a Government that Works Better and Costs Less: Report of the National Performance Review.* Darby, PA: Diane Books Publishing Company

Gore C (2000) The rise and fall of the Washington consensus as a paradigm for developing countries. *World Development* 28(5):789–804

Gowan P (1999) *The Global Gamble: Washington's Faustian Bid for World Dominance.* London: Verso Press

GTC (1981) *The Environmentalist* 1(2):168

Hart G (2006) Post-apartheid developments in comparative and historical perspective. In V Padayachee (ed) *The Development Decade? Economic and Social Change in South Africa 1994–2004* (pp 13–32). Pretoria: HSRC Press

Harvey D (2005) *A Brief History of Neoliberalism.* New York: Oxford University Press

Heynen N, McCarthy J, Prudham S and Robbins P (eds) (2007) *Neoliberal Environments: False Promises and Unnatural Consequences.* New York: Routledge

Hutton J, Adams W M and Murombedzi J C (2005) Back to the barriers? Changing narratives in biodiversity conservation. *NUPI* Forum for Development Studies: No. 2-2005

ICC (2008) International Conservation Caucus: Members. http://www.royce.house.gov/internationalconservation/members.htm (last accessed 1 May 2008)

ICCF (2007a) *ICCF Partners in Conservation.* Washington DC

ICCF (2007b) International Conservation Caucus Foundation overview: Mission. http://www.iccfoundation.us/aboutus.htm (last accessed 19 December 2007)

Igoe J and Brockington D (2007) Neoliberal conservation: A brief overview. *Conservation and Society* 5(4):432–449

Jepson P (2005) Governance and accountability of environmental NGOs. *Environmental Science & Policy* 8:515–524

King B (2009) Commercializing conservation in South Africa. *Environment and Planning A* 41:407–424

Kux M (1991) Linking rural development with biological conservation. In M Olfield and J Alcorn (eds) *Biodiversity: Culture, Conservation and Eco-Development* (pp 297–316). Boulder: Westview Press

Lancaster C (1999) *Aid to Africa: So Little Done, So Much To Do*. Chicago: University of Chicago Press

Lancaster C (2007) *Foreign Aid: Diplomacy, Development, Domestic Politics*. Chicago: The University of Chicago Press

Lewis D, Bebbington A, Batterbury S, Shah A, Olson E, Siddiqi M S and Duvall S (2003) Practice, power and meaning: Frameworks for studying organizational culture in multi-agency rural development projects. *Journal of International Development* 15:541–557

Lewis D and Mosse D (eds) (2006) *Development Brokers and Translators: The Ethnography of Aid Agencies*. Bloomfield, CT: Kumarian Press

Mawdsley E (2007) The Millennium Challenge Account: Neo-liberalism, poverty and security. *Review of International Political Economy* 14(3):487–509

McAfee K (1999) Selling nature to save it? Biodiversity and green developmentalism. *Environmental Planning D: Society and Space* 17:133–154

MCC (2007) Millennium Challenge Corporation selection criteria. http://www.mcc.gov/selection/index.php (last accessed 19 December 2007)

McCarthy J (2005) Devolution in the woods: Community forestry as hybrid neoliberalism. *Environment and Planning A* 37:995–1014

McCarthy J and Prudham S (2004) Neoliberal nature and the nature of neoliberalism. *Geoforum* 35:275–283

Mohan G and Stokke K (2000) Participatory development and empowerment: the dangers of localism. *Third World Quarterly* 21(2):247–268

Mosse D (2005) *Cultivating Development: An Ethnography of Aid Policy and Practice*. London: Pluto Press

Nowels L (2005) *Millennium Challenge Account: Implementation of a New US Foreign Aid Initiative*. Report for Congress. Congressional Research Service. Updated 25 May

Peck J and Tickell A (2002) Neoliberalizing space. *Antipode* 34(3):380–404

Raco M (2005) Sustainable development, rolled-out neoliberalism and sustainable communities. *Antipode* 7(2):325–347

Redclift M R (1987) *Sustainable Development: Exploring the Contradictions*. New York: Methuen

Rich B (1994) *Mortgaging the Earth: The World Bank, Environmental Impoverishment, and the Crisis of Development*. Boston: Beacon Press

Rodríguez J P, Taber A B, Daszak P, Sukumar R, Valladares-Padua C, Padua S, Aguirre L F, Medellín R A, Acosta M, Aguirre A A, Bonacic C, Bordino P, Bruschini J, Buchori D, González S, Mathew T, Méndez M, Mugica L, Pacheco L F, Dobson A P and Pearl M (2007) Globalization of conservation: A view from the South. *Science* 317:755–756

Scherr J (1978) Statement on behalf of the National Audubon Society, the Natural Resources Defense Council, the Nature Conservancy and the Sierra Club on S. 2646 The International Development Assistance Act of 1978 Regarding Environment,

Natural Resources and Development to the Senate Foreign Relations Committee, 4 May

Shaffer M L, Satterson K A, Carr A and Stuart S (1987) The biological diversity program of the US Agency for International Development. *Conservation Biology* 1(4):280–283

Sirica J J (1975) *Environmental Defense Fund, Inc et al, (plaintiffs) v. United States Agency for International Development, et al, (defendants)*. Order with attached stipulation, Civil Action No. 75-0500. 1 December

Smillie I (1997) NGOs and development assistance: A change in mind-set? *Third World Quarterly* 18(3):563–577

The White House (2002) *National Security Strategy of the United States of America*. September

UN (1987) *Our Common Future*. Report of the Brundtland Commission, A/42/427, General Assembly, Development and International Co-operation: Environment.

USAID (1988) *USAID Policy Paper: Environment and Natural Resources*. Report No. PN-AAV-464. Bureau for Program and Policy Coordination, Washington DC, April

USAID (1995) *Core Report of the New Partnerships Initiative*. Report No. PN-ACA-951. Washington DC, draft 21 July

USAID (2003) *The Global Development Alliance: Expanding the Impact of Foreign Assistance through Public–Private Alliances*. Washington DC

USAID (2004) *White Paper: US Foreign Aid: Meeting the Challenges of the Twenty-first Century*. Bureau for Policy and Program Coordination, US Agency for International Development, Washington DC, January

US Congress (1977) International Development and Food Assistance Act. *Title I, International Development Assistance, of a bill to amend the Foreign Assistance Act of 1961 to authorize development assistance programs for fiscal year 1978, to amend the Agricultural Trade Development and Assistance Act of 1954 to make certain changes in the authorities of that Act*. Public Law 95-88. 3 August

US Congress (1981) International Security and Development Cooperation Act. *Title III, Development Assistance, of an original bill to amend the Foreign Assistance Act of 1961 and the Arms Export Control Act to authorize appropriations for development and security assistance programs for fiscal year 1982, to authorize appropriations for the Peace Corps for the fiscal year 1982, to provide authorities for the Overseas Private Investment Corporation, and for other purposes*. Public Law 97-113. 29 December

US Congress (1983) International Environment Protection Act. *Title VII, International Environmental Protection, of a bill to authorize appropriations for fiscal years 1984 and 1985 for the Department of State, the United States Information Agency, the Board for International Broadcasting, the Inter-American Foundation, and the Asia Foundation, to establish the National Endowment for Democracy, and for other purposes*. Public Law 98-164. 22 November

US Congress (1986a) Senate Report 99-443 to accompany S.2824. *Foreign Assistance and Related Programs Appropriations Act, 1987*. Public Law 99-59. 16 September

US Congress (1986b) Special Foreign Assistance Act. *Title III, Protecting Tropical Forests and Biological Diversity in Developing Countries, of a bill to amend the Foreign Assistance Act of 1961 to provide assistance to promote immunization and oral rehydration, and for other purposes*. Public Law 99-529. 24 October

US Congress (1995) Senate Report 104-143 to accompany HR1868. *Foreign Operations, Export Financing, and Related Programs Appropriations Act 1996*. Public Law 104-107. September 5

US Congress (1999) House Conference Report 106-479 to accompany HR 3194. *Making consolidated appropriations for the fiscal year ending September 30, 2000, and for other purposes.* Public Law 106-113. 17 November

US Congress (2002) Foreign Operations, Export Financing, and Related Programs Appropriations Act. *Making appropriations for foreign operations, export financing, and related programs for the fiscal year ending September 30, 2002, and for other purposes.* Public Law 107-115. 10 January

US Congress (2003) Foreign Operations, Export Financing, and Related Programs Appropriations Act. *Division E, Foreign Operations, Export Financing, and Related Programs Appropriations, of joint resolution making consolidated appropriations for the fiscal year ending September 20, 2003 and for other purposes.* Public Law 108-7. 20 February

US Congress (2007) Consolidated Appropriations Act, 2008. *Making appropriations for the Department of State, foreign operations, and related programs for the fiscal year ending September 30, 2008, and for other purposes.* Public Law 110-161. 26 December

US Department of State (2007) *Supplemental Reference: Foreign Assistance Standardized Program Structure and Definitions.* 15 October

Vincent R M (1991) Biological diversity and Third World development: A study of the transformation of an ecological concept into natural resource policy. Unpublished PhD thesis, Oregon State University

Wade R (1997) Greening the bank: The struggle over the environment. In D Kapur, J Lewis and R Webb (eds) *The World Bank: Its First Half Century* (pp 611–734). Washington DC: The Brookings Institution

Watts M (2001) Development ethnographies. *Ethnography* 2(2):283–300

West P and Brockington D (2006) An anthropological perspective on some unexpected consequences of protected areas. *Conservation Biology* 20(3):609–616

Wilshusen P R, Brechin S R, Fortwangler C L and West P C (2002) Reinventing a square wheel: Critique of a resurgent "protection paradigm" in international biodiversity conservation. *Society and Natural Resources* 15:17–40

Wilson E O (ed) (1988) *Biodiversity.* Washington DC: National Academy Press

Wolmer W (2003) Transboundary protected area governance: Tensions and paradoxes. Paper presented at the Transboundary Protected Areas in the Governance Stream of the 5th World Parks Congress, Durban, South Africa, September

Zaidi S A (1999) NGO failure and the need to bring back the state. *Journal of International Development* 11:259–271

Zeller S (2004) On the work force roller coaster at USAID. *Foreign Service Journal* April:33–39

Chapter 5

Disconnected Nature: The Scaling Up of African Wildlife Foundation and its Impacts on Biodiversity Conservation and Local Livelihoods

Hassanali T. Sachedina

Introduction

> One of the things I'm a real believer in is what I call "T-shirt diplomacy". You produce T-shirts focusing on certain key flagship species, then you distribute them widely in the communities immediately adjacent to the protected area...and when people are wearing a shirt about a particular species, they start thinking about it. T-shirt diplomacy is one of the best tools for conservation education [Russell Mittermeier, President, Conservation International (CI), in the documentary feature "Hotspots"].

Conservation non-governmental organizations (NGOs) strongly influence conservation funding, wildlife policy and rural livelihoods throughout the world. Despite increasing academic and popular critiques of these organizations (Chapin 2004; Dowie 2005; Igoe and Kelsall 2005), and an extensive literature on governance of development NGOs, there is little empirical information about the financial and programmatic accountability of conservation NGOs (Edwards and Hulme 1996; Salamon and Geller 2005). The basic problem here is a lack of the good data, experience and close observation necessary to explore how NGOs perform in different circumstances, and how they respond to the pressures on them. NGOs and their staff can be reluctant to talk about problems.

This study describes some of the challenges of operating in the highly bureaucratic world of modern conservation where there are often significant and fundamental differences between the values of donors and funding agencies and those people whose lives are influenced by the ensuing conservation interventions. Conservation NGOs may cast themselves as rural development agents in their pursuit of conservation. The "T-shirt diplomacy" quote by the President of CI, one of the world's largest conservation NGOs, moments before departing from a rural Malagasy community in an expensive helicopter charter, illustrates that conservation agencies are potentially out of touch with the needs and aspirations of rural populations. The irony of a wealthy organization, traveling expensively, and hoping to enhance conservation outcomes with T-shirts, illustrates the well-intentioned but often disconnected nature of conservation NGO interactions with people. In this regard, CI is not necessarily alone. This chapter attempts to provide empirical information regarding the financial and programmatic accountability of another large NGO, African Wildlife Foundation (AWF). It is based on my 6 years' work with AWF as a fundraiser and 3 years PhD research in the Maasai Steppe, one of AWF's core regions of activity.

The Maasai Steppe, in northern Tanzania, is a significant place for conservation interests. It is considered by conservation and development agencies as a site of global biodiversity, primarily because of its significance due to its population of large migratory mammals. The Maasai Steppe includes lucrative national parks such as Manyara and Tarangire and includes part of the famed Ngorongoro Conservation Area (Figure 1).

My PhD research was conducted outside Tarangire national park in Simanjiro District. The Simanjiro Plains are critical to Tarangire national park in that large numbers of zebra and wildebeest migrate to the plains for approximately half the year. Crucial to any understanding of conservation dilemmas in the area is the fact that the health of wildlife populations in these ecosystems depends on their access to lands outside existing national parks and game reserves. Existing protected areas, although large, are not sufficient to provide for wildlife. To survive they have to move onto neighboring village lands, following the rains and the mineral phosphorus, which is deficient in Tarangire's soils. Populations of most large mammals in the Tarangire ecosystem (the park and its dispersal areas) declined by at least 50% over a 10-year period due to poaching and habitat change, primarily outside of the park. To advance conservation goals, therefore, conservation organizations have to work in rural areas on village land and attempt to make these often problematic wildlife valuable to villagers. Crucially, in these predominantly pastoral areas, this means discouraging cultivation and supporting pastoralism, because the extensive livestock husbandry pastoralists practice allows more space for wildlife than does crop cultivation.

The challenge that conservationists face in that respect is that for many decades rural villagers have been in conflict with the state over conservation in this region. Some have suffered eviction,

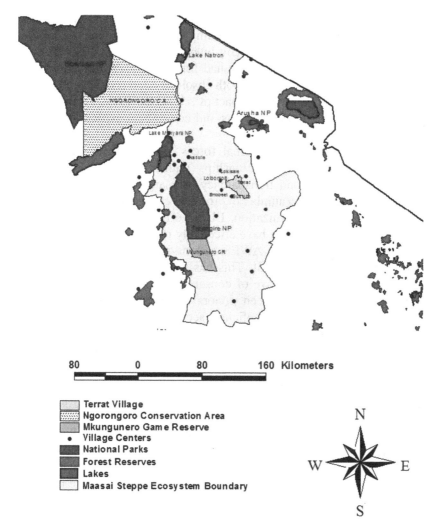

Figure 1: A map of the Maasai Steppe

others restrictions on their land use activities which have caused
impoverishment (Brockington 2002; Igoe 2004; Igoe and Brockington
1999; Homewood and Rodgers 1991). Advancing conservation goals
requires overcoming this legacy of discontent and suspicion. In doing
so, more than trust needs to be built; people need to experience positive
aspects of conservation such as poverty alleviation and empowerment.

I will examine AWF's work in the region in the context of its remarkable growth financially and as a major player in the conservation NGO
scene in Africa. I will argue that AWF has successfully transformed itself
from an important regional player into an organization with continental

ambition, with relationships with donors and government scaling up in proportion to its goals and funding needs. I document AWF's successful transformation into a Big International NGO (BINGO), and the consequences this growth has had for the choices and actions of the organization. AWF's growth both absolutely and relatively makes it an excellent case to study the impact of scaling up—not just increased size—and the structural constraints and contradictions that this could entail.

I examine the local and global forces shaping NGO values and accountability and how relationships with donors and the State affect accountability and transparency. My argument is that AWF's transformation has had a number of profound consequences for the work and practices of the organization. I believe that its relationships with donors and the government have effectively reduced AWF's grassroots accountability. As a result, AWF's actions increasingly contribute to poverty and disempowerment. This has heightened, not reduced, the suspicions that villagers have of conservation initiatives. I argue that existing arrangements between donors and grantees present serious challenges to notions of "beneficiary accountability", which suggests that to be effective as empowerment agents, NGOs should have some "downward" accountability to their beneficiaries (Kilby 2006).

Methods

This chapter is summarized from my PhD thesis. This largely took the form of traditional biological anthropological research of agro-pastoral livelihoods. I spent about 2 years in Emboreet village (in Simanjiro District, Manyara Region), in which several thousand people live on the western edge of Tarangire national park (Figure 1). I conducted two livelihood surveys comprising one extensive survey of a large number of households (n = 226), and another intensive repeat round survey of a smaller group (n = 37). I also attended village government meetings, ceremonies and NGO meetings and talked with diverse groups and individuals about their lives, histories and environmental interactions on a more informal basis. This work was conducted in Kiswahili, with translation into Maa by local interpreters where necessary. I read local, regional and national archives, interviewed tourism operators in the region about their interactions and payments to village government officials, and followed some Maasai migrants to the nearby tanzanite mines where they worked as brokers. This work gave me some insights into the needs, politics and dynamics of rural life and the fears, needs and tensions that it comprised.

To fund my stay in Tanzania, I worked part-time for AWF in Arusha as a Senior Program Design Officer. This role was primarily related to fundraising from government agencies and professional foundations. I had worked for AWF for 2 years prior to beginning my doctorate, and for other organizations in a similar capacity, and had been the lead or sole writer of successful conservation funding applications worth over $12 million over 6 years. I took the job with AWF against the wishes of my academic supervisors and in order to position myself well for a career in conservation after my thesis. I needed the experience and the money. And besides I am Kenyan born and raised, and have worked all my life in Kenya and Tanzania on different conservation projects. I did not view "fieldwork" as a foreign experience that is endured. I was coming home and thought nothing of taking part-time work.

To my surprise and disquiet, AWF's performance became a central part of the story my thesis had to tell. I found myself, unwittingly, in a position both to observe its fundraising success, and the consequences of its expenditure decisions. By working both at a village level and in the central offices I was able to form a unique perspective on the organization's work and achievements.

AWF's Growth and Development

AWF began as the African Wildlife Leadership Foundation (AWLF) and was incorporated in 1961 in Washington DC. The founding vision of AWLF was to train African Protected Area managers to replace European colonial-era wardens. AWLF's patriarchs shared several values: a love for African big-game hunting, a conviction that Africans could destroy wildlife in newly independent countries, and close connections to the US political establishment.[1]

AWF remained a relatively minor player until the late 1980s; thereafter two factors became important in its growth. First there was the ivory ban campaign in 1989. AWF is credited with catalyzing the marketing campaign to end the ivory trade. Its profile grew and membership grew to unprecedented levels—to over 100,000 members in 1990—the highest figure in its history (cf Bonner 1993:53). It proved a valuable source of unrestricted funds.[2]

Second, restricted funds have become particularly important to AWF since the 1990s. In particular, US Agency for International Development (USAID) support transformed the AWF Tanzania program, which in turn transformed the outlook and operations of AWF. In 2004, US government support to AWF totaled 74% of the total organizational budget, although it steadily slid to 55% by 2006 (Sachedina 2008). USAID supported a US$3.5 million project over 7 years beginning in 1989 in Tanzania and a further 8-year project worth US$10.5 million in 1998. AWF's relationship with USAID was privileged; no other organization in Tanzania initially benefited from the levels of support

AWF obtained. It resulted in close working relationships with the Tanzanian government.

USAID support helped to metamorphose AWF into one of the wealthiest and more powerful conservation organizations working in Tanzania. The success with USAID owed much to the skills and influence of Patrick Bergin, who was rising rapidly through the ranks of the AWF at the time to become the President and CEO based in Washington DC in 2001. Bergin was the prime mover behind the transformation of the vision and sphere of operations of the AWF. In 1998, 37 years after its inception, AWF was limited to East Africa. Internally, senior management strongly believed it needed to grow. They recognized the huge opportunity that government-based funding, especially from the USA, represented for AWF. However, AWF needed a new program—one compelling enough to attract increased funding. As then Vice President for African Operations, Bergin envisioned and championed the "Heartlands" program—a number of landscape level programs to reach across Africa.[3] AWF pursued, and was rewarded with, significant funding from USAID and other US government agencies to establish Heartlands in other African countries.

AWF also introduced a proactive approach to its branding. It sought the advice of an image consultant in the USA, who was able to use their extensive knowledge of American values to brand AWF. For example, the term "Heartland" was selected as it alludes to the geo-political and culturally significant Heartlands of the American Midwest. Another criterion was "inspirational" value; this means, at least in part, the ability of a landscape to attract funding.[4] Note, however, that a general aspect of the Heartlands program is that AWF branded some landscapes as Heartlands before the organization had much of a presence there, such as in the Samburu Heartland in Kenya and Limpopo in Southern Africa. At the outset of the Heartlands program, AWF needed compelling descriptions to attract support for the Heartlands program, but in a number of Heartlands AWF operated with a skeleton budget and staff. The Heartlands programme could sometimes be more evident as an aspirational vision than a programme of activities.

By 2004, AWF had become increasingly dependent on US government sources of funding, with 40% of the annual organizational budget contributed from this source alone, and 78% of AWF's restricted budget. As a result, AWF values began to be intertwined more fully with US foreign policy goals. AWF needed to integrate its funding, budget and policy to mirror its primary donor, summed up by the CEO as: "Basically, AWF is becoming an extension arm of USAID in Africa".[5] AWF's enterprise programs were designed to appeal to USAID's models of business development, institution building and democratization.

The importance of this relationship, however, has begun to change. The amount of aid from US government sources to AWF's total budget declined steadily to 29% of the total annual budget by 2006. A primary catalyst in shifting donor flows from US government sources was a new "Strategic Framework for US Foreign Assistance" following September 11, 2001 and related to the "War on Terror" (cf Corson this volume). Formerly an independent agency, USAID was absorbed into the Department of State to "ensure that our foreign policy and development programs are fully aligned to advance the National Security Strategy of the United States" (USAID 2003), generally reflecting the increasing militarization of US foreign policy in Africa (Donnelly 2006; Garamone 2007; Pincus 2007).

The restructuring of USAID reduced the amount of funding available for biodiversity in the post 9/11 geo-political environment (Elliott 2006). Non-environment related levels of US foreign aid to Tanzania significantly increased, primarily driven by Tanzania's geo-political importance in the "War on Terror". For example, Tanzania received an unprecedented aid package worth US$885 million for roads, power and water projects (*The Guardian* 2007), in addition to increased investments in HIV/AIDS and the health sector. While USAID reduced biodiversity funding, it concurrently increased funding to the HIV/AIDS sector. Overnight, AWF tried to style itself as a competent health education delivery agency in order to access funding to keep staff employed and operations going (AWF 2004).[6] The bid was unsuccessful.

This has precipitated another metamorphosis in AWF. AWF's new fundraising strategy to counter the drop in US government funds was to focus on private individuals, foundations and European multilateral funding (AWF 2007: 2).[7] Sustaining existing sources of support and finding new ones were vital for the organization's ambition. In 2006, AWF announced a major partnership with Starbucks Coffee.[8] AWF's 2006 *Annual Report* illustrates this shift towards private fundraising with a US$100 million capital campaign announcement that aimed to raise US$65 million from individuals, corporations and private foundations, and US$35 million in official aid (AWF 2006a:3).

The results of all this activity have been, by most assessments, remarkable. AWF has grown considerably (Table 1). It is now the fourth largest conservation NGO working in Africa (see Brockington and Scholfield this volume and Table 2). Globally, AWF belongs to a second tier of NGOs, below the four main "BINGOs" (Table 3). Within Africa, however, it is a key player. AWF is one of the main partners of The Nature Conservancy (the world's wealthiest conservation NGO) as the latter sought to begin operations in Africa. AWF's rapid recent growth makes it a particularly interesting case study of the nature of conservation work produced in neoliberal policy environments.

Chasing the Burn Rate: The Consequences of Financial Success

The more recent trend (driven partly by the World Parks Congress) of linking international conservation funding to development has clearly had a big influence on many "aspirational" NGOs such as AWF. The growth of funding, of donor support and of a closer relationship with government, has had a number of consequences for the financial and operational cultures of AWF. Not all of these were conducive to effective, or just, conservation practice.

USAID refers to expenditure of its funding by NGOs as the "burn rate". Organizations are required to report to USAID on a quarterly level regarding whether their "burn rate" is being met. If not, they are required to explain why expenditure does not meet prescribed targets. The very term "burn rate" does not psychologically lend itself to effective performance in terms of prioritizing expenditure. The emphasis is on spending money, not spending it well. The aid agency incentive system is such that opportunities for prestige and promotions are greater when aid agencies conduit more money (Gibson et al 2005).

As a decentralized organization, each AWF office managed its own finances with oversight from Washington. Independent auditors conducted an annual organizational audit in Washington where consolidated program expenses were examined based on information received from the field. Donor reporting policies called for the receipt of a spreadsheet of aggregated expenses and a slick narrative report; financial auditing was not conducted at a country office level as part of this annual audit. The external auditor noted the limitations of the organizational audit:

> Accordingly, we do not express an opinion on the effectiveness of AWF's internal control over financial reporting...Our consideration...would not necessarily identify all deficiencies in internal control that might be significant deficiencies or material weaknesses (Raffa 2007:16).

In the Arusha office, where I worked, the pressure to achieve a high burn rate, coupled with a lack of program staff and financial management capacity, resulted in weak financial management. Following a USAID budget cut to AWF in 2004, an internal analysis of AWF Tanzania's financial management—in which I took part—revealed a quagmire of budgeting and accounting problems, over-expenditure, and wrongly allocated finances.

AWF's relationship with its donors could have been more conducive to spending money more effectively in the field. Despite some internal resistance, AWF adopted USAID's travel allowance rates (known as per diems) to compensate its own staff when they travelled.[9] In rural

Africa, where subsistence costs were often only a few dollars at most per day, USAID's per diem rate of US$52 per day was a lucrative source of personal income for NGO staff and government partners. Per diems have become a general phenomenon in Tanzania's development sector. NGOs in Tanzania promoted the per diem system and made it a pillar of the development culture. Few NGO meetings will attract government attendance unless attendees are paid per diems. AWF adopted the widespread use of high per diems based on AWF experience in Tanzania where per diems had been a necessary part of AWF gaining access to, and legitimacy with, government bureaucrats. An AWF senior manager referred bluntly to per diems in managing community and district officials as "...basically we are bribing them".[10]

As AWF income grew from 2001 to 2006, so did its indirect cost rate: from 11% to 18.75% by 2006. Indirect costs are costs that cannot be identified with a single contract or grant. Each grant has the indirect cost rate deducted to fund activities of the organization. Examples of indirect costs are office space rental, utilities, and clerical and managerial staff salaries (USAID 1992:1). The increase in indirect cost harvesting illustrated the increasing centralization of AWF, for example, more managers paid international salaries and expenses of larger offices. The most important expenditure in the offices with which I was familiar were operational funds—labour and benefits, professional services, institutional overheads, office expenses, travel allowances, workshops and vehicle running. Table 4 provides AWF Tanzania's financial year 2007 budget with an example of funding allocations grouped by AWF designated budget lines. It shows that operational funds consumed 63% of the total budget.

It was not clear to me and several other senior staff how well conservation objectives were advanced by expenditure of these operational funds.[11] Conservation audits were not conducted. The highest allocation of AWF Tanzania's budget, by far, was staff salaries and benefits. This was the same model for AWF operations across Africa.[12] However, very few of these staff were actually based in a project area. In Tanzania, all staff resided in the city of Arusha and travelled to the field intermittently for work. The organization effectively spent more on its presence in urban and policy circles than in villages where conservation issues were being played out on a day-to-day basis. For example, consider the spread of AWF offices. In 2007, AWF purchased a house in Nairobi in a million dollar project (AWF 2007:42) as "An organization that seeks to ... position itself as a primary champion of Africa's great conservation and tourism landscapes needs to look the part" (AWF 2007: 2). In 2006, the CEO announced the establishment of a significant presence in Johannesburg (Bergin 2006). Impressive urban offices lent an aura of legitimacy to donors and government but they are also ineffective instruments for building good relations with people in rural areas. Aside from using up funds, offices distant from actual Heartlands, and heavily invested with staff and infrastructure, often kept AWF far from the realities of conservation on the ground. Furthermore AWF's fundraising model resulted in most program staff working largely or exclusively as relationship managers for donors and government bureaucrats. The growth strategy skewed AWF's capacity towards urban grant managers rather than rural extension agents, a core capacity needed for engaging local communities in conservation and livelihood improvement.

A rural focus is vital because that is the main arena where AWF's conservation goals are played out. AWF has to contribute to effective conservation practices on community and private lands beyond the national parks. However, most Heartlands were staffed with only one or two "community conservation officers". Community conservation officers were usually junior in hierarchy and disempowered, under-capitalized and marginalized from AWF's management. It was virtually unheard of for community officers to attend AWF program or annual meetings, yet these fora were where AWF's thinking and conservation learning took place. The pressure to raise more resources while achieving the quarterly burn rate refocused even well intentioned staff towards donor and state relationship management, and away from rural activities. AWF Arusha employed two symbolic community conservation officers, both urban based.

AWF's urban focus contributed to the problems of engaging with conservation problems at the village level. Most people living and working in urban areas, and urban-based conservation offices, do not appreciate the dilemmas and problems of rural work. I certainly did not while I was there, or at least initially. My colleagues and I believed wholeheartedly in the work and rhetoric of the organization. That is what motivated us. It was only as I spent more time in the villages (partly at my PhD supervisor's insistence) that I came to appreciate the problems of an urban focus. It is particularly important to consider how AWF's performance fared at the village level, where the conservation problems it sought to address in its Heartlands programme were found.

AWF at Village Level
In rural areas AWF's performance, despite its fundraising success, presented a number of problems. AWF's approach was to work through government to enhance its legitimacy, networks of power, and donor relations. However, this approach compromised its ability to function as an independent civil society organization, and failed to challenge the power of the central state or the corruption of local government.

AWF found itself promoting policies that either disadvantaged rural villagers or supported poor local institutional performance, or both. As we shall see, this added to a history of conflict between rural people and conservation authorities that has had the effect of making conservation acutely unpopular in precisely the places where it needs to be valued for conservation goals to be advanced.

As part of a broader strategy to promote wildlife's survival outside protected areas, the Tanzanian government is supporting the establishment of wildlife management areas (WMAs) in which villages can set aside land for tourist use in partnership with tour companies and benefit from the resulting revenues. All WMAs are to be facilitated by different NGOs that work in different parts of the country. The WMAs are AWF's primary community conservation implementation strategy in Tanzania. But this approach had weaknesses due to AWF's poor rural extension strategies and the fact that the WMA mechanism is so often contested at village levels.

The Burunge WMA in Babati District, which was facilitated by AWF, was among the first to be gazetted in Tanzania. Different accounts report significant internal conflicts within the WMA; two villages, Minjingu and Vilima Vitatu, claimed they never accepted a WMA (Igoe and Croucher 2007; Nelson, Sulle and Ndoipo 2006). Igoe and Croucher reported that the Babati District Game Officer (DGO) was responsible for evictions of families within the Burunge WMA, that beacons marking village natural resource management areas were placed without Village Assembly agreement, and that villagers alleged that the DGO had bribed village leaders (Igoe and Croucher 2007). AWF did not get involved in the unsavoury business of displacements, but was aware of them in Burunge and continued to fund the work of DGOs in various districts of the Maasai Steppe (AWF 2000). AWF is also responsible for facilitating the Enduimet WMA in Longido District in which Sinya village is found. Local people complained of attempts by the DGO to "force" them to accede to the WMA (Nelson, Sulle and Ndoipo 2006). A lack of understanding about the WMA has led to the defacement of several Enduimet WMA beacons and local level calls for the withdrawal of villages from the WMA (Nelson 2007).

The "Sinya battles" and AWF's failure to challenge the government reveal the depth of the problems posed to AWF's conservation and development objectives. Since the late 1990s, AWF had worked with communities in the West Kilimanjaro area. Sinya Village in Longido District entered into partnership with Tanganyika Wilderness Camps to establish a photographic tourism camp starting around 2000. By 2004, Sinya may have been generating up to US$40,000 per year from tourism (Nelson, Sulle and Ndoipo 2006: 22). Tanganyika Wilderness Camps also established a tourism camp adjacent to Tarangire in Minjingu

Village in the Burunge WMA. Both Sinya and Minjingu villages were located within tourism hunting blocks allocated to Northern Hunting Enterprises Ltd. Northern Hunting sued TWC for violating its use rights in village lands allocated through the Ministry of Natural Resources and Tourism. Tanganyika Wilderness Camps subsequently ceased operations in both villages.

The basic background to these conflicts is that villages control land and make land use decisions while the Wildlife Division (the government agency charged with managing wildlife in game reserves and on communal land throughout Tanzania) administers hunting concessions on village lands. The Wildlife Division has strong instrumental interests in the perpetuation of this system, as do the hunting companies that buy the concessions and engage in a lucrative industry. The Wildlife Division had earlier mounted its own bid to evict Tanganyika Wilderness Camps, based on the Tourist Hunting Regulations of 2000, and consistently supported Northern Hunting's case to exclusive control over the area. Tour operators engaged in community-based tourism, and local NGOs keenly followed the proceedings of the Tanganyika Wilderness Camps–Northern Hunting court case. They were concerned that if Northern Hunting was successful, that it could set a legal precedent for hunting companies and the Wildlife Division to evict photographic tourism operators from villages in tourism hunting blocks.

Northern Hunting won the case and Tanganyika Wilderness Camps relocated their camps to other villages. Villagers felt disempowered losing control of their revenue source and land tenure. The case highlighted critical legal issues between land jurisdiction granted by the Wildlife Conservation Act of 2004 over game controlled areas (many of which fall on village land) and jurisdiction granted by the Village Land Act (1999). The Commercial Court showed a disregard for village rights to enter into contracts and manage their lands as a corporate entity, rights established in the local government and Village Land Acts (Tanzania Natural Resource Forum—TNRF 2005).

To the disappointment of villagers, private sector and local NGOs, AWF failed to support the villagers' cause against the government. Although they felt that, as an international organization, AWF should have been more creative with the State, AWF, from its perspective, did not feel that supporting community interests was worth jeopardizing its legitimacy with the state by openly becoming a Wildlife Division adversary.[13] Through its disengagement from contentious community conservation rights issues in pastoral landscapes and from civil society, AWF inadvertently undermined policy reform and sources of tourism revenue to local communities.

AWF is not in an easy position. If it adopts a strong line with the Tanzanian government it risks losing its access to power. A shot was fired across AWF's bows when the Wildlife Division instigated an audit of AWF in 2003. In its pursuit for funds, AWF had apparently failed to adequately sub-grant funding to the Wildlife Division through heavy investments in Manyara Ranch, a private landholding, and to fund AWF's central machine. A high-level government fact-finding delegation reinforced the role that the state expected AWF to play and threatened to deregister AWF if it did not support its policies.[14] Organizations like "Savannas Forever" and "HakiElimu", which have displeased the government, have placed their whole operations in jeopardy. The Tanzanian Ministry of Education and Culture attempted to ban the widely respected NGO "HakiElimu" for allegedly disparaging the image of the Tanzanian education system by providing critical data in relation to the education sector's performance (URT 2005). But this incident shows that a more robust response to this sort of sabre rattling is possible. Ninety-six national and international NGOs— including local and international conservation organizations—signed a statement challenging the legal basis for the ban and supporting the role of civil society to participate in national policy processes. AWF was among several NGOs that did not endorse this statement, illustrating its isolation from the wider civil society movement, engendered by its donor and State ties. AWF might have been able to circumvent some of its organizational constraints by better partnering with and fostering Tanzanian civil society, especially local conservation NGOs.

Furthermore, the increasing importance of USAID funding has had important implications for its relationship with African governments. In order to achieve growth, AWF needed to position itself as a close government partner in order to gain legitimacy, influence and funding. USAID funds are actually given to NGOs with a strong "partnership" with the Tanzanian state. Therefore, AWF had no choice but to pursue close relations with the Tanzanian government, and especially with the Wildlife Division. Not surprisingly, AWF has since appeared wary of criticizing the Wildlife Division for slow implementation of WMAs and other policy reforms. By not placing pressure on government to reform the WMA process and provide greater support for community rights, AWF's support for WMAs is effectively dysfunctional. AWF's networks of power help to prop up a centralized and non-transparent state apparatus that, almost 10 years after WMAs were legislated in the Wildlife Policy of 1998, had failed to gazette a single WMA, delaying opportunities for poverty alleviation and enhanced resource rights. AWF is contributing to slowing down reform in the wildlife sector.

In addition to its lack of engagement with the central government, there were also problems with the organization's engagement in regional and local government. In one glaring case, AWF actively promoted economic displacement in communal land in Simanjiro District. An

AWF newsletter in April–July 2006 self-congratulated: "a process of zoning and management planning...has recently received support from the Regional Government who issued a moratorium on land allocations to agriculture" (AWF 2006b:5). This statement reflected an extraordinary sequence of events that involved poor science, ignorance of socio-economic processes occurring at the village level, and networks of power. The Simanjiro Plains are an AWF "priority intervention"— meaning that it requires focused investment on an annual basis due to its conservation value. Priority interventions are about quick wins—there was pressure to report progress to donors each year in what were often very complex projects. After years of community resistance to formal conservation initiatives in Simanjiro, AWF saw an opportunity to use its government influence to limit farming in the Simanjiro Plains. It reported an imminent threat to tourism revenues using inaccurate rates of agricultural change in the Plains. AWF did not conduct its own survey, consult researchers or remotely sensed imagery prior; but it reported:

> There is increased unplanned agriculture in the Simanjiro Plains (about 50% cultivated) threatening wildlife and livestock ecosystems in the Simanjiro. They will have adverse affects on migratory wildlife of the core protected areas of Tarangire and L. Manyara which greatly use the Simanjiro Plains for calving (USAID/Tanzania 2005:1).

AWF obtained two non-empirical studies from the internet to support its claims on land use change and submitted them to the regional authorities: a 1998–1999 case study from Monduli District (Conroy no date);[15] and a cursory 1996 consultancy report (Sikar 1996).[16] Manyara regional authorities accepted this outdated, non-spatially specific work. The Manyara Regional Commissioner then issued an edict directing the Simanjiro District Council to immediately stop land allocations for agriculture in the Simanjiro Plains until village Land Use Plans were completed. The edict directly equated land use change with a decline of over 50% in wildebeest and zebra populations in Simanjiro (Manyara Region 2005). In response to the Regional Commissioner's letter, the Simanjiro District Council (SDC) forbade villages in the Simanjiro Plains to allocate land for agriculture until Land Use Plans were completed. The edict threatened:

> For those of you who have already been allocated agricultural land to stop all agricultural activities in the plains until the plan mentioned above is completed. Anyone who disobeys this directive will be prosecuted (SDC 2006).

This is exactly what Simanjiro villagers fear from external conservation interests. Village and district leaders were reluctant to share or discuss the Regional Commissioner's letter with their fellow villagers. I was given a copy by the acting Simanjiro District Executive Director with a warning not to show it to villagers due to its inflammatory nature. Even the District Lands Officer had not heard about the Regional Commissioner's letter until I asked him about it in May 2006. The edict had the potential to affect thousands of people. Not all villagers of Emboreet, Terat and Sukuro farmed in the plains, but almost all were inextricably tied to the overall agro-pastoral economy.

Village elites profited from the lack of transparency by ward and village leaders regarding the letter to interpret its message in their own way. In Emboreet, village leaders surveyed each of the sub-villages for external farmers using the letter as an opportunity to expel outsiders, and potentially increase local land holdings for their own networks. As word slowly leaked out regarding the true nature of the letter, grassroots concern flared. Although no NGO is mentioned in the letter, people suspected NGOs of being behind the Regional Commissioner's letter. They also associated it with an address on national radio by President Kikwete calling for cattle de-stocking and preservation at a village level to protect the environment. Confusion reigned at a local level: villagers were confused why white commercial farmers were still permitted to farm in Simanjiro; they thought that the Regional Commissioner's letter was forged or illegal in the face of the Village Land Act (1999); or it was simply a ploy for Tarangire to expand. Why did it state the land was unsuitable for farming when people continued to farm the same plots year after year? The letter spoke of the wildebeest population; but why did it not reference growing human populations and their needs? Village leaders attempted to get a Village Assembly to obtain communal agreement to abide by the Regional Commissioner's letter, without actually sharing a copy of the letter. Villagers reported that it felt like the colonial treaty that removed people from Mkomazi and Ngorongoro. One informant reported that large areas of Emboreet remained unfarmed in 2007 in compliance with the Regional Commissioner's directive: "all those big shambas of Waarusha and all other invaders are left the way they are, no tenant is allowed to farm in Emboreet".[17]

Emboreet was already dependent upon food aid. In the dry season, Red Cross vehicles escorted trucks of food aid to Emboreet and poorer pastoral families depended on these deliveries. Many households were unable to depend on their livestock for their subsistence needs; 80% of households did not own a minimum of five tropical livestock units per adult unit (TLU/AU). These people were required to farm in order to survive. Restricting farming in this village was a direct threat to pastoral food security, in addition to the impoverishment effects of losing a revenue stream through crop sales. Crop sale values in Emboreet were close to US$100,000 per year in 2003 and 2004, not including the value of crop gifts and subsistence consumption. To put it mildly, the multiplier effects of reduced revenue from crops would have a massive effect on household economies.

These effects were a direct result of AWF interference in local and regional government. AWF's ability to influence the state into a ban on agriculture was the dividend of years of investment in developing good relationships with the State. However, AWF's use of its power to lobby for the regulation of pastoral livelihood activities conflicts with its rhetoric of empowerment, human rights and poverty alleviation. Not consulting scientific data or researchers and not involving local people risked causing hunger.

Finally rural distrust and suspicion of AWF was furthered by the problem of displacement. AWF's rhetoric towards the unacceptability of displacement in conservation practice is underscored by AWF's CEO:

> we must place a far greater priority in setting aside and protecting significant pieces of land to ensure the survival of any species, and secondly, the livelihoods and well being of African people must not be excluded by any wildlife conservation objective on the continent.[18]

But AWF has in fact sought opportunities to increase the number of large land units it controls or owns, specifically targeting underperforming government ranches or private lands which were occupied. AWF encouraged the removal of people living in these areas, but the removal was actually undertaken by other agencies, as AWF did not want to be associated with evictions. In addition, AWF promised to fundraise internationally for land management but based on the caveat that evictions were completed before AWF assumed management, such as the example of Saburi Estate adjacent to Manyara Ranch. Former employees, who had allegedly not been paid for several years, were labeled "squatters". The term "squatter" was problematic, implying illegal occupation and trespass. Brought to work there, and now destitute, it did not seem appropriate to label them trespassers. Likewise, Makuyuni JKT, a 9000-acre estate bordering Manyara ranch, was encouraged by AWF to relocate a district secondary school with several hundred students. Near Moshi, AWF sought to acquire the 75,000-acre West Kilimanjaro Ranch where there were legal proceedings underway to determine the fate of squatters there.[19] In other cases, AWF raised money for protected areas only to withhold it until evictions were complete, as in the case of Mkungunero Game Reserve.[20] The reality is that an NGO's financial clout can increase the incentive for displacement, such that government agents wanting to tap into its financial resources may feel compelled to deliver a wilderness devoid of people and legal liabilities.

Conclusion

As an American NGO, AWF became a preferential grantee for US government funding and a tool for exporting US models of democracy and enterprise to rural Africa. This suited AWF's ambitions to scale up to a continental level. AWF's transformation was successful in this regard, but it inevitably had consequences for the structure of the organization.

This chapter has sought to emphasize the dilemmas and contradictions that AWF faces. Choices taken at the time of scaling up (to expand, to pursue donors and government) often seem reasonable and the consequences of these choices may not have been clear at the time. In some ways, AWF is not entirely free in the choices it can make. AWF's dependence on US government funding and on Tanzanian government approval place it in a difficult situation.

AWF managers may have become locked into certain ways of behaving and being from which it was difficult to escape. It is not clear whether less reliance on donors and less concentration on government relations would have been a more beneficial approach for AWF. Adopting different practices may have brought about different costs and benefits. AWF thus experienced the challenges faced by organizations whose growth becomes a constraint when not properly managed.

AWF's organizational culture centralized power and muted transparency and accountability in ways that excluded local people from the conservation process and isolated AWF conservation staff in urban offices and well heeled lodges. Instead of increasing popular support for conservation and AWF in my study region, relations have worsened. The decisions that AWF took to work closely with government and donors have variously limited, handicapped, and sabotaged its desires to work with villagers, and sometimes meant it is actively working against them. Tragically, it contributed to the impoverishment of local people and stifled wildlife reform debate linked to empowerment of local peoples. Tens of thousands of dollars which tourists bring to villages like Emboreet are not effectively distributed among the rural poor because of failures in local government practice, which AWF has been unwilling to challenge. Attitudes to conservation are marked by their hostility and Tarangire's neighbours are even deliberately ploughing land in order to make it less attractive to conservation interests and to strengthen their own claims to it.

The case of AWF in Tanzania illustrates what happens when the pursuit of wealth and certain networks of power (such as with the state and donors) conflict with programmatic delivery of an NGO's mission. The people who end up mattering most to international conservation NGOs are donors in the west and African government elites, not poor communities marketed as key partners in glossy communications materials. However, while this account, and my data, emphasize the role of AWF, it is vital to recognize that this is the outcome of the joint activities of donors and grantees. USAID did not take opportunities to manage the growth of its grantee effectively. The success and failures of the AWF are not the property of one institution; they are general to a system of aid and giving.

Postscript

Working for AWF had been the fulfillment of many years' ambition for me. Since the age of 17, and while growing up in Kenya, I had played an active role in conservation activities. It was as a 17 year old that I joined the Kenya Wildlife Service's rhino capture unit as an ordinary hand and helped move 35 animals threatened by poachers. I had no intention of studying AWF for my PhD. I worked there while pursuing the thesis because I needed the money, wanted the experience, and believed in the values and work of the organization.

As my fieldwork came to a close I had a rather different perspective on the achievements of AWF, the consequences of its fundraising success and its contribution to conservation goals. I had communicated my fears to my managers, but did not feel that the information was being acted on. At the end of my fieldwork I was offered a lucrative position in the AWF Washington DC office. Such a position had been my goal for over a decade. I was not convinced, however, that my work there would be as useful to the conservation goals I believed in as it could be. I declined the offer, resigned from AWF and returned to the UK (without income) to complete my studies. Making that decision and living with the break it entailed from the conservation community which had been my world for so many years was one of the hardest things I have done.

I do not regret it. The thesis that this chapter has been based on came as a surprise to the AWF, but they have not contested the data it presents. Moreover, a significant number of people have broadly and enthusiastically welcomed it in the East African conservation community. I have also been able to discuss its findings and implications with AWF staff. I remain, and always have been, an ardent conservationist. My hope is that my findings will become useful to AWF, and the broader conservation movement.

Acknowledgements

My thanks are above all extended to the villagers of Emboreet who permitted me access to the community for 3 years. I owe a large debt of thanks to Dr Daniel Brockington for his academic supervision. Professor Jim Igoe read earlier versions of this manuscript. I am grateful to the Government of Tanzania for permitting me to conduct research in Tanzania, colleagues at AWF who provided feedback on this analysis, and to the University of Oxford, which helped to fund this research.

Endnotes

[1] AWLF was conceived by Judge Russell E. Train, Kermit Roosevelt and other members of the Washington DC chapter of Safari Club International (Garland 2006). Train was the head of the US Environmental Protection Agency (1973–1977) and later President (1978–1985) and then Chairman of the World Wide Fund for Nature (WWF)-US (Bonner 1993). He is recognized as one of the chief architects of the contemporary conservation movement (Bonner 1993; Flippen 2006). Kermit Roosevelt, grandson of former US President Theodore Roosevelt, is also well known for orchestrating a Central Intelligence Agency operation that toppled the Government of Iran in 1953 (Bonner 1993; Perkins 2005).

[2] Restricted funding comprised bi-lateral donor and professional foundation funding that was restricted to specific activities. Unrestricted funding, in contrast, is raised from private donors who do not specify how the money should be spent.

[3] This was part of a trend of international conservation organizations establishing landscape-scale approaches, but in separate places (Redford et al 2003) such as CI's "Hotspots", World Conservation Strategy (WCS)'s "Living Landscapes", WWF's "Ecoregions" and transnational corporation's "Last Great Places".

[4] E-mail, P. Bergin to AWF staff members, 6 November 2006.

[5] P. Bergin, pers. comm. (2002, 2003).

[6] Known as the "tail wagging the dog" phenomenon in fundraising circles.

[7] A key constituency for AWF is the 1 million estimated individuals who contribute to wildlife conservation in America (AWF 2007:21).

[8] http://awf.org/content/solution/detail/3372/ (last accessed 5 March 2007). This was the same month that the company was accused of using its might to block an attempt by Ethiopia's farmers to trademark their most famous coffee bean types http://www.guardian.co.uk/frontpage/story/0,,1931675,00.html (last accessed 5 March 2007).

[9] Per diem refers to travel and subsistence costs on business travel, meals, laundry and ad hoc expenditures. At the time, USAID had a federally granted annual budget of $9 billion; AWF's budget was roughly $9 million from disparate sources.

[10] Discussion, AWF senior manager, Arusha, 16 June 2006.

[11] Discussions, senior AWF staff, Arusha, 2005; discussion, senior AWF manager, Kenya, 2006.

[12] As part of the seven-person technical design fundraising team charged with raising about 50% of AWF's funding from bilateral sources, I was intimately aware of budget needs, donor grants and funding forecasts in different AWF Heartlands.

[13] Discussion, NGO employee, Arusha, 2004.

[14] Interview, Coordinator, Arusha, 16 June 2006.

[15] http://www.lead.virtualcentre.org/en/enl/vol2n1/maasai.htm (last accessed 9 March 2007).

[16] http://www.fao.org/docrep/x0271e/x0271e00.HTM (last accessed 9 March 2007).

[17] E-mail, Emboreet villager to H. Sachedina, 20 February 2007. *Waarusha* are a Maa speaking agro-pastoral ethnic group predominately from the Arusha region. Waarusha constituted the majority of immigrant farmers to the plains as land shortages became acute around Arusha. This was a source of conflict with Simanjiro Maasai.

[18] http://awf.org/content/headline/detail/1182/ (last accessed 7 March 2007).

[19] The case seemed to go in favour of the squatters and resulted in plans to reduce the size of the ranch.

[20] Discussion, Coordinator, Arusha, 2004.

References

AWF (2000) *The African Wildlife Foundation Support to Wildlife Conservation in the Tarangire-Manyara Heartland*. Arusha, Tanzania: African Wildlife Foundation

AWF (2004) *HIV, Environment and Livelihoods (HEAL) Project: Enhanced HIV Mitigation Strategies in the Maasai Steppe Heartland in Northern Tanzania*. Arusha, Tanzania: African Wildlife Foundation

AWF (2006a) *Annual Report*. Washington, DC: African Wildlife Foundation

AWF (2006b) *Putting the Landscape Back Together in Maasai Steppe*. Nairobi: African Wildlife Foundation

AWF (2007) *Board Book of the Annual Meeting of the Board of Trustees*. Johannesburg: African Wildlife Foundation

Bergin P J (2006) President's report. Board of Trustees Briefing Book, Fall Meeting 2006. Nairobi: African Wildlife Foundation

Bonner R (1993) *At the Hand of Man: Peril and Hope for Africa's Wildlife*. London: Simon and Schuster

Brockington D (2002) *Fortress Conservation: The Preservation of the Mkomazi Game Reserve, Tanzania*. London: James Currey

Brockington D and Scholfield K (2010) The conservationist mode of production and conservation NGOs in sub-Saharan Africa. *Antipode* 42(3):551–575

Chapin M (2004) A challenge to conservationists. *World Watch* November/December:17–31

Conroy A (no date) Maasai agriculture and land use change. *Livestock, Environment and Development Initiative (LEAD) of the FAO Electronic Newsletter* 5(2):1

Corson C (2010) Shifting environmental governance in a neoliberal world: US AID for conservation. *Antipode* 42(3):576–602

Donnelly S (2006) The pentagon plans for an African Command. *Time* 24 August

Dowie M (2005) Conservation refugees: When protecting nature means kicking people out. *Orion* November/December:1–9

Edwards M and Hulme D (1996) Too close for comfort? The impact of official aid on nongovernmental organisations. *World Development* 24(6):961–973

Elliott J (2006) Vice President for Technical Design report to the Board of Trustees Fall 2006. AWF Board of Trustees Briefing Book, Fall Meeting 2006. Nairobi: African Wildlife Foundation

Garamone J (2007) DoD establishing U.S. Africa Command. *American Forces Information Service Press Articles*

Gibson C C, Andersson K, Ostrom E and Shivakumar S (2005) *The Samaritan's Dilemma: The Political Economy of Development Aid*. Oxford: Oxford University Press

Homewood K and Rodgers A (1991) *Maasailand Ecology: Pastoralist Development and Wildlife Conservation*. Cambridge: Cambridge University Press

Igoe J (2004) *Conservation and Globalization—A Study of National Parks and Indigenous Communities from East Africa to South Dakota*. Denver: University of Colorado

Igoe J and Brockington D (1999) Pastoral land tenure and community conservation. *International Institute for the Environment and Development, Pastoral Land Tenure Series* 11:1–103

Igoe J and Croucher B (2007) Conservation, commerce, and communities. *Conservation and Society* 5(4):534–561

Igoe J and Kelsall T (2005) *Between a Rock and a Hard Place: African NGOs, Donors and the State*. Durham: Carolina Academic Press

Kilby P (2006) Accountability for empowerment: Dilemmas facing non-governmental organizations. *World Development* 34(6):951–963

Manyara Region (2005) *Circular of the Regional Commissioner No. 1 of the Year 2005.* Colonel (Rtd) Anatoli Tarimo, Manyara Regional Commissioner to Simanjiro District Commissioner. Babati, Tanzania: Manyara Region

Nelson F (2007) *Emergent or Illusory? Community Wildlife Management in Tanzania.* International Institute for Environment and Development

Nelson F, Sulle E and Ndoipo P (2006) *Wildlife Management Areas in Tanzania: A Status Report and Interim Evaluation.* Arusha: Tanzania Natural Resources Forum

Perkins J (2005) *Confessions of an Economic Hit Man.* London: Ebury Press *Washington Post* (2007) U.S. Africa Command brings new concerns. 28 May

Raffa P (2007) *African Wildlife Foundation Financial Statements and Report Thereon for the Year Ended June 30, 2007—Independent Auditor's Report.* African Wildlife Foundation

Sachedina H T (2008). *Wildlife is Our Oil: Conservation, Livelihoods and NGOs in the Tarangire Ecosystem, Tanzania.* Oxford: Oxford University Centre for the Environment

Salamon L M and Geller S L (2005) *Nonprofit Governance and Accountability,* http://alliance1.org/Research/Listening_Post/Communique4_Oct05.pdf (last accessed 1 September 2007).

SDC (2006) *Stoppage of Land Allocations for Any Activities in the Simanjiro Plains of Manyara Region.* Dr Karaine Kunei, Simanjiro District Executive Director to Village Executive Officers—Emboreet, Kimotorok, Loiborsirret, Narakauwo, Terat, Loiborsoit and Nadonjukini. Orkesumet, Tanzania: Simanjiro District

Sikar T O (1996) Conflicts over natural resources in Maasai District of Simanjiro, Tanzania. *Food and Agriculture Organisation of the United Nations Newsletter* 30

The Guardian (2007) Kikwete secures 1trn/- US grant. 1 August

TNRF (2005) *Meeting Minutes: January 25th, 2005.* Arusha, Tanzania: Tanzania Natural Resource Forum

URT (United Republic of Tanzania) (2005) Circular No. 5, 8 September. Tanzania Ministry of Education and Culture

USAID (1992) *CIB 92-17 Indirect Cost Rates, Memorandum for All Contracting Officers and Negotiators.* Washington DC: United States Agency for International Development

USAID (2003) *Strategic Plan: Fiscal Years 2004–2009.* Washington DC: United States Agency for International Development

USAID/Tanzania (2005) *Minutes of the 4th Oversight Steering Committee of the USAID/TZ SO2.* Bagamoyo, Tanzania: United States Agency for International Development

Chapter 6
The Rich, the Powerful and the Endangered: Conservation Elites, Networks and the Dominican Republic

George Holmes

The Dominican Republic is often held up as a paragon of environmentalism, an example of excellent conservation of tropical forests in the global South. It has an extremely extensive network of protected areas, with 21.5% of the country in national parks and scientific reserves (International Union for Conservation of Nature—IUCN category I and II protected areas—the strictest levels of protection), the fourth highest percentage of any country in the world (UNEP-IUCN 2006). It is lauded because it has achieved this despite its relative poverty and high population density. Yet there are two other important aspects to Dominican conservation that have led geographer Jared Diamond to use it in his best selling book "Collapse" (2005) as an example to illustrate his argument that societies chose their ecological destiny and ultimately their ability to survive. Firstly, the Dominican Republic shares the island of Hispaniola with Haiti, but the two countries have stark social, political and environmental differences. Diamond argues that the Dominican Republic has chosen to strictly protect much of its forests, creating a relatively stable society and a reasonably prosperous economy, whereas Haiti has chosen not to protect its forest, trapping itself in a cycle of environmental devastation, political instability and underdevelopment. Such a comparison makes the policies of the Dominican Republic look even more enlightened, progressive and effective. Secondly, Diamond argues that these policies are even more remarkable because they were created by Dominicans, not imposed or brought in by outside actors. He views them as the result of a "vigorous indigenous conservation movement" (p 332) consisting of Dominican NGOs and activists, and the legacy of a repressive but environmentally minded president.[1]

The Dominican Republic provides an excellent case for understanding issues of environment and development, particularly because it is held up by conservationists as an example to follow. This chapter is concerned with how the Dominican Republic came to be so conserved, how this "vigorous indigenous" movement achieved so much. The case is important not just for the reasons stated above, but because of how it interacts with two trends in conservation's relationship with capitalism.

Firstly, it adds to a growing literature that shows how conservation policy results not from popular movements, but from well-connected elites. Alongside others in this collection, this chapter explores how specific, identifiable networks can be shown to drive conservation. Membership of these networks is often facilitated by wealth, and they often include corporations and their directors. The capitalist influence on conservation comes not merely through the presence of these actors in networks, but through their ability to shape how conservation happens and what form it takes, its ideologies and its practices, in particular in ensuring that conservation does not critique capitalism. Elites are important when considering conservation's relationship with capitalism. This chapter presents an account of global and Dominican conservation elites, the similarities they share, the complex relationships between them, and their relationship with capitalism.

Secondly, it provides a counterpoint to a related body of literature critiquing the influence of large international conservation NGOs, who are seen as too powerful, able to act with impunity, often causing harm whilst pursuing biodiversity protection. These NGOs are often considered prime conduits by which forms of contemporary capitalism affect conservation practices. Their seemingly excessive influence over conservation in the global South is considered damaging to local environmentalist politics. The Dominican Republic shows a case where their actions are restricted, where they cannot act as they might like, identifying where the limits to this trend may lie.

At the heart of this chapter is the idea of networks of elites in conservation, and capitalism's place within this. It discusses different elites, the tensions between them, yet it shows how elite networks sharing a particular form are present throughout conservation, beginning with an exploration of the structures and workings of global conservation elites, particularly the role of large conservation NGOs. It then chronicles the growth of Dominican conservation, exploring the networks of influence within this. There is a particular focus on the struggles for control of protected areas between globally powerful and nationally powerful conservation actors. It shows the Dominican conservation elite as a scaled-down version of the transnational counterpart, using the same methods to be successful, while at the same time rejecting the actors (although not necessarily the ideologies and practices) of the global elite. Both elites share a reluctance to critique capitalism. This chapter concludes by showing how and why elites may have had such an influence on conservation both in the Dominican Republic and globally.

Elites and Conservation

This chapter uses Woods' (1998) idea of elites, defined by the disproportionate influence they have in comparison with the rest of society, based on ideas of societal networks and connections. Certain individuals have more influence than others in particular areas, and they form the elites for that particular area. In contrast to early conceptualisations of elites, membership in this model is partially detached from social structures—for example, although parliamentarians may be seen as the quintessential elites, television presenters and academics can also be highly influential within society, albeit in a different way. There can be multiple elites in the same field, working at different spatial levels.

Elites in conservation are not a new phenomenon. Adams (2004) and Mackenzie (1988) show how a mix of aristocrats, wealthy hunters and naturalists and colonial bureaucrats drove the creation of protected areas in British colonies in Africa, using their strong social connections to important figures in government as their primary asset, creating the first international conservation NGOs in the process. Gadgil and Guha (1992) demonstrate a similar process occurring simultaneously in British India, built on a foundation of earlier Mughal elite hunting regulations (Rangarajan 2001). The emergence of US protected areas in the late nineteenth and early twentieth centuries was driven by the formal and informal networks of wealthy urbanites, such as hunting or social clubs, and their contacts in government, most notably Theodore Roosevelt (Jacoby 2001). Well-connected elites have continued to drive the expansion of many US protected areas (Fortwangler 2007; Walker 2007). What is new in recent decades is the emergence of a strong global elite in conservation, the strong position of a few NGOs within this, and the opportunities this offers for transnational capitalism to affect conservation.

Recent analyses of global transnational capitalism have offered an insight into the structure of global elites, based around personal interactions and social networking (eg Rothkopf 2008; Sklair 2000), and these provide a good model for understanding global conservation. This transnational capitalist class, as Sklair calls it, consists not just of the directors of large companies who have an obvious impact on the shape of global capitalism, but also the politicians who provide the political structures that allow these companies to be successful, the professionals and thinkers whose ideas vitalise and reinforce capitalism and the media and trendsetters who promote concepts such as consumerism that ultimately underpin neoliberal capitalism. The elite works through the movement of individuals between different sectors, face-to-face interactions and personal contacts between members. Rothkopf highlights both the varied makeup of this elites and the importance of personal contacts by analysing the form and mix of attendees at the annual Davos summits. This structure resembles that of global conservation, which has become dominated by a networked elite made of individuals from NGOs, governments, corporations, science and the media, working through personal contacts built up as individuals move between these sectors and meet socially or at formal spaces such as conferences. Interactions allow the exchange of ideas and lead to mutually beneficial projects across different sectors. While each organisation and actor have their own interests, these exchanges encourage relatively homogenous and dominant discourses and practices in global conservation to emerge, creating the global scale in conservation. There is considerable heterogeneity in conservation overall, but at a global scale there is a much more homogenous set of actors, discourses and practices. As later sections show, similar elites can form at the national scale, and interact with transnational elites in interesting ways.

The transnational conservation elite is driven and dominated by the directors and senior staff of a few large conservation NGOs. NGOs are the heart of global conservation: they have been globalised since their inception (see Adams 2004), they are virtually the only organisations working on environmental issues with global aims and manifestos (younger bodies such as the UN environment programme excepted), and they have a long history of dominating global conservation issues. Their position means that NGOs are often seen as the most capable actors in global conservation and they are given a prominent position in treaties and negotiations on the environment (Princen and Finger 1994). NGOs are at the forefront of conservation thinking, employing large numbers of conservation scientists who are arguably producing the most innovative conservation research (da Fonseca 2003). These two points put NGOs at the heart of a conservation epistemic community, allowing them to frame the terms of debates. Conservation NGOs have benefited in recent decades from a political climate that has been favourable to NGOs generally (Mitlin, Hickey and Bebbington 2007). Globally, conservation is dominated by a few, very large organisations with billions of dollars of assets and excellent political contacts, around whom power and money is increasingly being centred at the expense of smaller organisations (Chapin 2004). Organisations such as The Nature Conservancy, World Wildlife Fund, Conservation International and Wildlife Conservation Society can be distinguished from the rest of the conservation movement by their size, influence and global ambitions, and it is these who form the heart of the transnational conservation elite, dominating how conservation is talked about and practiced at a global scale.

The work of NGOs is aided by politicians and bureaucrats working for national governments or international structures such as the World Bank or UN, who provide structures and policies which support and promote the work of NGOs. These range from formal inclusion of protected areas and NGOs into international development plans and international treaties, such as the UN convention on biodiversity, to debt-for-nature swaps, whereby part of a nation's foreign debt is cancelled in exchange for pledging to create new protected areas. Under US versions of this, large NGOs work closely with the US Agency for International Development to plan, fund and facilitate these swaps, and end up administering the new protected area (Lewis 1999). NGOs extensively lobbied for the creation of such legislation, and swaps are seen as a means by which NGOs have used their contacts to facilitate international expansion (Bernau 2006), as demonstrated in the Dominican Republic. Conservation NGOs have deliberately forged close links to the World Bank and aid agencies, through exchanges of personnel and integrating conservation into major development programmes as a way of increasing their influence (Goldman 2001).

Conservation NGOs have developed close ties to corporations, particularly as corporate directors join NGO boards—between 62% and 72% of the board members of the biggest three NGOs have also had board membership of large corporations,[2] greatly outnumbering those with scientific or technical backgrounds. This provides funding but more importantly brings corporate strategies and management structures into conservation (Birchard 2005). Many of these strategies, such as expansion of NGOs by taking over or merging with others, directly mirror those of large corporations, and have allowed NGOs to rapidly expand.

Global conservation is supported by a group of thinkers and professionals—the scientists, consultants and academics who research, write and advise on conservation—who create and promote forms of conserving biodiversity which are taken up and implemented by other actors. Media actors are very influential, controlling how biodiversity issues are portrayed in the media, endorsing conservation as an intrinsically good thing which cannot be challenged (Mitmann 1999). Repetition of images of wilderness landscapes devoid of their human occupants promotes the exclusionary forms of conservation, such as protected areas, upon which NGOs thrive. Complex realities of animal behaviour or ecology are brushed over to present a simpler portrait that is closer to viewers' preconceptions. Brockington (2009) highlights the importance of "conservation celebrities", individuals who use their fame and influence to promote conservation.

There is considerable cross-over and movement between these factions of the conservation elite: a prominent scientist may also work for an NGO or a bureaucracy, as well as becoming a conservation celebrity by presenting a nature programme on television. Interactions and networking taking place in formal meetings, such as the World Parks Congress and other conferences and workshops, or through informal social contacts such as cocktail parties, allow ideas to circulate and influence to be asserted, and the key tools are access and personal relationships. We can argue for the existence of a transnational conservationist elite, consisting of NGO directors, scientists and thinkers, corporate directors, bureaucrats, celebrities and individuals working in the media who all combine to promote the notion and practice of conservation across the world.

There are two important consequences of the transnational conservation elite. Firstly, it has produced a set of dominant practices and attitudes in global conservation, notwithstanding diversity in conservation at other scales. The transnational conservation elite determine how biodiversity is thought about, discussed and practised by those outside of the elite in a way that reinforces their dominant position. Certain things become taken for granted and others are beyond the terms of the discussion. Debates within global conservation are centred on what form of protected area is best and which type of NGO is most effective (see Adams and Hulme 2001; Adams and McShane 1992; Brandon, Redford and Sanderson 1998; Terbourgh 1999; Western 1997), yet whether NGOs and protected areas—the central actor and key policy recommendation of the transnational conservation elite—are the best ways of protecting biodiversity is beyond the terms of the discussion. By remaining unchallengeable, their dominance continues: protected areas continue to expand worldwide (West, Igoe and Brockington 2006), media representations continue to portray conservation and NGOs as good things, worthy of support (Brockington 2009). Additionally, criteria for assessing success and failures of conservation are set by these same NGOs (Bundell 2006), and by defining concepts and measurements, they can set the terms of debate in a way that serves their own interests and reinforces their status. While there is diversity in conservation practices, the NGOs in the transnational conservation elite have tended to promote exclusionary forms of conservation that have had significant social impacts on local populations, aided by media representations of biodiversity occupying spaces free of human impact and of NGOs as unchallengeably good actors. Importantly the elite's connection with capitalism, through funding and more importantly corporate involvement in NGO boards, makes it unwilling or unable to provide a bulwark to capitalism's impact on the environment—within global conservation, capitalism's environmental impact is beyond the terms of discussion. Not only do protected areas and other projects which they espouse fail to environmentally critique neoliberal capitalism, they allow "greenwashing" (McAfee 1999) of the corporations and organisations that sponsor conservation, as some of the papers in this edition show (see also Chapin 2004; Dowie 1995; Goldman 2001; Rothkopf 2008; Sklair 2000).

Secondly, recent decades have seen the rise of a few large conservation NGOs, whose financial and political resources have expanded massively—The Nature Conservancy grew from annual incomes of US$110 million in 1990 to over $800 million by 2005 (Birchard 2005), and Conservation International, founded in 1987, now has annual incomes in excess of US$200 million (MacDonald 2008). Critics have argued that their dominance of global conservation means they are now too powerful, particularly when these global actors become involved in national scale politics in the global South (Chapin 2004). One way of understanding this is through what Harrison (2004) terms governance states. Following recent post-conditional turns in global governance, the World Bank and aid agencies have been pushing governments in the global South to engage civil society such as NGOs as stakeholders or partners. Governance states form as banks, agencies and civil society actors become so integrated into the workings of the state that they compete with it for sovereignty. Global conservation NGOs have become partners on biodiversity issues, but their ability to lobby powerful organisations such as the World Bank and aid agencies— the result of consciously forged close links—means NGOs have strong influence over the policies of governments in the global South. They now compete with states for sovereignty over natural resources and protected areas. Perhaps the most extreme case of an environmental governance state is Madagascar, explored by Duffy (2006), where conservation NGOs sit on the Donor Consortium which determines future funding priorities and policies for Madagascar, from where they compete with the Malagasy state to shape national environmental policies. As part of governance states, these NGOs are less inclined to criticise the environmental impacts of World Bank driven neoliberalisation and increased capitalisation of nature.

These concepts explain how NGOs, as part of the transnational conservation elite, can become involved in environmental management in the global South. Governance states literature argues that states compete with NGOs for control of policy. The case of the Dominican Republic provides a useful counterpoint, giving a different perspective to theories of weak Southern states and powerful global NGOs. Here there was another set of actors, a vibrant, powerful national conservation elite, with whom the global elite had to compete for sovereignty, for control over the running of protected areas. This shows where the some of the limits to the transnational conservation elite may lie. The next section describes the growth of Dominican conservation and the forces behind it, showing the similarities between transnational and national elites, and the significant tensions that have emerged between the two.

Protected Areas in the Dominican Republic

Conservation in the Dominican Republic has been linked to elites since its inception. The first protected areas were forest reserves created in the 1920s by the occupying US forces to protect the watersheds that fed the tobacco and sugar producing valleys, products destined for US markets (Moya Pons 1995; Turits 2003). These reserves later expanded and became national parks during the Trujillo dictatorship (1930–1961). This was not a conservation strategy, but instead aimed to expropriate land and appropriate timber resources for the personal benefit of Trujillo, who owned a monopoly on timber and sawmills (Bolay 1997). Trujillo's dictatorial rule has been described as "a regime of total plunder organized to furnish him with total control of every economic enterprise existing in the country" (Moya Pons 1995:359).

The main periods of expansion for protected areas and environmental legislation occurred during the rule of Joaquin Balaguer. Originally Trujillo's protégé and puppet president during the last few years of the Trujillo regime, he was elected[3] in 1966 following a tumultuous period of civil war, coups and US invasion that followed Trujillo's assassination in 1961. Balaguer's government was dominated by a small body of advisors, largely made up of military personnel, with a highly centralised presidency as a way of retaining power (Ferguson 1992). As I show later, Balaguer's style of government was of great importance in allowing elite-driven conservation measures. Balaguer's environmental legislation, particularly regarding protected areas, was characterised by deep social impacts and highly centralised planning. In 1967, Balaguer passed law 206-67, which banned lumbering and made all trees, including those on private land, property of the state, rules ruthlessly enforced by the newly created military-run forest police. It significantly disrupted peasant life until its repeal in 1997: Roth (2001) and Rocheleau and Ross (1995) show that this changed how Dominican peasants engaged with trees and forests by removing any incentive they previously had to plant trees and created a disincentive to have trees on their land. As is shown below, such a strategy of crude regulation rather than sustainability or livelihood issues is a consistent feature of Dominican environmentalism. A number of protected areas were created during Balaguer's first period of power (1966–1978), although like the tree law of 1967, they were implemented with minimal planning or consultation, and local peasantry were evicted and excluded without compensation.[4]

Balaguer enacted far more environmental legislation than other presidents, and most Dominican national parks and all scientific reserves were created during his presidencies. In the first year of his second period (1986–1996), he enacted operation *Selva Negra* (Black Forest), a high-profile paramilitary crackdown on charcoal burners and shifting cultivators, and created four new protected areas that increased area under protection by 96% (Roth 2001). In the penultimate year of his rule (when, due to a history of vote rigging and political violence, the Clinton administration forced him not to stand for re-election) his administration created 32 new protected areas totalling 1129.9 km[2]. In August 1996, just 12 days before the handover of power to the new president, Balaguer created another 37 new areas and expanded existing ones to a total of 4932 km[2].

The outburst of environmental legislation at the very beginning of Balaguer's rule was in small part due to outside influences: reports by the Organisation of American States into infrastructure development persuasively argued for watershed protection for hydroelectricity.[5] Yet the large proportion of protected areas in the Dominican Republic is the result of Balaguer's drive to create them, rather than any external pressure. Although his motives are not certain, there is no suggestion that international actors or agendas affecting his thinking. Nor was it due to broad public pressure, as there is no widespread culture of visiting protected areas among Dominicans—foreign tourists visiting beach resorts constitute the vast majority of visitors to Dominican national parks, and the largest park receives fewer than 3000 Dominican visitors annually (ABT Associates 2002). One academic argues that his policies were part of his Machiavellian system of control rather than environmental convictions—park boundaries were created so that enemies could be evicted, or the eviction compensation scheme manipulated to reward supporters.[6] Balaguer's land reforms were a clientelist form of maintaining support (Ferguson 1992; Moya Pons 1995), where favoured groups received large compensation payments, and perceived enemies nothing. Diamond (2005) speculates, among other things, about the influence of his family and his childhood in forming his opinions and desire to protect the environment. Others consider Balaguer's concept of development, particularly his desire to make the Dominican Republic the polar opposite of neighbouring Haiti, socially, culturally, economically and environmentally. Balaguer's virulently anti-Haitian views strongly influenced his politics, and the idea of protecting the environment to benefit hydroelectricity and agriculture projects provides a stark contrast to Haiti's notorious deforestation, erosion, undeveloped agriculture and poverty

(Balaguer 1983; Moya Pons 1995). Anti-Haitian sentiment is widespread in Dominican society, and is manifested largely in "othering" of Haitians as African, black, poor and uncivilised in contrast to a European, white(r), developed, civilised Dominican people (Ferguson 1992; Howard 2001; Wucker 1999). The idea of preventing their country from become like Haiti is a constant theme among Dominican conservationists.[7] As one NGO director wrote:

> Several years ago, I dared to speak in public ... about the state of natural resources in our country. On that occasion, I presented the dramatic situation of Haiti as a mirror of what our future could become if we kept on following the same path ... the country would soon be classified as a "red alert" zone in the world, or even worse—God forbid—as an "ecological catastrophe", just one step removed from being declared as "wasteland" (Armenteros 1989a:10).

Whatever the roots of Balaguer's environmental views, the protected areas policies of his administrations were the result of a particular conservation elite, uniquely placed in Balaguer's style of government. Many analysts describe Balaguer as a *caudillo*, a reference to the quasi-feudal warlords who dominated nineteenth century Dominican politics, using coercion, patronage and a strong public persona to maintain power (Ferguson 1992; Moya Pons 1995). Balaguer operated in a very small circle of selected associates, mainly senior military officials, while remaining as a very visible personality and wielding strong, centralised individual power as president, refusing to delegate decisions to ministers (Moya Pons 1995; Wucker 1999). Despite Balaguer's strongly held environmental views, he created no environment ministry, keeping close control over environment issues; the forestry service was military and thus part of his political inner circle, and the protected areas directorate was part of the office of the presidency. According to foreign aid agency staff, Balaguer's concern combined with the centralised power made liaising with the state on environmental matters very efficient.[8]

Within Balaguer's close circle of loyal associates was Eleuterio Martinez, a connected member of the Santo Domingo elite well known for his environmental concern and rhetoric. A biology professor at the state university, he wrote eloquent newspaper pieces on environmental threats, was a member of the elite Academy of Sciences, and had family and social ties to important senators. Like Balaguer, there was minimal international influence in fomenting his environmentalism—he was educated and has worked solely in the Dominican Republic, always for the Dominican state or Dominican NGOs. His articles for the *Listin Diario* over the last 20 years have evoked impeding threats and the necessity of urgent action:

In 1979, a United Nations official predicted that Haiti could very well become the first real desert in the Caribbean. In 1980, a report by the government of the United States warned that, unless current trends be halted, Haiti would be an ecological wasteland by the year 2000. We can certainly not remain indifferent to the Haitian case. That country is our Siamese twin, a binding condition for very existence on this island ... Can we remain indifferent while our country is being converted into a desert? (Martinez 1988, quoted in Armenteros 1989b:3).

He became very close to Balaguer, who appointed him as head of the forestry service, and he became the main architect of Balaguer's environmental laws in later years, particularly the protected area system. He was influential in shaping conservation because of his close relationship with the president,[9] his academic status and his media profile. There are other cases where closeness to a head of state has allowed conservationists to wield a very large amount of influence (Boza 1993; Garland 2006). Martinez's vision, based on strict protection of areas with no human use allowed, dominated. Category I and II protected areas became the method selected for protecting the Dominican Republic's environment, rather than sustainable development focused policies.

If he [Eleuterio Martinez] had his way, the whole country would be a park (academic and activist).

[Martinez] has a focus that is extremely protectionist. He has little social vision of conservation problems, seeing parks as isolated islands and not part of a sea that is social, political ... far from being good it caused harm to the system of protected areas (protected areas consultant).

Having a centralising presidency with a small number of key advisors creates problems where one person is advising politicians on areas in which the politicians do not have technical expertise, and where no one else is being consulted. Legislation was pushed through on the will of Balaguer and Martinez, without prior studies of the social and economic makeup of the area, run on an ad-hoc basis without long-term planning or management plans. Currently out of a total of 69 terrestrial protected areas, only 13 have management plans and 32 have full-time staff. The few management plans that exist largely contain description of ecosystems, with minimal or no socio-economic data. Many protected areas had ill-defined boundaries, did not consider how local people used the resources contained within them, and did not plan for dealing with the aftermath of evictions and displacement of peasants from newly created parks (Geisler and Barton 1997). Protected areas created substantial hardships for neighbouring communities, resulting in sustained opposition to them:

None of these parks that were created in this other period, in none was the community consulted, and many ended up displacing sizeable populations that were inside or next to the parks that were created, without studies, without consulting, without compensation, without any of this. So if you go to the countryside there remains a negative attitude towards what is the protection of natural resources, protected areas, national parks (protected areas operations director).

NGO workers, consultants and national parks directorate employees all complained of the massive task of reforming the protected area system so that it takes social factors into account.[10] Observers say this process has been "hijacked" by large international tourism companies allegedly aided by corrupt politicians, as subsequent laws aimed at changing park boundaries were altered, removing key beaches from coastal parks, allegedly so they could be developed for resorts.[11]

Conservation Elites in the Dominican Republic since 1996

The end of the Balaguer regime in 1996 represented a new phase in Dominican politics, society and conservation. Environmental issues became a lower political priority: rather than increasing environmental regulation and expanding protected areas, many Balaguer era laws were repealed and national parks shrunk. An environment ministry was created in 2000, but investment in it has waned. The National Parks Directorate, which had reported directly to the office of the presidency during the Balaguer era, became the Sub-secretariat for protected areas as part of the Ministry of the Environment. The ministry has become a means to reward loyal party supporters with jobs, bloated with unqualified staff who take resources away from infrastructure investment, and who are replaced after every election, losing expertise and institutional memory.[12] There has been some devolution of environmental governance, although centralising tendencies are still strong, to the annoyance of environmental and aid NGOs. Some limited protected area co-management projects have been created with community groups, and some other larger projects set up in conjunction with Dominican environmental NGOs. The Dominican government gives little support to NGOs generally, except NGOs directed by politicians used as a front for co-opting state funds to buy political support. Although environmental NGOs almost never receive money from the state, they remain highly politicised along party lines; ruling parties choose to work only with NGOs who support that party. This provides an element of temporariness to membership of the Dominican conservation elite—if the NGO's favoured political party is in power, this gives them the connections to be influential, but they lose this when the party looses.

Another feature of Dominican society now manifest in the conservation sector is the almost oligarchic power of a few well-connected families which dominate the economy. These are descended from the families who controlled trade during the Spanish colonial period and post-independence, particularly industries such as sugar and tobacco. These positions were reworked and reinforced by Trujillo, who rewarded loyalty by granting monopolies to his supporters (Moya Pons 1995). The elites emerged with sufficient power from the Trujillo era to play the central part in the overthrow of Juan Bosch as president in 1962, and continue to dominate politics and the economy (Ferguson 1992; Rosario 1988). These families have also entered conservation, and head up many of the key NGOs. Fundación Progressio, one of the most successful NGOs in terms of influence, was founded by one of the country's leading bankers. The vice-president is the extremely influential Cardinal Nicolas de Jesus Lopez Rodriguez and other board members include oligarchs from tobacco and manufacturing. The board has also included newspaper directors, university rectors and an ex-president of the Supreme Court (Armenteros 1991).These elite board members and their personal contacts have been central to the success of Fundación Progressio since its inception in 1982. As one of their employees put it "[our directors] can phone up the [presidential] palace if they need to". In a rare example of Balaguer era decentralisation, in 1989 the directors of Fundación Progressio used their contacts with the president and successfully lobbied for the organisation to be granted total autonomy to administer a small state-owned scientific reserve. This lobbying was helped by the personal friendship between their founder and Eleuterio Martinez, who intervened on their behalf. In later years, they lobbied subsequent presidents to overrule senators who wanted to downgrade or abolish the reserve, allegedly to take advantage of its timber resources for their own gain. They have also been able to survive sustained resistance from subsistence farmers evicted to make way for the reserve because they have the political and financial backing to oppress and ignore them.

Like the global conservation elite, Dominican NGOs have rivalries, competition and alliances. There are a few official fora where NGO workers meet together, such as workshops hosted by the US Agency for International Development, DED (German Development Bank), the Ministry of the Environment, and The Nature Conservancy, the only large international conservation NGO working in the Dominican Republic. Relationships between NGOs are not always good—some NGOs have formed following a split from another, accusations of corruption fly between groups; rivalries are such that some conservationists refuse to be in the same room as each other. NGOs' relationships with different political parties, which have implications for their financial and political resources, are also important. Yet despite the passing of the Balaguer regime, Eleuterio Martinez's influence remains strong. This is due to his polemical books and articles in the national press, strong personal contacts with key political figures, positions in civil society such as the state university and the Academy of Sciences, and because of a paucity of other commentators. He dominates all media discussions of the environment, giving him unique ability to influence public perceptions of conservation. Most recently, he was reappointed as director of protected areas in September 2008 following a political crisis and reshuffle of the environment ministry. The ruling party which brought him back into government is not the party of Balaguer which first nurtured him, reflecting his status across the political elite. One of his first acts in office was to send 1200 soldiers into Los Haitises National Park to evict 100 "squatter" farmers, a return to the Balaguer era strict protectionist policies.

Just as having a small networked elite in global scale conservation has led to complete domination of certain actors and concepts, the same has occurred in Dominican conservation. Amongst the civil servants, NGOs, oligarchs and experts in Dominican conservation, there is an undisputed acceptance that protected areas should be at the heart of environmental policy, while other issues such as pollution and urban ecology have been neglected by both government and the environmental movement. The town of Haina, 15 km from Santo Domingo, was the site of a highly polluting lead battery recycling plant. Although inadequate regulation resulted in massive pollution—over 90% of residents suffer lead poisoning, and the town is considered one of the top 10 most polluted cities in the world (Caravanos and Fuller 2006; Kaul et al 1999)—it has received minimal attention from government or Dominican NGOs. The Dominican conservation elite have not paid attention to commercial agriculture nor mining, both of which are large economic sectors with significant environmental impacts. They have been unable or unwilling to critique the consequences of capitalism on the environment, nor discuss issues of sustainable development. Instead, Dominican environmentalists debate how protected areas should be run, by whom and where exactly their boundaries should be, while other issues remain beyond the terms of discussion. One academic interviewed argued this was because protected areas and biodiversity issues were far more politically "easy", with fewer difficult questions than sustainable development or critiquing urban pollution, particularly as the oligarchy who dominate the Dominican economy are involved in conservation NGOs. Yet while there are battles within the Dominican conservation elite, it has been successful in uniting to repel external competition from international conservation NGOs.

International Conservation NGOs and the Dominican Republic

The Dominican Republic is praised not just for the extent of its conservation policies, but for their indigeneity, in contrast to much of Latin America. Large conservation organisations are criticised for imposing foreign ideas about land use in the global South, harming the rural poor and indigenous peoples, while remaining too powerful and well connected to be held accountable (Brockington and Schmidt-Soldau 2004; Chapin 2004). They are also accused of bringing particular neoliberal, capitalist forms of conservation (Igoe and Brockington 2007). Yet the Dominican Republic appears to challenge the notion that international conservation NGOs can involve themselves as they would like throughout Latin America.

The only international conservation NGO with a permanent presence in the Dominican Republic is The Nature Conservancy, the largest conservation NGO in the world, with assets worth US$4 billion (Birchard 2005), and a central actor in the transnational conservation elite. It works extensively in Latin American protected areas, first getting involved in the Dominican Republic in the late 1980s, working with the US Agency for International Development on "debt for nature swaps". This proposed creating new protected areas, which The Nature Conservancy would run, in return for reductions in Dominican foreign debt. Most recent work has been part of The Nature Conservancy's pan-American Parks in Peril programme, partly funded by USAID. This aims to build capacity in "paper parks"—protected areas that exist in law but which have very little protection in reality—so they become effective in biodiversity conservation (Brandon, Redford and Sanderson 1998). The Dominican Republic has many "paper parks" and The Nature Conservancy began targeting three regions,[13] investing US$1.6 million in this programme alone between 1990 and 1997.[14] To administer this, The Nature Conservancy has nine permanent staff in the Dominican Republic, working in a fashionable suburb of Santo Domingo. In terms of expenditure and numbers of full-time staff, it is certainly one of the biggest conservation NGOs in the Dominican Republic, if not the largest overall, and has unique connections to globally powerful institutions such as USAID.

Under the Parks-in-Peril programme, The Nature Conservancy's policy throughout the Americas is to partner a pre-existing local NGO (Brandon, Redford and Sanderson 1998), which in the Dominican Republic has been the Moscoso Puello Foundation. Created in 1988, this conservation NGO has a particular focus on the central highlands, and the two organisations began working together in 1994, developing close links. In 2002 the director of Moscoso Puello became country director of the Nature Conservancy for the Dominican Republic, while holding his old job, and a number of other staff members transferred between the two organisations. It is through this process of partnering local NGOs and using their staff that The Nature Conservancy has been able to ensure that most of its personnel are Dominican. Moscoso Puello was hit by scandal in 2006 when it was embezzled by its accountant, and subsequently collapsed, and The Nature Conservancy formally took over its projects and staff. Many in Dominican conservation commented that Moscoso Puello had become subsumed into The Nature Conservancy, and that other Dominican NGOs reportedly do not want to work with it for fear of being taken over.[15]

Yet the Nature Conservancy's large, sustained presence and connections to USAID have not translated into significant involvement in Dominican conservation. Despite repeated attempts, it has not been able to get involved in conservation practice or to enact its programmes as it would like. The Nature Conservancy faces strong resistance from Dominican NGOs, the state and the media, who are very strongly opposed to its involvement in protected areas administration. This is based on a form of anti-US sentiment, part of wider Dominican opposition to US ownership of land: giving a US NGO a role in running a protected area is seen as a damaging attack on sovereignty. Twice there have been attempts by Dominican presidents to sell or lease parts of the country to the USA, and both were met with fervent opposition, in the 1860s (when opposition led to the overthrow of the president) and in 1972, which was a rare case of open opposition to Balaguer's rule (Moya Pons 1995). Criticisms of debt-for-nature swaps as attacks on national sovereignty are widespread throughout Latin America (Lewis 1999), sentiments mobilised in opposition to The Nature Conservancy's four attempts in the late 1980s and early 1990s to set up debt-for-nature swaps in the Dominican Republic (see Chantada 1992). Leading figures in the conservation elite put pressure on the government to abandon the swaps, through the media and newspaper articles, and by lobbying senators to whom they were connected.[16] As a direct result, in three cases the plans were abandoned, and one altered so that a Dominican NGO inherited the protected area rather than The Nature Conservancy. This opposition goes beyond debt-for-nature swaps: attempts by The Nature Conservancy to instigate a co-management project with the Dominican government in a major park in 2002 under the Parks in Peril programme sparked similar opposition and lobbying, and it too was cancelled. Whereas in the rest of Latin America The Nature Conservancy has used debt-for-nature and Parks in Peril to get involved in the management and everyday running of protected areas, it has no such involvement in the Dominican Republic. This opposition was a rejection of an actor, voiced in terms of sovereignty and land ownership, not on competing conservation ideologies—indeed foreign consultants criticised both Dominican NGOs and The Nature Conservancy for sharing the same vision of people-free protected areas as the basis of environmental protection (a concept that was dominant in the Dominican Republic long before The Nature Conservancy began working there).[17]

The Nature Conservancy admits it faces opposition to what it would like to do in the Dominican Republic, particularly purchasing land or running a protected area, and that it faces considerable political constraints. The stated goal of The Nature Conservancy is to get involved in protected area management,[18] not to ensure that an extensive protected area network and a dynamic conservation politics exists, and in this it has failed. Instead of withdrawing in the face of this opposition to its involvement in protected areas, it has worked on different priorities. It contrasts its work in the rest of Latin America, where it gets involved in a few large projects with a hands-on role in the day-to-day management of protected areas, with the Dominican Republic, where it works across many different small issues, producing policy documents and reports.[19] Its country report for 2005, which details the 34 projects it has been working on, shows that all relate to the production or dissemination of data, and that none involved a hands-on role in protected areas (Marte 2006). A number of different parts of the Dominican conservation sector argue that the choice of The Nature Conservancy to take a hands-off, document-producing role is an attempt to keep attracting funding while not addressing its inability to have a role in running protected areas nor the opposition it faces. One consultant described how The Nature Conservancy was doing unnecessary and methodologically flawed studies in order to appear to its funders to be doing something. A manager in the National Parks Directorate described the activities of The Nature Conservancy as follows: "They make their big noise to sell themselves abroad and to have resources, but for the protected areas, nothing ... These institutions are for finding financial resources for themselves, for their employees." The Nature Conservancy claims that its ideas still filter into Dominican conservation practice through its publications and through the workshops it hosts for local NGOs, bureaucrats and aid agencies, yet it is difficult to assess what impact this has had. Other actors attend, and maintain cordial relationships with The Nature Conservancy, despite opposition to its programmes, while denying that they are influenced.

Rather than being excluded from the Dominican Republic because they are not needed, as Diamond (2005) insinuates, international conservationists have been actively resisted, their attempts to gain sovereignty over protected areas out-competed. Preventing a powerful organisation like The Nature Conservancy from acting as it likes in the

Dominican Republic is remarkable given the perceived dominance of large NGOs in the rest of Latin America, and only happened through the locally powerful and well connected conservation elite.

The Nature Conservancy could be seen to be a failure in the Dominican Republic, yet paradoxically it has also been a great success, if considered using different criteria which argue that the primary goal of NGOs is to sustain their funding and continue their existence (Jeffrey 2007). Despite local opposition from government and other conservationists, and despite not meeting their stated goal of direct involvement in protected area management, The Nature Conservancy has been able to sustain a large, multi-million dollar presence in the Dominican Republic. It has successfully adapted its strategy away from direct involvement in protected area management and into the production of documents and reports, retaining the support and financing of both The Nature Conservancy head office in the USA and USAID. This enduring presence despite strong political challenges represents a great success.

The Dominican Republic, Conservation Elites and Capitalism

The case of the Dominican Republic tells us several things about elite networks in conservation. Firstly, as with other case studies and with current global scale patterns, it shows that a well connected elite can be very successful, creating one of the most extensive protected area systems in the world despite the pressures of working in a densely populated island. This was for many years driven by the personal wishes and centralising power of a quasi-dictator, shaped by the environmentalist who had unique access to him. Later, it became dominated by a network of social elites, political elites and NGO directors, with access and personal contacts as key commodities in the network.

The Dominican conservation elite has been successful because its networks have allowed it privileged access to political power, and the ability to shape public and policy discourses on the environment. This is the same as the transnational conservation elite, who are able to influence how conservation happens at a global scale and to make it successful (as measured by the extent of protected areas globally and the size of transnational conservation NGOs) through similar strategies of using personal contacts to access wealth, political power and influence. The form of operating of the transnational conservation elite is present in the Dominican conservation elite.

Furthermore, both elites show that NGOs and other organisations who are outside of the elite network lose out because they cannot use contacts successfully. As Chapin (2004) shows, while funding for conservation overall may be slowing, the financial resources of the very largest organisations who are at the heart of the transnational conservation elite are expanding. Other case studies have shown how large conservation NGOs deprive their smaller counterparts of resources and influence because they are better able to use their connections (MacDonald 2008). In the Dominican Republic, those NGOs who have temporarily been excluded from the elite because they are linked to politicians temporarily out of power find themselves unable to have any influence. This demonstrates the dominant power of each elite in conservation at a particular scale—to have influence, one must be part of the network.

More importantly, the global and the national elite both share the same non-critique of global and national capitalism. Conservation elites are interesting because they, with their links to corporations and trends in neoliberalism, seem unable or unwilling to engage with the environmental consequences of capitalism. Global scale conservation actors have been heavily criticised in recent years for becoming complicit in the destruction of the environment by large corporations, particularly by refusing to criticise corporations who have provided them with money or board members (Chapin 2004; Dowie 1995; McAfee 1999; Rothkopf 2008).The ever strengthening links between these conservation actors and neoliberal trends in development (Goldmann 2001; Igoe and Brockington 2007) allows them to extend into new geographical and political areas, yet in doing so, they introduce a form of environmentalism that is weak in its critique of the environmental impacts of the expansion of neoliberal development and accompanying capitalism. The Dominican elite shares this non-critique of capitalism. By almost entirely concentrating on protected areas, they are neglecting other important issues such as urban pollution, mining and agriculture that occur in the 78.5% of the country which is not in a protected area. Indeed, the lack of information on forest cover and erosion rates means it is difficult to find data to support claims that protected areas have been effective in protecting the environment, preventing the Dominican Republic becoming another Haiti. As with the global scale, capitalist elites in the Dominican Republic are deeply involved in the conservation elite, and they appear unable or unwilling to mount a critique of the environmental impacts of capitalism. It remains beyond the terms of discussion. The reluctance of the transnational conservation elite to critique capitalism is also present in the Dominican Republic, demonstrating the importance of addressing the networks that shape conservation when considering how conservation should relate to capitalism.

Yet although the Dominican conservation elite appears to be a scaled-down version of the transnational conservation elite, with the same form, workings and ideology, the actors central to the latter have been repeatedly rejected. This is in contrast to literature that argues that large international NGOs are seen as being able to dominate conservation in countries of the global South, involving themselves in protected areas as they like, marginalising local conservationists. Yet although the actors have been rejected, it would be difficult to also state that the ideology and practices of the transnational conservation elite have also been refused, and so the question of whether the Dominican Republic represents a limit to hegemony of international NGOs remains unanswered.

Under the idea of governance states, large conservation NGOs able to compete with Southern states for sovereignty to control conservation policy, using their connections to donor agencies and international financial institutions. Yet in the Dominican Republic, The Nature Conservancy competes not with the state for sovereignty, but with a local elite consisting of NGOs, industrialists, politicians and media actors. They have their own access to power and influence that allows them to out-compete The Nature Conservancy's bid to get involved in conservation practice. Governance state theory assumes that local actors are much weaker than international actors in accessing political power. The exceptional case of the Dominican Republic points to the importance of national elites exercising particular forms of nationalism to rebuff international actors, using their superior access to political power at the national scale.

Acknowledgements

The author would like to acknowledge the organisers and participants at the Conservation and Capitalism conference, held at the University of Manchester, 9–10 September 2008, where an earlier version of this chapter was presented, as well as the willing interviewees in the Dominican Republic. Fieldwork was funded by an ESRC studentship.

Endnotes

[1] This chapter is based on two research trips to the Dominican Republic, in April–May 2005 and October 2006–June 2007. Although much of the research was an ethnographic study of a community living near a protected area, this chapter is based on transcripts of 50 face-to-face interviews conducted with civil servants, international and national NGOs, academics, activists, aid agency officials, politicians and development and conservation consultants. Many were selected because I had an interest in their institutions, but others came from recommendations from previous interviewees. Given the relatively small number of people involved in Dominican conservation, this represents a significant proportion of the sector.

[2] The NGOs in question are the Nature Conservancy, World Wildlife Fund for Nature and Conservation International. Data from http://www.nature.org, http://www.conservation.org and http://www.worldwildlife.org, all accessed 4 January 2009.

[3] Although technically elected for a total of six terms, none of these elections could be considered free and fair, involving outright vote rigging, manipulations of media, use of military force to intimidate and murder opponents and a multitude of other dirty tricks. This was successful up until the 1994 election, when loyal senior military figures stopped the election count early and declared Balaguer the winner. Following intense diplomatic pressure from the USA, Balaguer promised to limit his term to 2 years and never run again.

[4] Interview, protected area consultant.

[5] Interview, protected areas consultant.

[6] Interview, academic.

[7] Interview, NGO director.

[8] Interview, worker for development NGO.

[9] Interview, protected area consultant, former protected areas director.

[10] Interviews with protected area consultant, national parks operations director, NGO worker, national parks planner.

[11] Interviews with protected area consulted, national parks operations director, NGO worker, national parks planner.

[12] Interview, development NGO worker, USAID employee.

[13] These are: Del Este national park in the east, Jaragua national park in the southwest, and a group of protected areas in the Cordillera Central consisting of Armando Bermudez, Valle Nuevo, Carmen Ramirez, Humeadora and Nalga de Maco national parks, as well as the Ebano Verde scientific reserve. The Nature Conservancy refers to this as *Madre de las Aguas* (Mother of the waters) in reference to the role the region plays in feeding the island's most important rivers.

[14] Source, The Nature Conservancy US press office.

[15] Interview, former protected areas director, protected areas consultant.

[16] Interview, protected areas consultant, academic.

[17] Interview, protected areas consultant, academic.

[18] Interview, The Nature Conservancy employee.

[19] Interview, The Nature Conservancy employee.

References

ABT Associates (2002) Publicacion Especial Sobre el Proyecto de Reforma de las Politicas Nacionales de Medio Ambiente [Special publication on the project to reform the national policy on the environment]. Santo Domingo

Adams J and McShane T (1992) *The Myth of Wild Africa: Conservation Without Illusion.* London: Norton

Adams W (2004) *Against Extinction: The Past and Future of Conservation.* London: Earthscan

Adams W and Hulme D (2001) If community conservation is the answer in Africa, what is the question? *Oryx* 35 (3):193–200

Armenteros Rius E (1989a) *The Crossword Puzzle of Dominican Forests: Concerns, Suggestions and Hopes.* Santo Domingo: Progressio

Armenteros Rius E (1989b) *Forestry's Zero Hour in the Dominican Republic.* Santo Domingo: Progressio

Armenteros Rius E (1991) *Impacto del Sector Empresarial en la Conservacion del Ambiente.* Santo Domingo: Progressio

Balaguer J (1983) *La Isla Al Reves: Haiti y el Destino Dominicano.* Santo Domingo: Fundación Jose Antonio Caro

Bernau B (2006) Help for hotspots: NGO participation in the preservation of worldwide biodiversity. *Indiana journal of global legal studies* 13(2):617–643

Birchard B (2005) *Nature's Keepers: The Remarkable Story of How the Nature Conservancy Became the Largest Environmental Group in the World*. New York: Jossey Bass

Bolay E (1997) *The Dominican Republic: A Country between Rain Forest and Desert*. Wiekersheim: Magraf Verlag

Boza M (1993) Conservation in action: Past, present, and future of the national park system of Costa Rica. *Conservation Biology* 7(2):239–247

Brandon K, Redford K and Sanderson S (1998) *Parks in Peril: People, Politics and Protected Areas*. Covelo: Island Press

Brockington D (2009) *Celebrity and the Environment: Fame, Wealth and Power in Conservation*. London: Zed Books

Brockington D and Schmidt-Soltau K (2004) The social and environmental impacts of wilderness and development. *Oryx* 38(2):140–142

Bundell J (2006) *Debating NGO accountability*. Geneva: UN Non-Governmental Liason Service

Caravanos J and Fuller R (2006) *Polluted Places—Initial Site Assessment*. New York: Blacksmith Institute

Chantada A (1992) Los canjes de deuda por naturaleza. El caso Dominicano [Debt-for-nature swaps: The Dominican case]. *Nueva Sociedad* 122:164–175

Chapin M (2004) A challenge to conservationists. *World Watch*. Nov/Dec 17–31

da Fonseca G (2003) Conservation science and NGOs. *Conservation Biology* 17(2):345–347

Diamond J (2005) *Collapse: How Societies Choose to Fail or Succeed*. New York: Viking

Dowie M (1995) *Losing Ground: American Environmentalism at the Close of the Twentieth Century*. Boston: MIT Press

Duffy R (2006) Non-governmental organisations and governance states: The impact of transnational environmental management networks in Madagascar. *Environmental Politics* 15(5):731–749

Ferguson J (1992) *Dominican Republic: Beyond the Lighthouse*. London: Latin American Bureau

Fortwangler C L (2007) Friends with money: Private support for a national park in the U.S. Virgin Islands. *Conservation & Society* 5(4):504–533

Gadgil M and Guha R (1992) *This fissured land: An Ecological History of India*. Berkeley: University of California Press

Garland E (2006) "State of Nature: Colonial power, neoliberal capital and wildlife management in Tanzania." Unpublished PhD Thesis, University of Chicago

Geisler and Barton (1997) The wandering commons: a conservation conundrum in the Dominican Republic. *Agriculture and Human Values* 14:325–335

Goldman M (2001) Constructing an environmental state: Eco-governmentality and other transnational practices of a "green" World Bank. *Social Problems* 48(4):499–523

Harrison G (2004) *The World Bank and Africa. The Construction of Governance States*. New York: Routledge

Howard D (2001) *Colouring the Nation: Race and Ethnicity in the Dominican Republic*. Oxford: Signal

Igoe J and Brockington D (2007) Neoliberal conservation: A brief introduction. *Conservation and Society* 5(4):432–449

Jacoby K (2001) *Crimes Against Nature: Squatters, Poachers, Thieves and the Hidden History of American Conservation*. London: University of California Press

Jeffrey A (2007) The geopolitical framing of localized struggles: NGOs in Bosnia and Herzegovina. *Development and Change* 38(2):251–274

Kaul B, Sandhu R, Depratt C and Reyes F (1999) Follow-up screening of lead-poisoned children near an auto battery recycling plant, Haina, Dominican Republic. *Environmental Health Perspectives* 107(11):917–920

Lewis A (1999) The evolving process of swapping for nature. *Colorado Journal of International Environmental Law and Policy* 10(2):431–467

MacDonald C (2008) *Green, Inc.* Guildford: Lyons

Mackenzie J (1988) *The Empire of Nature: Hunting, Conservation, and British Imperialism.* Manchester: Manchester University Press

Marte D (2006) The Dominican Republic Country Plan, Financial Year 2005. Santo Domingo: The Nature Conservancy

McAfee K (1999) Selling nature to save it? Biodiversity and green developmentalism. *Environment and planning D: Society and Space* 17:133–154

Mitlin D, Hickey S and Bebbington A (2007) Reclaiming development? NGOs and the challenge of alternatives. *World Development* 35(10):1699–1720

Mitmann (1999) *Reel Nature: America's romance with Wildlife on Film.* Cambridge: Harvard University Press

Moya Pons F (1995) *The Dominican Republic: A National History.* New York: Marcus Weiner

Princen T and Finger M (1994) *Environmental NGOs in World Politics: Linking the Local and the Global.* London: Routledge

Rangarajan M (2001) *India's Wildlife History.* Delhi: Permanent Black

Rocheleau D and Ross L (1995) Trees as tools, trees as text—Struggles over resources in Zambrana-Chacuey, Dominican-Republic. *Antipode* 27(4):407

Rosario E (1988) *Los Duenos de la Republica Dominicana.* Santo Domingo

Roth L (2001) Enemies of the trees? Subsistence farmers and perverse protection of tropical dry forest. *Journal of Forestry* 99(10):20–27

Rothkopf D (2008) *Superclass: The Global Power Elite and the World They are Making.* New York: Farrar, Straus and Giroux

Sklair L (2000) *The Transnational Capitalist Class.* London: Blackwell

Terbourgh J (1999) *Requiem for Nature.* Washington DC: Island

Turits R L (2003) *Foundations of Despotism: Peasants, the Trujillo Regime, and Modernity in Dominican History.* Stanford: Stanford University Press

UNEP-IUCN (2006) *World Database on Protected Areas.* Cambridge: United Nations Environment Programme–International Union for Conservation of Nature

Walker R (2007) *The Country in the City: The Greening of the San Francisco Bay Area.* Seattle: University of Washington Press

West P, Igoe J and Brockington D (2006) Parks and peoples: The social impact of protected areas. *Annual Review of Anthropology* 35:251–277

Western D (1997) *In the Dust of Kilimanjaro.* Washington DC: Shearwater.

Woods M (1998) Rethinking elites: Networks, space, and local politics. *Environment and planning A* 30:2101–2119

Wucker M (1999) *Why the Cocks Fight: Dominicans, Haitians, and the Struggle for Hispaniola.* New York: Hill and Wang

Chapter 7
Conservative Philanthropists, Royalty and Business Elites in Nature Conservation in Southern Africa

Marja Spierenburg and Harry Wels

Introduction

When German-born and globally networked Prince Bernhard of the Netherlands died at the age of 94 in December 2004, his old billionaire friend Anton Rupert—Afrikaner founder of the Rembrandt Group in South Africa and founder of the Peace Parks Foundation (PPF), the major lobby organisation for transfrontier conservation in southern Africa—organised a special memorial service for him at Stellenbosch University in South Africa. Prince Bernhard had served as president of the World Wide Fund for Nature (WWF) between 1962 and 1976, but even after he stepped down, had remained involved and committed to the organisation. At the memorial service Rupert claimed that "for me he was without doubt the most important person in nature conservation in the twentieth century. He has achieved a lot".[1]

Rupert himself passed away, aged 89, in January 2006, and in obituaries in South Africa he was primarily described as a philanthropist, who had greatly contributed to and lobbied for nature conservation in Africa, particularly the idea of creating Peace Parks or transfrontier conservation areas through the PPF. This Foundation was established in 1997 with an initial grant of Rand (R)1.2 million (US$260,000) from Anton Rupert, provided through the Rupert Nature Foundation (Hanks 2000).[2] The South African President at that time, Thabo Mbeki, "said Rupert will be remembered for his total devotion to nature and environmental conservation, adding he was a true philanthropist".[3] In the university magazine of the University of Pretoria Tukkie, where Rupert had studied and had been Chancellor between the mid 1980s and mid 1990s, he is also primarily remembered as "*n filantroop in wese*", a philanthropist by nature. The magazine proudly reported that in November 2003 Rupert was the first South African who was honoured with a medal by WWF for his continued efforts in nature conservation (Tukkie 2006:9–11).

Anton Rupert and Prince Bernhard, the main characters of this chapter, have indeed both contributed substantially to nature conservation. Perhaps even more importantly, they have worked together successfully to market conservation and raise considerable funds from the corporate sector. The public reactions upon their passing away suggest that, in popular memory, they are primarily remembered as philanthropists. The way they obtained their capital and connections that allowed them to play this role seem to be, if not obscured, then at least rendered acceptable in the light of their contributions to nature conservation. The fact that Anton Rupert and Prince Bernhard established and maintained their connection during South Africa's apartheid years, and that Rupert was for many years an influential member of the Afrikaner Broederbond (Afrikaner brotherhood, a partly secretive organisation that exerted a lot of influence on the apartheid era governments), hardly seems to matter anymore.

This contribution to the theme issue on capitalism and conservation focuses precisely on the background of the two protagonists, the development of Anton Rupert from humble beginnings to become a successful businessman and "oracle of trademarks" (Domisse 2005:15), and the increasing interconnectedness of Rupert's and Bernhard's networks which they used for their fundraising activities. The chapter explores the ways in which their background and social connections influenced the way they perceived nature and nature conservation, and how this in turn influenced nature conservation practices in southern Africa. The aim is not to "expose" the two men, but rather to increase our understanding of the way in which capitalism and conservation interact (see Brockington, Duffy and Igoe 2008), how philanthropy enhances corporate images and business, and how business (networks) influence(s) conservation.

Capitalism and Conservation, the Role of Philanthropy

Tracing the history of philanthropy, Lässig (2004) shows how at the end of the nineteenth century, members of the newly emerging bourgeoisie sought to integrate themselves in the leading social and political elite, still dominated by the aristocracy. Devoting part of their accumulated capital to philanthropy provided a way into the elite, allowed them to build up "social wealth" in order to gain socio-political influence (cf. Bourdieu 1983; see also Anheier and Leat 2006), but also became part of a process of identity formation of the bourgeoisie. As such, one could argue that philanthropy was linked to processes of social change. At the same time, however, philanthropy may also serve to prevent the advance of other groups in society, notably the poor, and maintain social inequalities. As Simmel (1965 [1908]:155) argued, giving is both a symptom and a necessity of socio-structural inequities: "to mitigate certain extreme manifestations of social differentiation, so that the social structure may continue to be based on this differentiation". Bornstein (2003) poses that when giving became institutionalised through the state—and to a certain extent through NGOs as well—what used to be philanthropy became reinterpreted as entitlements. However, in the neoliberal paradigm of the 1990s, she argues, the discourse of conservative (and) religious NGOs moved to the charity model again. Describing the activities of such NGOs in Zimbabwe, she states that by giving assistance to those suffering the consequences of the Economic Structural Adjustment Programme pushed by the IMF and the World Bank, they allowed the adjustments to be implemented without too much resistance. Similarly, a report by the US National Committee on Responsive Philanthropy concludes that in the USA conservative religious philanthropic foundations have consistently directed a majority of their funding to organisations and programs that pursue an overtly ideological agenda based on industrial and environmental deregulation, the privatisation of government services, reductions in federal anti-poverty spending and the transfer of authority and responsibility for social welfare from the national government to the charitable sector and state and local government (Covington 1997).

The importance of conservation NGOs in nature conservation is partly a reflection of neoliberalism, as well as a driver of an increasingly neoliberal agenda for nature conservation. Conservation organisations increasingly cooperate with and rely on funding from the corporate sector (Chapin 2004; see also chapters by MacDonald, and Brockington and Schofield in this volume). The links between capitalism and conservation have a direct impact on the way landscapes are changed and categorised; they are also changing attitudes towards wildlife and landscapes, introducing markets in nature conservation and commodifying nature (Brockington, Duffy and Igoe 2008). Brockington and colleagues (2008) argue that the global expansion and proliferation of protected areas and related conservation strategies can be directly related to these increasingly tight links. Protected areas are transformed from places to protect the elite's trophy animals (Beinart and Coates 1995) to places that can be marketed as "tourist habitat" (Brockington, Duffy and Igoe 2008). Reliance on environmental NGOs and the corporate sector in the funding, planning and implementation of nature conservation poses challenges to the powers of the nation-state. Even more so does the increasingly popular strategy of creating protected areas that straddle national boundaries, so-called transfrontier conservation areas (TFCAs). These TFCAs challenge the very boundaries of the nation-state and allow for an even greater influence of a myriad of non-state actors involved in the governance of such areas (Büscher 2009; Duffy 1997).

Although philanthropic contributions to and influence on nature conservation by business elites are not a new phenomenon (Ben Barka 2003; Coniff 2003), Sklair (2001) argues that at present the corporate sector is dominating the debates on strategies towards nature conservation to an extent we have not witnessed before (cf MacDonald this volume). He poses and analyses the existence of a transnational capitalist class—composed of corporate executives, merchants, professionals, bureaucrats and politicians—striving for global economic growth and a global adoption of a neoliberal economic model. It does so through, among other things, a profit-driven culture ideology of consumerism. This class, however, also tries to resolve the ecological crisis that arises from the drive towards global economic growth, but without compromising its economic goals. Linking up with the discussions initiated by the Brundtland report "Our common future", its focus is on sustainable development—albeit in a piecemeal fashion focusing on separate, "manageable" problems rather than a wholesale transformation of the corporate sector—arguing that consumption is not only reconcilable with sustainable development, but even a prerequisite. The marketing of protected areas as "tourist habitat" is directly connected to this (Brockington, Duffy and Igoe 2008).

Prince Bernhard and Anton Rupert were both active participants in a number of the corporate–policy networks that Carroll and Carson (2003) maintain were and are crucial to the policy influence of the transnational capitalist class, including the Bilderberg conferences that for many years were chaired by Prince Bernhard. The authors emphasise the cosmopolitan character of the most active members of these corporate–policy networks, but unlike Sklair (2001), who maintains that members of the transnational capitalist class strive for a complete denationalisation of economies, they do not exclude the possibility that their national identity and interests still play a role in their activities (see also Carroll and Fennema 2002).

As we aim to demonstrate below, Anton Rupert started out as an Afrikaner capitalist, actively promoting the development of an Afrikaner entrepreneurial class in South Africa to counter British domination of the South African economy. His early philanthropic contributions towards the preservation of historic buildings as well as nature conservation in South Africa can be linked to these Afrikaner nationalist sentiments (cf Carruthers 1994, 1995). His motivation for and impact of his international contacts—strengthened through his connection with Prince Bernhard—can be interpreted in different ways, however. Although he became increasingly critical of apartheid policies, which he came to consider as "bad for business", the interlocking of his business and nature conservation networks also offered possibilities for South African businessmen to network with international businesses despite the international economic boycott of South Africa. Rupert's promotion of transfrontier conservation during the last years of his life, with its explicit references to socio-economic benefits for local communities, could be interpreted as an attempt to "denationalise" nature conservation and as being directly linked to the dominance of the transnational capitalist class in the sustainable development debate. However, it could also be interpreted as an attempt by a member of the "old" South African elite to find and justify a place in the new South Africa (see Draper, Spierenburg and Wels 2004).

Anton Rupert and Prince Bernhard: Networking for Capitalism and Conservation

Prince Bernhard and Anton Rupert worked together closely to promote and market nature conservation. Anton Rupert joined the board of trustees of the World Wildlife Fund (WWF)[4] in 1968 when Prince Bernhard was president of the organisation. Together they developed the idea of the establishment of the "1001 Club". "The 'one' was Prince Bernhard. The other one thousand were wealthy individuals who could be persuaded to part with $ 10.000" (Bonner 1994:68). South African businessmen were also included in the list, at a time when South Africa was still officially boycotted by the world business community. On the 1989 list, "at least sixty individuals were from South Africa", many of them members of the highly politically influential Afrikaner Broederbond, including Rupert himself (Bonner 1994:68). Below we will describe how Prince Bernhard and Anton Rupert developed their networks that fed the WWF's 1001 Club.

The Development of Anton Rupert as an Afrikaner Capitalist

De Klerk (1975:292) describes Rupert as one of the two men [the other being Jan S. Marais, founder of the Federale Volksbeleggings (FVB), an Afrikaner investment bank] from the *platteland* of the Karroo that "epitomise" the rise of Afrikaner capitalism in South Africa:

> In the late [nineteen] thirties, Rupert had been a leader of the very nationally-minded Afrikaner students [the Afrikaanse Nasionale Studentebond, ANS] who had sat at the feet of Diederichs, Cronjé and Meyer. He had shared their [nationalist] sentiments deeply, and his own little tobacco venture had started as a direct result of the appeal of Kestell and others that a nation should save itself (De Klerk 1975:293).[5]

Rupert started his business when he was a member of the rather
secretive Afrikaner Broederbond (Brotherhood). His Rembrandt Group
started as Voorbrand Tobacco Corporation and its products were
distributed at Broederbond meetings. "'We [Broederbond members]
were asked to smoke and cough for volk and vaderland,' one of
them told us [the authors] with a chuckle" (Wilkins and Strydom
1978:428). The Broederbond was established in 1918 with the aim
"to protect the Afrikaner by means of an efficient organisation against
vilification, humiliation and oppression [from the British]" (Pelzer
1979:6). Its two first presidents, Ds J F Naudé and Ds L J Fourie,
were church ministers."This symbolised the inextricable link over
the years between the Bond and the Afrikaans churches" (Serfontein
1979:34; see also Moodie 1980). According to Serfontein (1979), by
1930 the Broederbond had consolidated its grip on Afrikaans culture.
Furthermore, the Broederbond exerted a profound influence at all levels
of South African politics, ranging from ideological legitimisation of
government policies to influencing political appointments at various
levels of government. "(I)t demands situating the Bond within the
broader context of Afrikaner Nationalism in South Africa" (O'Meara
1977:156–157) that developed in the 1920s and 1930s. Marx (1998)
argues that this Afrikaner Nationalism found inspiration in and support
for its anti-British sentiments in German National Socialism, which
developed about the same time. Marx, however, does not suggest a
simple "ideological diffusion" from National Socialism to Afrikaner
Nationalism, but does state that:

> (t)he sympathy with the Third Reich among the Afrikaans-speaking
> white population was widespread and for all to see. A distinction
> must be made, however, between a pro-German attitude—whose roots
> go back much further into the late nineteenth century—and a clearly
> pro-National Socialist one (Marx 1998:512).

During the 1930s there had been regular lecture visits by German
"cultural" figures, Afrikaner students had travelled to Germany on
special tours, and prominent young Afrikaner intellectuals, according
to Furlong (1986:7), were so deeply impressed by what they saw there
that their own works began to reflect National Socialist philosophy.
Influential people like Piet Meyer and the future State President,
Nico Diederichs, both at a later stage chairman of the Broederbond,
and close contacts of Rupert, were heavily influenced by German
National Socialist ideology (Furlong 1986; Marx 1998). Afrikaner
nationalists and German national socialists shared an ideology in which
mythical connections between the "Volk" (the people/nation) and a
certain type of landscape were central. In the case of the Afrikaners
these connections centred on the South African *veldt*, the landscape

they had conquered, tamed, and wanted to protect (see Carruthers 1995).

In terms of tracing the historical antecedents of patterns and inter-relationships between Afrikaner capitalism and conservation, one of the major myths and most persistent icons of the relationship between Afrikaners and nature is the story of the Voortrekkers, a sturdy group of religiously devoted Afrikaners who tried to escape British influences at the Cape by travelling by ox wagon out of the colony and settling deeper into the African continent (see for instance Becker 1985). They were seen as rugged individuals who lived close to the land (Sparks 1991). The Great Trek of 1836 and especially the ox-wagon symbolises this intimate relationship. In 1938 the same symbol of the ox-wagon was taken by Verwoerd to announce the Second Great Trek which was aimed to "Afrikanise the cities, and take a legitimate place in commerce and industry" (Verwoerd in Bloomberg 1989:121; see also O'Meara 1983). This Second Trek was explicitly placed in the context of a divinely willed destiny (Templin 1984). "The Afrikaner volk has been planted in this country by the Hand of God, destined to survive as a separate volk with his own calling" (Moleah 1993:328). This calling implied, among other things, that the Afrikaners had to preserve a landscape in its "pristine" state, with all animals that belonged to it, "just as the Voortrekkers saw it" (Reitz quoted in Beinart and Coates 1995:77). In the process, the image of a pristine African landscape became a commodity in itself that still serves to attract (Western) tourists to (South) Africa today (Brockington, Duffy and Igoe 2008; Wels 2004). Former Transvaal President Paul Kruger was portrayed as descending in a straight line from the Voortrekkers and is described in mythic proportions in an older tourist book about the Kruger National Park as follows:

> Already in 1884 the president of the Zuid Afrikaansche Republiek had recognised the beauty and value of this much-maligned bushveld and had urged in the Volksraad the need to preserve some of this land. In 1898 (26 March) after much opposition, he succeeded in officially proclaiming the establishment of a "Government Reserve" in the Transvaal's eastern Lowveld (Braack 1983:11).

Carruthers (1994) has convincingly argued that it was not so much Kruger's concern for the environment, but rather his political acumen and opportunism towards uniting the Afrikaners behind a shared myth that stimulated his position in declaring the reserve. In 1926 the park was named after Kruger to foster a sense of unity between the English and Afrikaners in the still precarious Union of South Africa, which included the Transvaal and Orange Free State (Carruthers 1994).

Rupert started his business with one of the first loans provided by the Afrikaner investment bank FVB (Giliomee 2003:438). He was active in stimulating emerging Afrikaner entrepreneurs and became leader of the small-business development unit of the Reddingsdaadbond (salvation union) (Domisse 2005:56). His business associates included a number of high-ranking Broederbond members (Domisse 2005). Once he decided to expand his business overseas, he changed the name of his company from the Afrikaner nationalist name "Voorbrand" to the Rembrandt Group. His business overseas grew considerably during the 1950s, and the Rembrandt Group became the owner of famous brands such as Rothmans, Montblanc and Cartier.

Parallel to his business development was the equally impressive growth of Rupert's philanthropic activities (De Klerk 1975:294). Rupert is hailed by De Klerk for his patronage to the arts and the fact that he restored so many of Old Cape houses in Stellenbosch and Graaff-Reinet, "many of which were restored to their *pristine beauty*, all due to the magnanimous interest shown by the Rupert Organization. *So, too*, did nature conservation" (1975:294, italics added). In his quite one-sided praise for Rupert's philanthropic giving, De Klerk goes so far as to suggest that, "nothing the Americans had ever done in the all-encompassing good works of Protestant capitalism could be said to have surpassed this amazing and sudden flowering of Afrikaner business efficiency and public spirit" (1975:294). Afrikaner business philanthropy is seemingly a league of its own where conservation and capitalism embrace each other and work hand in hand, brought together by an Afrikaner tycoon in luxury consumer goods.

Whether Anton Rupert maintained his connection with the Broederbond is a moot point. Domisse (2005:85) claims that from 1948 onwards—the year the Afrikaner National Party came to power—Rupert no longer considered there was a need for the Broederbond to defend Afrikaner interests, and that he let his membership lapse. Serfontein (1979:178) states that Rupert left the Bond in 1974, whereas Wilkins and Strydom (1978:428) claim that in 1974 he indeed intended to leave the bond, but was convinced by Piet Meyer to stay. The same ambiguity pertains to his attitude towards apartheid, which was endorsed and supported by the Broederbond. After the Sharpeville shootings in 1960, a number of Afrikaner business men, including Rupert, "felt that the costs of apartheid had become too steep" (Giliomee 2003:622). In his own publications, Rupert (1967, 1981) pleads in favour of partnerships and coexistence, however whether that would be on the basis of total equality is not entirely clear. He specifically mentions the need for increasing economic chances for black and coloured people, but remains rather silent about their political rights. According to his biographer Domisse (2005:128–129), he tried to set up a partnership with coloured people in the 1950s, but this attempt supposedly was blocked by government. He maintained the headquarters of his business in South Africa till the late 1980s. In 1988, at the instigation of his son Johann, the South African branches were separated from the international branch to protect the latter against the sanctions in place against South Africa (Domisse 2005:322–328). At the time of the hearings of the Truth and Reconciliation Commission, set up by the post-apartheid government, accusations were levelled that the Rembrandt Group had continued to support the National Party. Rupert told the TRC about his problems with Verwoerd, but according to Nattrass (1999:377) "was silent about the benefits which the company's close association with the National Party [through the Broederbond] may have delivered". Marx (1999:360) claims that Rupert maintained contacts with the ultra-right scene in South Africa at least till the 1960s.

Connecting the Networks of Anton Rupert and Prince Bernhard

Apart from funding the protection of Afrikaner cultural heritage, Rupert also became very active in nature conservation in South Africa. In the 1960s he established the South African Nature Foundation (SANF) over which he also presided, at the request of Prince Bernhard—then President of WWF International—as a national branch of WWF. From 1980 to 1989 Frans Stroebel was the chief executive of the SANF. Previously Stroebel had been a diplomat and private secretary to the then Foreign Minister of South Africa, Pik Botha (Ellis 1994:59–60). WWF was founded in 1961, intended initially as an international fundraising organisation for IUCN (International Union for the Conservation of Nature and Natural Resources) (Hey 1995:224). The SANF was, due to increasing international concern about apartheid, denied an official status in the IUCN.

> Despite the fact that the SA Nature Foundation had contributed over R200,000 to the administrative costs of WWF and IUCN, and had been urged repeatedly by their officials to apply for full membership of the IUCN, Council turned down the application in 1981 (Hey 1995:229).

·Rupert was angry and "withdrew (the) application" (1995:229).
This unofficial status in the IUCN, however, did not prevent close ties between SANF and WWF, in particular with Prince Bernhard of the Netherlands. Rupert was invited to become a member of the Board of Trustees of Prince Bernhard's WWF International, and from 1971 he served on its executive committee (Ellis 1994:59). Rupert and Prince Bernhard worked closely together in many ways; thinking about financial strategies to give WWF a kick-start was one of them.
It is not really clear how and when Anton Rupert and prince Bernhard first met. One of the more plausible suggestions is that this happened

during the first time Prince Bernhard visited South Africa, on a trade
mission for the Dutch government in 1954, through a mutual friend of
the two men, Francis (Freddie) de Guingand. Prince Bernhard knew De
Guingand from the Second World War when the latter was a military
intelligence agent for Great Britain. After the war De Guingand went to
southern Rhodesia to become a business man. He founded and chaired
the South Africa Foundation which became an important business lobby
organisation for South African companies. Some of Rupert's companies
joined this Foundation. In 1954 Rupert appointed De Guingand to the
board of Rothmans Canada, Rothmans being one of Rupert's latest
acquisitions at the time (Duineveld 2009:55–59).

It appears it was Rupert who came up with the idea of starting the
"1001-club". In this club he brought together his own skills in branding
(Domisse 2005), with his and Bernhard's extensive worldwide business
networks. Bernhard was especially well known for his extensive contacts
in business and with power bearers around the world. He participated in,
chaired and promoted several major business networks in his lifetime,
two of those being the Bilderberg Group (since 1954) and the Mars
and Mercury Group. What we try to argue here is not only that Prince
Bernhard was a skilled and accomplished networker and made good
use of this for obtaining funding for WWF, and later the PPF, but
we also want to explore the nature of these elite networks to show
how these contributed to fostering the links between conservation and
capitalism.

Prince Bernhard has been a strong networker throughout his whole
career. From 1935 to 1937, when he married Princess Juliana, he worked
for IG Farben in Paris. His networking skills made him a very valuable
employee of the organisation's intelligence bureau. Official sources even
suggest that his contacts with—and later marriage to—the Princess may
even have been a deliberate company strategy:

> IG Farben had a very effective intelligence bureau of its own: NW
> 7, headed by Max Ilgner . . . An approved (IG Farben) tactic was to
> parachute prominent employees into foreign countries where, in due
> course, they became naturalised. They had orders to infiltrate the
> highest social classes and were expected to become respected citizens
> in their new fatherland. Prince Bernhard zur Lippe-Biesterfeld, an
> employee of NW 7, who married Princess Juliana, the future queen of
> the Netherlands, was according to one of his biographers [Klinkenberg
> 1979] sent to the Netherlands for exactly this reason (Aalders and
> Wiebes 1996:22).[6]

Prince Bernhard seems to have been parachuted through German
business connections to become part of important business networks
that were acting globally at the time, long before the term globalisation

became fashionable. His later marriage to Princess Juliana gave him, in addition to his already existing networks, almost unlimited access to political power holders around the world. According to Hoffman (1979:122), Bernhard's wife Princess Juliana, "was concerned with the speed with which Bernhard had used his elevated position to insinuate himself to business and government affairs". The Bilderberg conferences could perhaps be called Bernhard's meeting place where all his government, intelligence and business networks eventually came together. The Bilderberg Group, named after Hotel Bilderberg where in 1954 the first meeting was held, brings together captains of industry and high-powered politicians from Western Europe and the USA for informal discussions on "the state of affairs in the world". The first meetings were organised at a time when the transatlantic bonds were cooling after the end of the Second World War and in which plans for a united Europe were not met with much enthusiasm—much to the dislike of the Americans, who wanted a united Western Europe as a shield against the threat of communism. The Bilderberg meetings were basically meant to bridge the divide and bring a united Europe closer to realisation. The results and agenda of the meetings were kept secret. Among the 75 invitees for the first meeting, for instance, were from the business side David Rockefeller and Giovanni Agnelli of Fiat. After the first meeting everybody agreed that Prince Bernhard with his charm and wit had done a wonderful job to bring the desired results.[7] Concerning the hidden political agenda of the conference, recent research by Gerard Aalders from NIOD[8] shows that the Central Intelligence Agency (CIA) has sponsored the first few meetings of the Bilderberg Group to promote the American agenda for Europe (Aalders 2007:33–35). In a newspaper interview, Aalders was quoted saying that he could not prove that Bernhard knew of this connection, but that it would be quite likely, as he was a good friend of two former CIA bosses.[9] In later years politicians like Henry Kissinger, Tony Blair and Bill Clinton were among the invitees to Bilderberg. What becomes clear from Aalders' (2007) book as well as from Caroll and Carson's (2003) article is the incredible powerful political and business network the Bilderberg Group represents. What is particularly striking about this network is the enormous influence of various American foundations, like the Ford Foundation, the Rockefeller Foundation and the Rockefellers Brothers Foundation. Aalders formulates it rather bluntly as follows:

> Both the Ford and the Rockefeller Foundation have been functioning as instruments for covert American foreign politics. Their Boards of Directors had close ties with American intelligence services or were even part of it. Charity foundations were most suitable to transfer

large sums of money to CIA-projects, without its happy recipients ever having the slightest notion that the money came from the American intelligence service (Aalders 2007:31).[10]

The highest officials from the world of intelligence, government and business were all brought together in the Bilderberghotel in the Netherlands, networking under the chairmanship of Prince Bernhard. It can without any doubt be concluded that Prince Bernhard was at least ideally positioned to "motivate" influential people from around the globe to give generously to, or otherwise facilitate, the plans he developed for nature conservation around the world, but particularly in southern Africa, together with his good friend Anton Rupert.

Bernhard's other network is less well known and is called Mars and Mercury, established in 1926 in Belgium, in order to commemorate the bonding and friendship between military officers during the First World War, *and* to promote the idea of a necessity to combine and cooperate in commercial interests. That is why the network was named after the Roman god of war (Mars) and the god of trade and commerce (Mercury). In the 1960s it invited other NATO partners to follow the example which led, among other things, to the installation of the *Commission de Liaison Internationale Mars et Mercure* (CLIMM). Four years later Prince Bernhard offered to become its patron.[11]

How did these business networks in which Prince Bernhard was involved contribute to finding suitable candidates for the 1001-Club during the WWF years? WWF refuses to give information about the members of the club, but journalist Kevin Dowling—who made a documentary on ivory poaching that had the support of WWF but who later fell out with the organisation over Operation Lock (see below)—discovered two lists, one for 1978 and one for 1987 (see http://www. isgp.eu) which suggest that some cross-fertilisation between the various networks took place. Ellis states that "(t)he identities of the 1001 members of the 1001 Club reflect quite closely Bernhard's own circle of acquaintances, as might be expected" (Ellis 1994:61). On these lists the names appear of a number of famous businessmen and bankers, such as David Rockefeller, Henry Ford II, several members of the De Rothschild family and the Agnelli family, to name but a few. Anton Rupert's name and that of his son Johann—who presently directs the Rembrandt/Richemont Group—appears on the list, and quite a number of their South African colleagues: Freddie de Guingand, Louis Luyt (the "fertiliser king" of South Africa), but also Rupert's greatest rival Harry F. Oppenheimer, owner of Anglo American and the De Beers company. Club members came together for receptions held at Prince Bernhard's palace in the Netherlands. The WWF organised international and national meetings for the members, as well as special trips to WWF projects all around the world (Duineveld 2009:83). Hence, the club offered a platform for South African businessmen to meet their colleagues even during the boycott of South Africa. Quite a number of the Dutch businessmen on the lists—connected no doubt through Prince Bernhard—have at one time or another been accused of sanction busting.

Anton Rupert cooperated closely with WWF International. WWF International was intensively cooperating with and intervening in South African nature conservation policies and via that route in the complexities of apartheid politics. An important issue here is the position of the Director General of the WWF between 1977 and 1993, Charles de Haes.[12] Between 1971 and 1977, before he was appointed Director General he was the personal assistant of Prince Bernhard, and "(o)ne of his tasks was to implement the 1001 Club project" (Bonner 1994:69). Before he became Prince Bernhard's assistant, he was attached to Rupert's Rembrandt Group. With Prince Bernhard writing the letters of introduction (with a royal letterhead) and knowing his business networks described above, it may not come as a surprise that, "it took De Haes only three years to find one thousand donors" (Bonner 1994:69). Everybody, friend and foe, described De Haes' fundraising skills as "brilliant" (p 70). Bernhard's networks and De Haes' skills were a perfect match. When De Haes became first joint Director General in 1975 and from 1977 Director General, Rupert paid his salary. "WWF never said at the time that Rupert was paying De Haes, and it still tries to conceal this fact . . . What this means, of course, is that de Haes was still employed by a South African corporation while working for WWF" (p 70). His leadership within WWF-International was certainly not uncontested (pp 74–75), but he did have the unwavering support of another royal figurehead of WWF, Prince Philip, the successor of John Loudon. Loudon had taken over the position of President of WWF-International when Prince Bernhard had to resign because of the Lockheed scandal in 1976.[13] This support helped De Haes to stay in power till 1994. During his reign, Charles de Haes had to deal with an issue that also bore witness to the close interconnection of at least WWF-International, and hence also Prince Bernhard, with South Africa: "Operation Lock", a joint operation initiated by Prince Bernhard, allegedly without the knowledge of Prince Philip.

The Fortress Approach to Conservation and Operation Lock
SANF was established at a time when the main focus of nature conservation in South Africa was on creating and expanding protected areas. As mentioned above, the drive to protect the South African landscape had certain nationalistic undertones (Carruthers 1994). The

National Parks Board of Trustees, the main South African conservation agency, was dominated by Afrikaners. In the 1960s, the apartheid government's spending on national parks steadily increased, and in 1975–1976 a considerable sum of money was allocated to fortify the border between Kruger National Park and Mozambique, as a reaction to the latter's recently won independence (Ramutsindela, Spierenburg and Wels forthcoming). That year, the Board presented its annual report with a picture of Paul Kruger's statue prominently on the cover (Ramutsindela, Spierenburg and Wels forthcoming). SANF's activities mainly supported these policies of expansion and fortification.

At the time, protected area management worldwide was dominated by an uncompromising attitude towards local communities referred to as "fortress" conservation (Anderson and Grove 1987; Brockington 2002), and in South Africa this overlapped with apartheid policies towards black communities. Evictions from protected areas occurred, and although Anton Rupert may have advocated economic partnerships with non-white communities, these did not pertain to conservation activities (yet). Members of local communities were stigmatised as trespassers and poachers and had to be kept at bay. The "war against poaching" was a dominant theme in nature conservation, a theme that was also high on the agenda of WWF-International and SANF (Bonner 1994).

In the 1980s, as the economic nationalisation of Afrikaans-speaking South Africa suffered from international sanctions, funding became an issue for the National Parks Board of Trustees. The Board finally agreed to a more important role of the private sector, and established the National Parks Trust Fund in 1985. Rupert's SANF shared equal representation with the Board in this Trust. The Trust's main activity was to raise funds for the expansion and maintenance of protected areas in South Africa (Ramutsindela, Spierenburg and Wels forthcoming). These increased linkages between the state and the private sector in combination with the war on poaching perhaps culminated in what became known as Operation Lock.

In the second half of the 1980s, by arrangement with WWF officials, KAS Enterprises trained anti-poaching units in Namibia—then still under the control of South Africa—and in Mozambique—where South Africa deployed a destabilising strategy. KAS Enterprises employed mercenaries who were former British commandos from the Special Air Services (SAS) founded by Sir David Stirling. The project was funded by Prince Bernhard.[14] In South Africa KAS implemented an operation to stop ivory and rhino horn trafficking. Various parts of the operation were exposed by a Reuters correspondent in Nairobi, Robert Powell, and later by Stephen Ellis (1994; see also Bonner 1994:79). Through KAS infiltration it became clear that the South African Defence Force

(SADF) was also heavily involved in poaching. SADF's involvement was such that it financed part of its operations to protect the apartheid regime through ivory and rhino horn smuggling. South African ports turned out to be logistical hubs for the smuggling of ivory and rhino horn. WWF "continued to remain silent on the matter" and its only reply to the question why South Africa was never targeted in the international campaign against the rhino horn trade was, "(t)he subject 'never came up'" (Ellis 1994:64). Operation Lock spun out of control and the mercenaries started to become involved in smuggling rackets themselves and in the process participated in anti-ANC activities which were part of the general strategy of apartheid's "general onslaught" (Ellis 1994; see also Breytenbach 1997; Potgieter 1995). When the operation was exposed, the embarrassment for WWF was complete and in an effort to distract attention from Prince Bernhard, Charles de Haes put all the blame on John Hanks, then head of the Africa Programme at WWF-International, saying that Hanks had initiated the project without approval from WWF officials. Loyal to WWF, Hanks took full responsibility for Operation Lock (Bonner 1994:78–81).[15] In 1995 Mandela ordered an investigation into WWF activities in South Africa during the apartheid years, led by Mark Kumleben. Reporting on its outcomes a Dutch journalist summarised the report as follows: "Kumleben found a wide and intertwined network of espionage and economic interests, where nature conservation was seemingly only used as cover" (Zwaap 2001:4, see also Zwaap 1997a, 1997b).[16] The South African case is an interesting one; while an often cited risk of privatisation of security in conservation areas is a certain loss of states' sovereignty (see Brockington, Duffy and Igoe 2008), in this case the South African state used it to defend its sovereignty in the face of increased resistance to its apartheid policies.

Adapting to the Winds of Change, Marketing Tourism as Community Development

In the 1990s things started to change in South Africa. Mandela was released from prison and in 1994 he became the first democratically elected president of South Africa. While earlier plans for cross border cooperation in nature conservation in southern Africa had always been suspected of being "primarily intended to reinforce white domination in the region" (Ramutsindela 2004:124), after 1994 both the regional and global political conditions for cross-border cooperation seemed right, with the end of the Cold War and the concomitant end of civil war in Mozambique, as well as the increasing global and regional dominance of neoliberalism. Anton Rupert emerged as a leading promoter and fundraiser for transfrontier conservation through his

Peace Parks Foundation (Ramutsindela 2004:124). Not surprisingly, Prince Bernhard became the foundation's patron; more surprising is the patronage of Nelson Mandela, who seemed to have forgotten the Kumleben investigation.[17] With Mandela's name linked to the project, Anton Rupert successfully sold his transfrontier conservation ideas to almost all of the regional leaders who subsequently were willing to become patrons of his foundation themselves (Ramutsindela 2004). In addition, Rupert started another club of companies and wealthy businessmen to finance his operations, this time called the "21-Club", with membership fees set at $1 million (Domisse 2005:385). The membership list is published on the foundation's website (http://www.peaceparks.org), and contains familiar names, some of which also appear on the 1001-Club lists that circulate. The official signing by the three founding patrons of the PPF, Nelson Mandela, Prince Bernhard and Anton Rupert, took place at the palace of the Prince in Soestdijk in June 2001 (Vermeulen 2006:230). In 2002 Mozambique honoured Prince Bernhard for his role in nature conservation in southern Africa, particularly for his support to the Great Limpopo Transfrontier Park, by issuing a series of postage stamps showing the Prince with his favourite animals (*NRC Handelsblad* 25 September 2002).

The PPF claims to endorse a different approach to nature conservation than the one pursued by the old SANF—now renamed WWF South Africa. Since the Brundtland report of 1987, where care for the environment and economic development were considered linked to each other, nature conservation has increasingly sought salvation in what is broadly described as a "people and parks-approach" in which the paradigm of Community Based Natural Resource Management (CBNRM) could flourish (Adams and Hulme 2001). Fortress conservation "had progressively been challenged by a new community conservation narrative which stresses the need not to exclude local people, either physically from protected areas or politically from the conservation policy processes" (Hutton, Adams and Murombedzi 2005:342). Emphasis was placed on the need to develop strategies for local communities to economically benefit from nature conservation in order to secure communities' support for and involvement in nature conservation. This approach was also adopted and captured by the corporate sector to foster its own interests, as it legitimated further commodification of nature and "conservation through consumption" (Brockington, Duffy and Igoe 2008; Sklair 2001).

In line with this, the PPF presents transfrontier conservation areas (TFCAs) as motors for economic development, especially through tourism (Wolmer 2003) and communities living in and adjacent to these areas are said to benefit through job creation and revenue sharing. The main vehicles for the promotion of tourism are public–private partnerships (Ramutsindela 2004), and the PPF emphasises the importance of creating investment opportunities in TFCAs. In a report on the "flagship" of transfrontier conservation, the Great Limpopo Transfrontier Park, on the PPF website the foundation reports that:

> The Great Limpopo joint management board developed a five-year integrated development and business plan that will provide a comprehensive package of business and investment opportunities. It will also guide the joint management board and management committee activities over the next five years.

Elsewhere, however, we have shown that it is quite difficult for communities to tap into the revenues generated by TFCA tourism developments and develop beneficial partnerships with the private sector (Spierenburg, Steenkamp and Wels 2006, 2008). Especially on the Mozambican side of the Great Limpopo, where the PPF is co-managing the newly established Limpopo National Park with Mozambican conservation authorities, community benefits are doubtful. Of the approximately 27,000 people living in the Limpopo park, about 7000 are living in an area that consultants hired by the PPF deemed most suitable for conservation and tourism development. Those 7000 people are now in the process of being resettled. The resettlement exercise is presented by the PPF and the Mozambican authorities as voluntary, and as a development project. Questions can be raised, however, about the extent to which the exercise is voluntary. Many of the residents complained that the translocation of wild animals and the restrictions placed on local agricultural production and livestock husbandry due to the National Park's regulations render life in the park increasingly difficult, so many people were "forced to accept" resettlement (Milgroom and Spierenburg 2008). Instead of creating opportunities for local communities, the extension of the Great Limpopo TFCA into Mozambique has resulted in a vast area being cleared of human habitation to make way for conservation and (mainly South African) investors in tourism.

The developments described above fit with what several authors have argued, namely that in fact, the whole transfrontier conservation movement is a disguised return to the barriers, a movement away from working with communities that are considered not such good ecological stewards after all (Chapin 2004; Dzingirai 2004; Hutton, Adams and Murombedzi 2005). Hutton, Adams and Murombedzi (2005) put the blame not only on disappointments with community-based conservation, but also and explicitly on the increased involvement of the private sector in conservation (see also Dzingirai 2003). Ramutsindela (2004) concurs that an important factor contributing to the acceptance of the idea of TFCAs in southern Africa has been the adoption by most countries in the region of a neoliberal economic development model, allowing a significant role for the private sector. He argues that conservationists promoting the establishment of TFCAs took advantage of the financial demands placed on the new democracies of South Africa and Mozambique. Given the high pressures put on these governments to redress the historically skewed distribution of resources and services, and address economic growth and poverty alleviation as top priorities, nature could only be conserved if it would "pay for itself". This condition created pressure for the privatisation of conservation and allowed the private sector to step in (Ramutsindela 2004:69). According to Hutton, Adams and Murombedzi (2006), as a result, the interests of private sector companies involved in conservation take precedence over those of local communities.

The PPF places strong emphasis on the need to develop opportunities for the private sector which, it maintains, will benefit local communities. Meanwhile, the technical committee on community involvement that used to advice the joint management board of the Great Limpopo has been abolished, since, the PPF argued, community participation is an issue of national governments (Makuleke 2007). This move is likely to further reduce the possibilities for local communities to participate in and benefit from natural resource management in conservation areas, and seems to undo what little progress had been made in that direction in the 1990s.

Concluding Remarks

Trying to trace the ideological influences of business philanthropy to nature conservation in southern Africa is a murky endeavour. Big money and big plans come with complex and intertwined global (financial, business and government) networks. When Ellis analysed the situation he started from the assumption that "(c)onservation requires *government* action to control land and the people and animals which occupy it. In other words, it requires the control of both natural and human resources, which is the stuff of power" (Ellis 1994:54, italics added). Nevertheless, it seems clear that in conservation there is quite a long history of governments relinquishing some public control to wealthy business elites. In recent years, the process of elite pacting, however, is playing an even more important role in nature conservation, especially in resource-poor countries in the South (Draper, Spierenburg and Wels 2004; Ramutsindela 2004; Seekings and Natrass 2006; Sklair 2001). With this pacting came the political, cultural and religious convictions that are determined by the various contexts of the wealthy individuals. The neoliberal economic wind that has been blowing globally, also very clearly in South Africa (see Hart 2002), helped to bring governments and private sector interests and networks together. It seemed to facilitate the combination of fierce capitalism with overwhelming philanthropic generosity.

The lifelong cooperation between Prince Bernhard and Anton Rupert brought capitalism and conservation together in a way that both needed each other. The latter not only, as Sklair (2001) argued, in the sense of a felt need to contribute to solving the ecological crises arising from the global spread of capitalism, but also in terms of improving their public images that might have been damaged by their involvement in apartheid and—in the case of Prince Bernhard—a bribery scandal. Through his promotion of transfrontier conservation and the Peace Parks, Anton Rupert managed to safeguard his position in the new South Africa, despite having been so prominently involved in "the old order", as well as perhaps the position of his old business partners, some of whom are now members of the PPF's 21-Club. Nevertheless, the ideological underpinnings and contextualisation of the relations between the two men—and between conservation and capitalism—at times had contradictory outcomes. Although their actions may have added to the amount of land under conservation, on some occasions they also endangered wildlife populations, and threatened regional political stability, as the example of Operation Lock showed. The example of the Great Limpopo Transfrontier Park, where a multiple-use approach under the guidance of the PPF turned into the creation of yet another park, with dire consequences for some of the local communities (see Spierenburg, Steenkamp and Wels 2006, 2008), also hints at the possible destabilising consequences of the marriage between capitalism and conservation, further feeding the antipathy of some local communities towards conservation efforts in southern Africa.

Acknowledgements

We thank the South Africa Netherlands Programme for Alternative Development (SANPAD) for its financial contribution to this study. We also thank Gerard Aalders, Stephen Ellis, the two anonymous reviewers of the original journal, and the participants of the European Environmental History conference (Amsterdam, June 2007), of the Anthropology of Elites conference (VU University Amsterdam, January 2008) and of the Colloquium on Conservation and Capitalism (Manchester University, September 2008), in particular Bernhard Giblsi, Jane Carruthers, Shirley Brooks, Malcolm Draper, Maano Ramutsindela, Rosaleen Duffy and Daniel Brockington, for their helpful comments on earlier versions of this chapter. We take full responsibility for any errors or misinterpretations.

Endnotes

[1] Translation by Marja Spierenburg and Harry Wels, original text: "(v)oor mij was hij zonder twijfel de belangrijkste figuur in natuurbehoud in de twintigste eeuw. Hij

heeft veel bereikt", http://www.refdag.nl/website/artikel.php?id=127682 (accessed 15 December 2004). For reasons of flow and readability of the text, references to websites and newspapers are put in footnotes.

[2] Rupert's overall donations to the PPF in the first years of the PPF, however, are rather difficult to trace in the annual reports of the PPF. We could not trace for instance the amount of R1.2 million that John Hanks mentions. Rupert's donations, however, can be roughly estimated on the basis of comparing the amounts of money generated by the fees of the Peace Park Club Members and through donations. For 1997 the donations totaled R555,004, while membership fees brought in R2,169,113 (Annual Report PPF 1998:19), so most money in 1997 was brought in by others than Rupert. For 1998 the Peace Parks Club also brought in most money (R4,820,090) compared to donations (R2,385,879) (Annual Report PPF 1999:20). The same applies to 1999 (Annual Report PPF 2000:20). In a way, one could conclude from this that it is not solely through the major financial contribution by Rupert that the PPF had a flying start, as some commentators would have it.

[3] *The Witness*, 20 January 2006, visionary and patriot.

[4] In 1986, the organisation changed its name to *World Wide Fund for Nature*, retaining the WWF initials, to better reflect the scope of its activities. However, it continues to operate under the original name in the United States and Canada.

[5] Pieter Johannes Meyer later worked for the Rembrandt Group from 1951 to 1959 as Chief Public Relations Officer (Wilkins and Strydom 1978:130).

[6] Dr Max Ilgner was present as a "guest of honour" at the wedding of Prince Bernhard and Princess Juliana, 7 January 1937 (Klinkenberg 1979:51).

[7] http://geschiedenis.vpro.nl/programmas/2899536/afleveringen/20182712/items/ 15868786/ (accessed 4 March 2007). See also Klinkenberg (1979:305–322).

[8] NIOD: Nederlands Instituut voor Oorlogsdocumentatie (Netherlands Institute for War Documentation).

[9] http://www.parool.nl/nieuws/2007/FEB/09/p3.html (accessed 4 March 2007). The article does not name the two CIA bosses. Probably they refer to Walter Bedell Smith who headed the CIA from 1950 to 1953 and his successor Allen W. Dulles (Aalders 2007:17).

[10] Translated by the authors, original text: "Zowel de Ford als de Rockefeller Foundation heeft gefungeerd als instrument voor 'covert' Amerikaanse buitenlandse politiek. Hun bestuursleden hadden nauwe banden met de Amerikaanse inlichtingendiensten of maakten er zelfs onderdeel van uit . . . Liefdadigheidsfondsen waren het meest geschikt om grote sommen geld door te sluizen naar CIA-projecten, zonder dat de gelukkige ontvangers ook maar het geringste benul hadden dat het geld van de Amerikaanse geheime dienst afkomstig was."

[11] http://www.mars-mercurius.nl/wie.asp (accessed 4 March 2007).

[12] http://www.panda.org/about_wwf/who_we_are/history/nineties/index.cfm (accessed 5 March 2007).

[13] Prince Bernhard was accused of having accepted slush money (1.1 million Dutch guilders = approx. €500.000) from this aircraft construction company. Prince Bernhard has always maintained that at least part of the money had been earmarked and paid to WWF for conservation purposes. Absolute and watertight evidence has never been found for the accusations. Nevertheless a constitutional crisis loomed as a result of this scandal, because of the danger of criminal prosecution of the Prince. The crisis was averted by not prosecuting Prince Bernhard, but he had to lay down all his functions in the private sector and his position as Inspector General of the Dutch military. Furthermore he was not allowed to ever wear a military uniform in public again (Van Wijnen 2000).

[14] See also Eveline Lubbers from Buro Jansen en Janssen on http://www.burojansen. nl/artikelen_item.php?id=204 (accessed 10 May 2007). Thanks to Stephen Ellis for bringing her name to our attention.

[15] John Hanks continued to be active in nature conservation, however, and in February 1997 was asked by Rupert to help him set up the PPF as its CEO, a post he held for $3^1/_2$ years until he was ousted in 2000 and replaced by Willem van Riet.

[16] Translated by the authors, original text: "Kumleben stuitte op een wijdvertakt netwerk van spionage en economische belangen, waar natuurbescherming alleen maar dienst leek te doen als dekmantel."

[17] "He [Mandela] is said to agree to connect his name to the organisation due to the promise of 500 000 guilders [now approx. € 200.000 MS/HW] per annum for his Children Fund. The 500,000 guilders come from the Dutch Postcode Lottery, which on their behalf, agreed to Rupert that they would become a major donor [of the PPF] if Mandela would come aboard" (Bosman 2007:36).

References

Aalders G (2007) *De Bilderberg Conferenties. Organisatie en Werkwijze van een Geheim Transatlantisch Netwerk* [The Bilderberg Conferences. Organisation and method of a secret transatlantic network]. Amsterdam: Uitgeverij van Praag

Aalders G and Wiebes C (1996) *The Art of Cloaking: The Case of Sweden*. Amsterdam: Amsterdam University Press

Adams W and Hulme D (2001) Conservation and community: Changing narratives, policies & practices in African conservation. In D Hulme and M Murphree (eds) *African Wildlife & Livelihoods* (pp 9–23). Oxford: James Currey

Anderson D and Grove R (eds) (1987) *Conservation in Africa*. Cambridge: Cambridge University Press

Anheier H K and Leat D (2006) *Creative Philanthropy*. London. New York: Routledge

Becker P (1985) *The Pathfinders. The Saga of Exploration in Southern Africa*. London: Penguin Books

Beinart W and Coates P (1995) *Environment and History*. London: Routledge

Ben Barka M (2003) Religion and environmental concern in the United States. In H Bak and W W Hölbling (eds) *"Nature's Nation" Revisited. American Concepts of Nature from Wonder to Ecological Crisis* (pp 281–293). Amsterdam: VU University Press

Bloomberg C (1989) *Christian Nationalism and the Rise of the Afrikaner Broederbond in South Africa, 1910–1948*. Bloomington and Indianapolis: Indiana University Press

Bonner R (1994) *At the Hand of Man. Peril and Hope for Africa's Wildlife*. New York: Vintage Books

Bornstein E (2003) *The Spirit of Development*. New York, London: Routledge

Bosman M (2007) "Partnership for a Southern Africa or 'Pre-colonial Measures'. A study on the Peace Parks Foundation and its Corporate Supporters." Unpublished MA Thesis, VU University Amsterdam

Bourdieu P (1983) The forms of capital. In R Kreckel (ed.) *Soziale Ungleichheiten* [Social inequalities] *(Soziale Welt, Sonderheft 2)* (pp 183–98). Goettingen: Otto Schartz & Co

Braack L E O (1983) *A Struik All-Colour Guide to the Kruger National Park*. Cape Town: C. Struik

Breytenbach J (1997) *Eden's Exiles*. Cape Town: Queillerie

Brockington D (2002) *Fortress Conservation: The Preservation of the Mkomazi Game Reserve, Tanzania*. Oxford: James Currey

Brockington D, Duffy R and Igoe J (2008) *Nature Unbound. Conservation, Capitalism and the Future of Protected Areas*. London: Earthscan

Brockington D and Scholfield K (2010) The conservationist mode of production and conservation NGOs in sub-Saharan Africa. *Antipode* 42(3):551–575

Büscher B (2009) "Struggles over Consensus, Anti-Politics and Marketing: Neoliberalism and Transfrontier Conservation and Development in Southern Africa." Unpublished PhD thesis, VU University Amsterdam

Carroll W K and Carson C (2003) Forging a new hegemony? The role of transnational policy groups in the network and discourses of global corporate governance. *Journal of World-Systems Research* 6(1):67–102

Carroll W K and Fennema M (2002) Is there a transnational business community? *International Sociology* 17(3):393–419

Carruthers J (1994) Dissecting the myth: Paul Kruger and the Kruger National Park. *Journal of Southern African Studies* 20(2):263–283

Carruthers J (1995) *The Kruger National Park*. Pietermaritzburg: University of Natal Press

Chapin M (2004). A challenge to conservationists. *World Watch Magazine* November/December, http://www.worldwatch.org (last accessed 20 September 2008)

Coniff R (2003) *The Natural History of the Rich: A Field Guide*. London: Arrow Books

Covington S (1997) Moving a public policy agenda: The strategic philanthropy of conservative foundations. A report by the National Committee for Responsive Philanthropy, http://www.commonwealinstitute.org/ncrp.covington.1.htm (last accessed 21 July 2009)

De Klerk W A (1975) *The Puritans in Africa*. London: Rex Collings

Domisse E (in cooperation with Willie Esterhuyse) (2005) *Anton Rupert: A Biography*. Cape Town: Tafelberg

Draper M, Spierenburg M and Wels H (2004) African dreams of cohesion: Elite pacting and community development in Transfrontier Conservation Areas in southern Africa. *Culture and Organization* 10(4):341–351

Duffy R (1997) The environmental challenge to the nation-state: Superparks and national parks policy in Zimbabwe. *Journal of Southern African Studies* 23(3):441–451

Duineveld W (2009) "Nature and networks. Prince Bernhard and Anton Rupert: Nature conservation activities in relation to business networks." Unpublished MA Thesis, VU University Amsterdam

Dzingirai V (2003) The new scramble for the African countryside. *Development and Change* 34(2):243–263

Dzingirai V (2004) *Disenfranchisement at Large: Transfrontier Zones, Conservation and Local Likelihoods*. Harare. IUCN ROSA

Ellis S (1994) Of elephant and men: Politics and nature conservation in South Africa. *Journal of Southern African Studies* 20(1):53–69

Furlong P J (1986) "Pro-Nazi subversion in southern Africa, 1939–1941." Paper presented at the twenty-ninth Annual Meeting of the African Studies Association, Los Angeles, USA, October–November

Giliomee H (2003) *The Afrikaners. Biography of a People*. Cape Town, Charlottesville: Tafelberg Publishers

Hanks J (2000) The role of TransFrontier Conservation Areas in southern Africa in the conservation of mammalian biodiversity. In A Entwistle and N Dunstone (eds) *Priorities for the Conservation of Mammalian Diversity—Has the Panda Had its Day?* (pp 239–256). Cambridge: Cambridge University Press

Hart G (2002) *Disabling Globalization. Places of Power in Post-Apartheid South Africa*. Pietermaritzburg: University of Natal Press

Hey D (1995) *A Nature Conservationist Looks Back*. Cape Town: Cape Nature Conservation

Hoffman W (1979) *Queen Juliana: The Story of the Richest Woman in the World*. New York: Harcourt Brace Jonanovich

Hutton J, Adams W M and Murombedzi J (2005) Back to the barriers? Changing narratives in biodiversity conservation. *Forum for Development Studies* 2:341–370

Klinkenberg W (1979) *Prins Bernhard: een Politieke Biografie: 1911–1979*. Amsterdam: Onze Tijd

Lässig S (2004) *Bürgerlichkeit*, patronage, and communal liberalism in Germany, 1871–1914. In T Adam (ed) *Philanthropy, Patronage and Civil Society. Experiences from Germany, Great Britain, and North America* (pp 198–218). Indiana: Indiana University Press

MacDonald K I (2010) The devil is in the bio(diversity): Private sector "engagement" and the restructuring of biodiversity conservation. *Antipode* 42(3):513–550

Makuleke L (2007) The power behind the scenes: A case study of the Makuleke community and their participation in the establishment of the Great Limpopo Transfrontier Parks (GLTP). Unpublished paper, Stellenbosch University

Marx C (1998) *Oxwagon Sentinel: Radical Afrikaner Nationalism and the History of the Ossewabrandwag*. Berlin: Lit Verlag

Milgroom J and Spierenburg M (2008) Induced volition: Resettlement from the Limpopo National Park, Mozambique. *Journal of Contemporary African Studies* 26(4):435–448

Moleah A T (1993) *South Africa, Colonialism, Apartheid and African Dispossession*. Delaware: Disa Press

Moodie D (1980) *The Rise of Afrikanerdom: Power, Apartheid, and the Afrikaner Civil Religion*. Berkeley, LA: University of California Press

Nattrass N (1999) The Truth and Reconciliation Commission on business and apartheid: A critical evaluation. *African Affairs* 98:373–391

NRC Handelsblad (2002) Prins op Afrikaanse wildzegels. 25 September

O'Meara D (1977) The Afrikaner Broederbond 1927–1948: class vanguards of Afrikaner Nationalism. *Journal of Southern African Studies* 3(2):156–186

O'Meara D (1983) *Volkskapitalisme: Class, Capital and Ideology in the Development of Afrikaner Nationalism, 1934–1948*. Cambridge: Cambridge University Press

Pelzer A N (1979) *Die Afrikaner-Broederbond: Eerste 50 Jaar*. Cape Town: Tafelberg Publishers

Potgieter De W (1995) *Contraband: South Africa and the International Trade in Ivory and Rhino Horn*. Cape Town: Queillerie

Ramutsindela M (2004) *Parks and People in Postcolonial Societies: Experiences in Southern Africa*. Dordrecht: Kluwer Academic Publishers

Ramutsindela M, Spierenburg M and Wels H (forthcoming) The nature of elites: Business philanthropy and networks in nature conservation in southern Africa. In B Gissibl, S Hoehler and P Kupper (eds) *Civilizing Nature: Towards a Global History of National Parks*

Rupert A (1967) *Progress through Partnership*. Cape Town: Nasionale Boekhandel

Rupert A (1981) *Priorities for Coexistence*. Cape Town: Tafelberg Publishers

Seekings J and Nattrass N (2006) *Class, Race and Inequality in South Africa*. Pietermaritzburg: University of KwaZulu Natal Press

Serfontein J H P (1979) *Brotherhood of Power: An Exposé of the Secret Afrikaner Broederbond*. London: Rex Collings

Simmel G (1965) [1908] The poor [translated by Claire Jacobson]. *Social Problems* 13(2):118–140

Sklair L (2001) *The Transnational Capitalist Class*. Oxford: Blackwell Publishers

Sparks A (1991) *The Mind of South Africa: The Rise and Fall of Apartheid*. London: Mandarin

Spierenburg M, Steenkamp C and Wels H (2006) Resistance of local communities against marginalization in the Great Limpopo Transfrontier Area. *Focaal, European Journal of Anthropology* 47:18–31

Spierenburg M, Steenkamp C and Wels H (2008) Enclosing the local for the global commons: Community land rights in the Great Limpopo Transfrontier Conservation Area. *Conservation and Society* 6(1):87–97

Templin J A (1984) *Ideology on a Frontier: The Theological Foundation of Afrikaner Nationalism, 1652–1910*. London: Greenwood Press

Tukkie (2006) 'n Man van waarde word wêreldwyd geag. May

Van Wijnen H A (2000) *De Macht van de Kroon*. Amsterdam: Uitgeverij balans

Vermeulen W (2006) *Operatie Natuur: Het Natuurbeschermingsleven van Zijne Koninklijke Hoogheid Prins Bernhard*. Zeist:Wereld Natuur Fonds

Wels H (2004) About romance and reality: Popular European imagery in postcolonial tourism in Africa. In C M Hall and H Tucker (eds) *Tourism and Postcolonialism* (pp 76–94). London: Routledge

Wilkins I and Strydom H (1978) *The Super-Afrikaners: Inside the Afrikaner Broederbond*. Johannesburg: Jonathan Ball Publishers

Wolmer W (2003) Transboundary conservation: The politics of ecological integrity in the Great Limpopo Transfrontier Park. *Journal of Southern African Studies* 29(1):261–278

World Commission on Environment and Development (WCED) (1987) *Cur Common Future*. Oxford: Oxford University Press

Zwaap R (1997a) Het wereld natuur leger. *De Groene Amsterdammer* 5 November, http://www.groene.nl/1997/45/het-wereld-natuur-leger

Zwaap R (1997b) Optellen en afschieten. *De Groene Amsterdammer* 17 December, http://www.groene.nl/1997/51/optellen-en-afschieten

Zwaap R (2001) Out of Africa. *De Groene Amsterdammer* 30 June

Chapter 8
Protecting the Environment the Natural Way: Ethical Consumption and Commodity Fetishism

James G. Carrier

One of the ways that people are encouraged to protect the natural environment is by assessing the objects that they buy in terms of the degree to which they meet environmentalist criteria. People are urged to buy the objects that meet these criteria and reject the others. This goes under the name of "ethical consumption", which is concerned with social as well as environmental issues. It is the current manifestation of a long history of social movements that have urged people to assess the objects that they confront not just in the classic economic terms of their cost and utility, but also in terms of the ways in which they are produced, processed and transported (Trentmann 2007).

Through ethical consumption, its advocates argue, people can do two things. Firstly, at a personal level, they can lead lives that are more moral. Secondly, at a public level, they can use their purchases to affect the larger world, by putting pressure on firms in a competitive market to change the ways that they do things. In both of these, though more obviously in its public aspect, ethical consumption marks a conjunction of capitalism and conservation, for it identifies people's market transactions, and market mechanisms generally, as the effective way to bring about protection of the environment.

In this chapter I want to look at this conjunction in terms of something that has long been part of the analysis of capitalism, Marx's (1867) notion of commodity fetishism. In doing so, I mean to point to some aspects of ethical consumption that reduce the likelihood that it can serve as a means by which ethical consumers can either lead personal lives that are more moral or influence the behaviour of firms in ways that they intend. What concerns me especially is the ways that the environment and conceptions of it are presented to would-be ethical consumers, presentations that are likely to be important in shaping the way people assess the conservationist merits of the various purchasing options available to them.

A visible sign of ethical consumption is objects that carry the mark of the Fairtrade Foundation, a certifying organisation concerned mostly with foodstuffs produced in tropical regions. An indication of the growth of ethical consumption in Britain is sales of items certified by Fairtrade. From 2005 to 2006 these increased by 49%, and rose about 80% in the following year, to a total of £493 million (Hickman and Attwood 2008). Fairtrade certification is available only to organisations that meet particular standards concerning the way in which items are produced and transacted. Coffee, for instance, can be sold as Fairtrade only if the merchant behaves in particular ways in relation to the producers and only if the producers are certain sorts of people who act in specific sorts of ways (Fairtrade Labelling Organizations International 2006). Although certification requirements vary for different items, broadly they enjoin a degree of social and economic co-operation and equality, and a loose, general concern to protect the natural environment.

Because it is concerned with a range of foodstuffs visible to shoppers, Fairtrade is probably the best known of the organisations that certify objects for sale in Britain in terms of ethical criteria. There are other such organisations, of course, ranging from the Soil Association, certifying foodstuffs, to the Forest Stewardship Council, certifying timber and its products. Equally, there are objects and forms of consumption that are not certified but that can be seen as part of ethical consumption, such as preferring fuel-efficient cars over conventional vehicles, or travelling to ecotourist destinations on holidays in preference to conventional resorts.

From a certain perspective, ethical consumption is the natural way to protect the environment. The most recent name for that perspective is "neoliberalism" but, like assessing the morality of what you buy, in one way or another it has been around for a long time (Carrier 1995a:157–166). From that perspective, ethical consumption is natural in two ways. Firstly, it reflects the importance of market transactions as a fundamental aspect of people's lives, what Adam Smith (1976 [1776]:17) famously described as an aspect of human nature, the "propensity to truck, barter, and exchange one thing for another". Secondly, it reflects the importance of the autonomous individual and of that individual's freedom. One aspect of this autonomy is the freedom to choose without constraint, including the constraint of government rules of the sort that could regulate commerce in ways that might achieve consumers' ethical goals (see Brown 1997). This freedom to choose is seen as important for human welfare, as it allows each person to satisfy his or her wants. In a more profound sense, it is said to be important because, without it, people would develop distorted tastes and values; only the free, in other words, can come to recognise and value the good (see Carrier 1995a:162).

While some claim, with Smith, that market transactions by autonomous actors are natural, it is perhaps more accurate to refer to them as "naturalised". That is because the natural basis and positive value assigned to autonomy and market freedom reflect a view of people and the world around them that is peculiar to some political-economic systems rather than others. That view tends to fetishise commodities, market transactions and, indeed, people themselves. As this invocation of fetishism and my earlier mention of Marx indicate, I will draw here on "The fetishism of commodities and the secret thereof" (Marx 1867) in *Capital*. Recall that Marx was concerned with how commodities tend to be presented and perceived in a peculiar way under capitalism, one that ignores or denies the labour time entailed in the processes involved in their production and their presentation to the would-be purchaser.

Here, I extend "fetishism" somewhat more broadly than did Marx. I use it to refer to the ignoring or denial of the background of objects. However, my concern is not restricted to what concerned him, the transmuting of the relationship between different amounts of human labour into a relationship between different commodities. Rather, I take from his argument a concern with the general tendency to obscure the people and processes, of which labour power is a component, that are part of creating an object and of bringing it to market. This tendency is an aspect of the abstraction of things from their practical contexts that is both widespread in modern capitalist societies and seen as natural and valuable (Carrier 2001). Of course, more is involved here than just material objects. For instance, seeing individuals as autonomous fetishises them. That is because it ignores or denies the people and processes in the past and present that shape them, that enable them to act in certain ways and that endow those actions with certain effects, people and processes that may be recognised and valued in other societies (eg Strathern 1988) and in subordinate parts of Western, market societies (Carrier 1995b:99–101). Instead, the autonomous individual's attributes, acts and effects are seen to spring, fetishistically, from within the person rather than from the interaction of the person and the contexts in which he or she exists.

I extend fetishism beyond Marx's original argument in another way as well. He was concerned with the production of things with the intention of selling them in a market transaction, commodities. I include things, whether material or not, that are not produced in the conventional sense but that can be appropriated and used for commercial gain, in the manner of Polanyi's (1957 [1944]:ch 6) fictitious commodities. For instance, a private nature reserve that charges admission fees does not produce the landscape or the living things within it in the way that a clothing company produces shirts or a law firm produces wills. However, if advertisements for the reserve urge people to come and see the landscape and animals for a fee, these things are commodities from my perspective. Somewhat more contentiously, a remote hotel in a spectacular setting that touts the view from the veranda to potential guests has turned the scenery into a commodity: it is part of what is being sold in market transactions even though the hotel did not produce the scenery.

I have sketched the battery of basic concepts that I will deploy. I use them to consider what is in this chapter's subtitle, ethical consumption and commodity fetishism. My purpose in this is not to present a detailed, empirical analysis of one or another sort of ethical consumption: I am not competent to do so. Rather, I mean to put forward a perspective on ethical consumption that raises questions about the ability of ethical consumers to pursue their moral goals through transactions in competitive, capitalist markets. In fact, I suggest that the nature of these markets in the current political economy encourages a fetishism of commodities that tends to subvert ethical consumption in ways that are likely to be invisible to ethical consumers and are likely to conflict with their values.

I will illustrate this commodity fetishism with two sorts of ethical consumption. One of these is the purchase of Fairtrade and similar items. I use Fairtrade because it is familiar to many people, because it illustrates well some aspects of fetishism and because Fairtrade certification requires practices that are at least loosely environmentalist. The other sort of ethical consumption I use is ecotourism. While it is not standardised and certified in the way of Fairtrade, it is more clearly concerned with the protection of people and nature. It is a form of tourism that entails travel to enjoy attractive and interesting surroundings, often identified as "natural", and attractive people and their activities, often identified as "indigenous" or "exotic". It is considered to be ethical because the purpose is not simply to enjoy these things. Rather, as Martha Honey (2003), then head of the International Ecotourism Society, explained, ecotourism "should: 1) protect and benefit conservation; 2) benefit, respect, and help empower local communities; and 3) educate as well as entertain tourists". Ecotourism may, in fact, fail to fulfil the promise that Honey identified (see, eg, Duffy 2002; Mowforth and Munt 1998), but the question of its efficacy is different from the issue I pursue here.

These two sorts of ethical consumption have different historical backgrounds, reflect different sets of concerns and operate in different ways. In spite of those differences, both illustrate the fetishism that concerns me. This is apparent in what follows, a description of three aspects of fetishisation in ethical consumption. These three aspects are the fetishisation of the object that is to be purchased, of the acquisition or consumption of that object, of the environment to be protected. Although I identify these three aspects as distinct, it will become clear that they are linked to each other to some degree.

The Object

The first aspect of fetishisation in ethical consumption that I want to consider is that of the object to be purchased. By this I mean the activities that remove it, to a greater or lesser degree, from the context in which it is produced, activities that range from the more symbolic to the more material. Because much ethical consumption is concerned with the social, political and environmental context of objects, this removal is not absolute: it is hard to see how it could be in a situation where the context of the object is part of what is advertised, sold and consumed. Rather, normally objects are presented in ways that manipulate that context, obscuring some aspects and featuring others. However, there is more at stake in this than just the fetishistic rendering of the object. In addition, this aspect of fetishism, like the others I will discuss, seems likely to influence what I call "ethicality", a notion that draws on Scott's (1998) idea of legibility, as it is elaborated by Errington and Gewertz (2001).

For Errington and Gewertz, as for Scott, making something legible means making it both visible and recognisable for purposes of survey and supervision by governments or other powerful bodies. These writers are concerned with how legibility entails identifying complex and variable realities in terms of "a common standard necessary for a synoptic view" (Scott 1998:2), and does so in terms of the interests and perspectives of the powerful. While legibility often has this political dimension, it involves a more general process as well. To make things legible is to render them as instances of a conceptual category: for Scott that might be "taxable income"; for Errington and Gewertz it is "traditional cultural practice". With the passage of time, however, it seems likely that the relationship between category and instance is reversed. What had been an instance of that category comes instead to define it, especially for those who are not particularly knowledgeable. It is this reversal that is pertinent for considering ethical consumption.

Ethical consumers want things that meet their moral criteria. An ecotourist wants a resort in a sound natural environment; an ethical consumer wants coffee produced in a way that does not exploit the growers. "Sound natural environment" and "production that does not exploit" are like "taxable income" and "traditional cultural practice" in that they are all conceptual categories, not things that exist independently of human thought and classification. Our ecotourist and ethical-coffee drinker need to be able to recognise a sound natural environment and non-exploitative coffee growing when they see them. These conceptual categories need to be legible, be visible and recognisable, if those ethical consumers are going to be able to assess the choices on offer.

Those who want to sell things to such people satisfy that need through images that encapsulate or manifest those conceptual categories, rendering them as instances of the categories. I think it likely that people who are not especially knowledgeable will come to take these images as defining a moral value, which is to say ethicality. So, for example, if photographs of people in exotic dress are used to represent indigenous people in general, then it is likely that people in exotic dress will come to define indigeneity, with the corollary that their absence will mark its absence. Similarly, to anticipate a point I develop later in this chapter, if colourful fish are used to represent healthy coastal waters, then environmental health will tend to be defined by the presence of those fish. Put more succinctly, images used to represent a state of affairs that satisfy ethical criteria make that satisfaction legible and come to define these criteria. They come to define ethicality. There is nothing very mysterious about this, nor does it imply that people are gullible. It is, after all, the same as people inferring the meaning of "cat" by seeing a hundred things that are called cats.

I will illustrate fetishisation of the object purchased with Peter Luetchford's (2008) study of a growers' co-operative in Costa Rica that produces Fairtrade coffee. Often, such coffee is sold in packages that include the picture of a co-operative member, and the Fairtrade web site regularly includes such pictures (Fairtrade Foundation 2008).[1] Typically, the picture is of a smiling man or woman of marked ethnic aspect presented as a small-holder working a few acres of land. However, because of the nature of coffee cultivation, presenting these growers in ways that suggest self-reliant peasantry fetishises the coffee that they grow. Tending a coffee small-holding is not terribly difficult most of the time. However, harvesting is so intense that the co-operative members that Luetchford studied frequently had to hire additional labour from their village and from further afield, including migrant workers from Nicaragua (a similar disjunction between an image of local small-holders and the use of hired labour is reported for a Fairtrade co-operative in Guatemala, in Lyon 2009).

In focusing on peasant small-holders, then, the picture on the bag of coffee elides the wage labour and the migrant workers that are an important aspect of the production, and so fetishises the product. Moreover, the pervasiveness of such images, coupled with the likely ignorance among ethical consumers of how coffee is grown, serves to shape ethicality. If I am correct, these images will come to define not just the form of coffee-growing that meets those consumers' moral values, but also those moral values themselves: "non-exploitative" will come to mean what ethical consumers imagine self-reliant, small-holding peasant production to be. Images of migrant workers living in barracks would be seen as indicating an unethical state of affairs, even though they are a part of Fairtrade coffee production.

This relationship of image and ethicality is clear in ecotourism, touristic consumption guided by environmentalist values. In the

Caribbean, the area that I know best, the most visible form of ecotourism is tourist divers enjoying the coastal waters, often in national parks. Even though these parks are not straightforward commercial firms in a competitive market, they resemble companies that sell Fairtrade coffee because they advertise. They do so because, like roasters handling Fairtrade coffee, they need the business; in their case, visiting ecotourist divers who will pay user fees (Carrier 2003:216–222). They need that business because the governments that made them parks do not give them enough money to operate, whether because the government is too poor, because it is constrained by bodies like the World Bank and the International Monetary Fund or because advocates assured the government that the proposed park would be self-supporting, and were wrong. These parks advertise themselves and the waters that they protect, and these advertisements contain images intended to attract visitors. Like the picture on the bag of Fairtrade coffee, these images deserve attention.

The images that concern me come from two parks in Jamaica that I have studied, at Montego Bay and Negril, though images from parks around the Caribbean are similar.[2] For each, the predominant objects portrayed are fish and coral growths, sometimes by themselves and sometimes with a diver in the picture. Such images make commercial sense: fish and coral growths are relatively easy to photograph and they are what a diver would want to see. And what a diver wants to see is important for those who run these parks. The director of the Montego Bay park repeatedly asked me for information about what made a "good dive", one that would attract tourist divers.

This concern with what attracts ecotourists is common in the region, where managers generally are concerned with increasing "the tourism product and recreational opportunities" of their parks (Geoghegan, Smith and Thacker 2001:10). Echoing this is a body of work that studies which aspects of nature attract paying tourists. Research in a Mexican terrestrial park concluded that "the presence of crocodiles . . . is the main attraction for visitors and that special emphasis needs to be put on their conservation" (Avila-Foucat and Eugenio-Martin 2004:13); research in the Turks and Caicos Islands found that "divers were willing to pay at least \$10 extra for dives during which more abundant Nassau groupers were observed" (Rudd and Tupper 2002:146).

While these attractive images make commercial sense, they fetishise the coastal waters that are being advertised to environmentally concerned divers. They do so when they focus on individual fish and coral growths as representing the state of those waters. For one thing, this focus ignores other aspects of those waters that may be more important ecologically, but are not portrayed because they are not likely to attract ecotourists. Prime among these are sea-grass beds. These are important in local nutrient cycles but appear to repel tourists, and hotels often remove beds from the waters where their guests are supposed to swim. As well, and to anticipate a point I will address at greater length below, these images represent individual creatures rather than populations, which has important consequences. One is that, in the context of environmentalist concern, this stress on individuals implies certain sorts of threats rather than others. Those are the threats that individual ecotourists can ameliorate by taking steps to avoid harming those individual fish and coral growths: they can forego spear-fishing, they can take care not to brush against a coral growth, they can make sure that the boats that take them on dives tie up at mooring buoys rather than using anchors.

I said that these sorts of representation do not only fetishise, they also serve to define ethicality. In saying this I am assuming two things. The first is that the vast majority of people who view these images are relatively ignorant of in-shore, tropical ecosystems and of the political-economic context of marine parks in Jamaica and the Caribbean more generally. The second is that most people see these images as produced by groups of knowledgeable and conscientious park staff concerned to protect the coastal waters, which endows those images with authority. The sheer repetition of such images, like the sheer repetition of photographs of co-operative members on bags of Fairtrade coffee, will, then, embody the ethical consumer's values, and so will define ethicality. Just as a smiling small-holder makes morally acceptable coffee growing legible, so pretty fish and colourful coral make a morally acceptable in-shore environment legible. Once we have seen the small-holder we can tell about the coffee; once we have seen the fish and the coral we can tell about the coastal waters and, indeed, know what a good environment is.

Jamaican marine parks and firms that sell Fairtrade coffee appeal to ethical consumers in different ways and in terms of different values. However, they operate in the same political-economic system, which constrains them to attract customers in the same way, by presenting appealing images that fetishise what they are selling. And as I have argued, these images are likely to be implicated in defining ethicality for ethical consumers who are ignorant of how these things are produced. Commercial pressure, then, and the commercial practices that it shapes, seem likely to shape the definition of a moral world, ethicality.

Purchase and Consumption

The second aspect of fetishisation that concerns me is the fetishisation of the means by which ethical consumers are able to purchase, and ultimately consume, an object. Those means are the commercial operations that bring the purchaser and object together, whether the result is the shopper confronting a bag of Fairtrade coffee on a supermarket shelf, a tourist settling in at an ecotourism resort or, as Neves (this volume) describes, a boatload of tourists confronting a whale in the Canaries.

When these institutions and processes are ignored, ethical consumers are prone to see the object to be purchased or consumed in terms of what are taken to be its properties, rather than the activities that brought it and the consumer together. The result is something like the "tourist bubble" (eg Cohen 1979; Graburn 1989), the carefully managed and mediated experiences that certain sorts of guided tours provide for tourists, experiences that hide the management and the mediation from view. The result is also rather like the conception of consumers in much anthropology of consumption and neoclassical economics (see Carrier 2006). In these, the focus is on the moment of consumer choice, rather than the contextual factors that shape that choice. So, relatively little attention is paid to the larger and longer-term processes by which consumers acquire their tastes or their utility functions, and by which the object of consumption comes into being and is brought to the presence of the consumer.

This fetishism appears in Fairtrade coffee. Such coffee is seen as allowing a more direct link between purchaser and grower, "cutting out the middle-man", enshrined in the picture of the co-operative member on the bag and in the name of a large British Fairtrade coffee company, CaféDirect. This imagery of a direct link ignores the roasters, shippers, wholesalers and retailers who stand between growers and drinkers of Fairtrade coffee, just as they do with ordinary coffee. In fact, with Fairtrade coffee they may be more numerous, as the Fairtrade organisation itself intervenes in the process. This fetishism appears as well in a recent exposition organised in Zaragosa, which sought to be "carbon neutral". Talking of a carbon-neutral exposition ignores the effects of the visitors travelling to the city to see the exposition and going back home afterwards: organisers were hoping to attract seven million of them (Ferren 2008).

This same sort of fetishisation is readily apparent in ecotourism. Its existence among ecotourists became especially apparent to me when I was communicating with a woman who said that she was concerned for the environment. As evidence of this, she said that she was careful not to step on moss when she visited Antarctica, because it was so fragile. She could claim to be ethical environmentally only if she viewed her consumption, her being in the Antarctic, fetishistically. Doing so meant that she could focus on her footsteps in the snow and ignore the operations that got her from Washington DC, where she lived, to the continent and back again.

The most obvious of these operations is air travel, and Stefan Gössling (1999) calculated that, on average, getting a visitor from a First World country to a tropical ecotourism destination and back again uses 205 kg of aircraft fuel and generates about 650 kg of CO_2 emissions (for estimates of emissions produced getting travellers to New Zealand, a growing ecotourist destination, see Becken 2002). There is, of course, increasing public concern about CO_2 emissions by aircraft: British Airways informs potential customers that "British Airways was the first airline to introduce a voluntary passenger carbon offsetting scheme. The money raised funds emission reduction projects such as hydro-electric power plants and wind farms" (British Airways no date).[3]

However, a range of other practices and events remain obscured, among them those that frame the nature visible at the ecotourist destination. Rosaleen Duffy illustrates this in her description of ecotourism in Belize. The wreck where tourists dive to see nature, in the form of sharks and rays, was not the result of a ship accidentally foundering on the coast. It was bought and purposely sunk by a local dive shop (Duffy 2002:29). More basically, the tropical beach in front of the tourist hotel was not natural, but the result of human engineering: the clearing of mangrove stands and sea-grass beds and the addition of sand (2002:45).

A different sort of event involved in a different sort of framing is described in Paige West's story of an ecotourist lodge in the village she calls "Maimafu", in the Crater Mountain Wildlife Management Area in the Highlands of Papua New Guinea (West and Carrier 2004:488–491). Although ecotourists are concerned with protecting and respecting local cultures as well as nature, this lodge was not an expression of local culture in any straightforward sense. In order to spread the benefits of the lodge and so keep tensions among land-owning clans as low as possible, villagers decided to build the lodge on a particular piece of land where members of many clans lived. It took 2 years to get all the clans to agree to that location. The decision was rejected by expatriate project managers in favour of a location of their own choosing that, they said, would be more attractive to tourists. Similarly, it took months to get villagers to agree on how the lodge should be run, to make its operation as equitable as possible among local clans. Expatriate managers rejected that decision in favour of arrangements that they preferred, again on commercial grounds. Just as ecotourists in Belize do not see the intentional sinking of a ship and construction of a beach, visitors to this lodge do not see the local decisions that were made and the managers who overrode them. These are the things that are lost to view when the purchase and consumption of the object are fetishised, presented and viewed in ways that ignore the activities and relationships that bring together the ethical consumer and the thing consumed.

This fetishisation is not the accidental result of a mass of random, local decisions. Rather, it reflects the commercial pressures that influence all organisations that want tourist money, whether commercial ecotourism companies running resorts or under-funded national parks and reserves. And as a consequence it fetishises ethical consumers themselves. They are portrayed, at least implicitly, as individuals motivated by their moral concerns. This hides the way that the expressions of that concern are shaped by commercial pressure, as well as the way that ethical consumers are a market that organisations seek to attract, even though many of the values of those consumers seem opposed to capitalist market rationality. The dive shop that bought an old ship and sank it off the coast of Belize was creating a product that, it thought, would attract ecotourists, as did those project managers who imposed their own commercial judgements when they rejected the decisions of Maimafu villagers about where to locate their lodge and how to run it.

In practice, then, ethical consumption extends the realm of just those commercial values and practices that this fetishisation hides from view, in the same way that Igoe, Neves and Brockington (this volume) say that environmentalist consumption extends the realm of capitalist exploitation into realms that are supposed to be protected from it. An instance of this extension in ecotourism is ECOSERV (Khan 2003), a questionnaire that elicits "the service level that [ecotourist] customers believe they 'should get' from the service provider" (2003:112). Use of the questionnaire reveals an important expectation that ecotourists have of hotels, that they have staff who are courteous, attentive and friendly (2003:117). Commercial pressure in a sector with increasing competition (2003:109–110) obliges hotels seeking to attract ecotourists to meet these expectations. Doing so means, among other things, employing people who will behave in a friendly and courteous way to strangers simply because it is part of their job. This commodification of affect and emotion (Hochschild 1983) may be common in the places where ecotourists come from. However, it is rare in many of the places where they go, and their expectation that they will experience it serves to extend commodification in spite of ecotourists' goal of respecting local cultures.

I have described how the fetishisation of purchase and consumption obscures the machinery that brings object and consumer together. As I have argued, this serves to restrict attention to what is seen as the inherent properties of the object. As a result, it hides from view not just the machinery but also the competitive capitalist way in which it operates, and indeed in which it must operate if ethical consumers are to be able to consume ethically. The head of CaféDirect put it this way: "If you want to change the trading system you've got to be on the same terms as the conventional system. You need to make a profit" (Martinson 2007).

The Environment

The last aspect of fetishisation that I want to describe is that of the environment itself. By this I mean the ways that the natural environment tends to be presented, which remove salient features of it from the larger context in which they exist.

I have already mentioned something that is pertinent here, the need of bodies like national parks to raise money through user fees, analogous to the need of commercial operators to raise money through sales to customers. I have noted that this need can reflect a number of factors, especially the relative poverty of governments in the tropical countries where ecotourists often go. In addition, however, it is reinforced by international funding agencies. In Jamaica, for instance, the Montego Bay park regularly received money from the United States Agency for International Development. However, Agency rules required that the funding be used for special projects, not to fund the basic operation of the park. Much of it was spent, in fact, on producing management and business plans as part of "capacity building". This was intended to put the park on a sound commercial basis by identifying possible sources of revenue, which turned out to be different aspects of tourism. This is indicated by the 1998 park management plan (Montego Bay Marine Park 1998), the bulk of which is devoted to revenue.

The situation in Negril was similar, although the source of funding was the EU rather than the USA. The EU gave a substantial amount of money for the Negril park, and explained its goals this way:

> The project aims at protecting the coral reef ecosystem and at achieving long-term financial sustainability of the Negril Marine Park and Protected Area through income generating measures. The project will contribute to the overall objectives of the Jamaican government to stabilize its economy by maintaining and increasing foreign exchange earnings through tourism, while at the same time protecting the fragile coastal environment of Negril, including the entire coral reef ecosystem (EU 2002).

Whatever the source of the pressure to treat parks as, in effect, tourism businesses, it is pervasive. Shortly after Jamaica began to undergo structural adjustment (see Bartilow 1997:ch 2–3; Payne 1994:ch 7), some long-term environmental activists in Montego Bay produced a report funded by the Organization of American States. That report evaluated a number of sites in Jamaica in terms of their suitability as the country's first marine national park. The crucial criterion was commercial, the ability to make money from tourists. The authors put it this way:

> marine parks can be pretty much self-supporting through a number of activities: snorkeling, SCUBA diving, glass bottom boat tours arranged

for a fee. Usually the marine park organization will leave most of these activities to commercial diver operators and watersport centers. In that case, however, substantial revenues may be obtained from concessions (O'Callaghan, Woodley and Aiken 1988:37).

On these grounds, Montego Bay was the best site. It was the main point of entry for visitors to Jamaica and by far the largest tourist destination in the country. When the Jamaican government declared the first two national parks, Montego Bay was the marine park of the pair, which prompted a number of environmentalists in the country to complain that the tourism that made the site attractive financially had so damaged the bay that a park there would do no real good.

This is the commercial orientation and financial pressure that underlay the pictures of fish and coral on the park web sites, which I mentioned previously. I want to return to those images.

I already noted one aspect of the fetishisation that they entail. I said that because they portrayed individual fish and coral growths, they implicitly identified threats to those things, and so implicitly identified environmentally ethical behaviour for ecotourists: do not take a fish or damage a coral. In the context of Jamaica's coastal waters, they implicitly point to another threat: the inshore artisanal fishers who are common around Jamaica's coast (see Garner 2009). Unlike ethical ecotourists, those fishers anchor their boats rather than using mooring buoys, and they spear and trap fish. Intentionally or otherwise, then, those images of individual fish and coral fetishise the nature that they portray in such a way that they reinforce the interests of one set of users of the coastal waters relative to another. Responsible tourists who spend money are welcome; irresponsible poor Jamaican fishers are not wanted. Here, then, is the political dimension of legibility that concerned Errington and Gewertz, and before them Scott; here too is the political dimension of ethicality.

There is a further aspect to this fetishisation, also political. In portraying individuals, these images imply that an ethical concern to protect nature is satisfied by reducing threats to individuals. However, the environmental state of the coastal waters is indicated not so much by the fate of individuals as by the state of populations, influenced by the context in which they exist. It would be difficult to portray populations in ways that would attract the ecotourists that these parks and companies need. However, in portraying the coastal waters in terms of individuals abstracted from their context, these images fail to portray populations, and so fail to imply that an ethical concern to protect these waters requires attending to threats to these populations. These are very different from threats to an individual fish or coral growth, which are matters of individual action: catching a fish or dropping an anchor. Rather, they are matters of the context that sustains these populations, which is to say the ecological system and the activities that can harm it. And unlike activities that harm individuals, these activities commonly take place on land.

In Montego Bay and Negril, the most important of these are the activities that generate the run-off that feeds into the coastal waters that these parks and ethical tourists seek to protect. Much of this pollution, and many of these activities, are a direct or indirect consequence of tourism. Directly, three-fifths of the waste water produced by hotels in Jamaica is inadequately treated or not treated at all (Burke 2005:11); the average tourist in Jamaica generates about four times the solid waste that the average Jamaican generates (Thomas-Hope and Jardine-Comrie 2005:3); these feed into the surroundings and, in these coastal areas, end up washing into the sea.

Tourism also has had indirect effects in these two sites. The establishment and expansion of tourism in Montego Bay and Negril has led to the destruction of natural features that are important for the in-shore ecology: the filling of swamps, destruction of mangrove stands, dredging of harbours and re-alignment of beaches. As well, Jamaicans have moved to these towns in search of work in the tourism sector. In 40 years the city of Montego Bay has tripled in size, to about 100,000, with hotels having about 8000 rooms servicing over 400,000 visitors in 2003; Negril has gone from a fishing camp in a cane-growing area to a town of about 20,000, with hotels having a total number of rooms of a little under 6000 servicing about 275,000 visitors in the same year (Bakker and Phillip 2005). This growth in population has not been matched by an expansion of urban infrastructure: houses are not connected to what sewer system there is and solid waste is not collected.

I have described the last of the three fetishisations that concern me, that of the environment consumed by ethical consumers. I have argued that this fetishisation, like the others I have considered, is shaped by the capitalist political economy that constrains the parks and companies that seek to attract ethical tourists. Like those others, this fetishisation does not only elide the context of the thing fetishised, it does so in a way that shapes ethicality. In this case it tends to define environmental values in terms of individual creatures, while ignoring the contextual processes that shape the populations of which they are a part. As a result, it encourages ecotourists to focus on the effects of what they do in the coastal waters, and to ignore the effects of the tourism sector that their ethical consumption supports.

Conclusion

What I have presented here is a polemic, contentious not least because it assumes more than it demonstrates. I have described a set of images and representations without knowing how ethical consumers in fact respond to them. Equally, I have asserted that those images shape understandings of the world and ethicality in ways that make ethical consumption a dubious vehicle for challenging what many see as the ill effects of capitalism, and again, I have done so without knowing how ethical consumers respond to those images. These are empirical questions, and if there are studies of those responses I do not know of them. The sorts of images that I have described are, however, quite common, which suggests that they are not there by accident. Rather, they appear to be there for a reason, to attract potential purchasers. Moreover, their persistent visibility, their dull repetition in everyday life, is likely to have an effect on ethicality, as they come to define morally acceptable coffee and coastal waters, and hence moral good itself.

I have argued that these representations fetishise the world that they portray. Reflecting the general tendency within ethical consumption to seek to place objects in their context, these images do not fetishise in the classical Marxian sense: they do not ignore context altogether. However, as I have shown, their presentation of some aspects of context is accompanied by ignoring many others, such as the flight to the ecotourist destination, the labour used to harvest the coffee and the decisions that lead to the location of the ecotourist lodge.

One could, of course, retort that the sort of images I have described are self-evidently condensations rather than thorough renderings, so that some ignoring of context is inevitable. However, what I have described involves a particular sort of ignoring. The eliding of contextual events and processes serves repeatedly to present the object at issue in ways that conform to what appear to be ethical consumers' moral values. For Fairtrade coffee this means eliding the roasters, shippers and merchants who stand between grower and purchaser. For ecotourists this means eliding the dubious processes that generate the bit of nature to be consumed, frame it and make it available to the consumer.

These elisions and the ethicality that they help define have two consequences that are worth mention. The first of these concerns the claim that ethical consumption is supposed to make visible the ways that things get into our lives. Doubtless in some ways it does so. At the same time, however, the fetishisation that I have described continues to direct our attention in certain directions rather than others. In doing so, it continues and even strengthens the mystification of objects of consumption, though now it is the mystification of what is stressed in appeals to ethical consumers, the context that is an important part of what is being consumed.

The second, less visible consequence of fetishisation in ethical consumption is that it strengthens the assumption that personal consumption decisions by autonomous market actors are an appropriate and effective vehicle for correcting what are seen to be the ill effects of a system of capitalist production and commerce. We need to buy the right coffee, relate more directly to the growers, not step on the moss or kill a coral or catch a fish. In a way, this assumption completes a logical circle, for it is not only coffee and coastal waters that fetishistically are stripped of the contexts that create them and in which they exist. Ethical consumers are too, in the focus on their moral choices rather than on the contexts that shape the ethicality people seek and the objects and mechanisms through which they are encouraged to seek it.

Endnotes

[1] Of course, these are not straightforward photographs of real people, any more than the reports of coffee buyers that West (this volume) discusses are straightforward descriptions of real people and events. That is because those who present them and those who perceive them endow these images with meanings that are not inherent in them and are likely to be alien to those who are presented.

[2] For Negril, see, eg, http://www.negril.com/ncrps/ncrps_files/page0004.htm; for Montego Bay, see, eg, http://www.mbmp.org/snorkeling.htm; other parks in the region include the Saba Marine Park, in the Netherlands Antilles (http://www.sabapark.org/marine.html) and the Hol Chan Marine Reserve, in Belize (http://www.holchanbelize.org) (all sites accessed 15 July 2009.)

[3] Such claims may make people feel better, but asserting the existence of a carbon-offset scheme is not the same as asserting its efficacy. In the United States, the phrase "carbon offset" has no legal meaning, and hence can be used without fear of legal challenge. Early in 2008, the Federal Trade Commission announced that it was concerned about the misleading use of the phrase and was going to conduct hearings to investigate the extent to which offset schemes actually did what they claimed to do (Story 2008; see also Fahrenthold 2008).

References

Avila-Foucat V S and Eugenio-Martin J L (2004) "Modelling potential repetition of a visit to value environmental quality change of a single site." Paper presented at the 10th biennial conference of the International Association for the Study of Common Property, Oaxaca, Mexico, August

Bakker M and Phillip S (2005) *Travel and Tourism—Jamaica—February 2005*. London: Mintel International Group Ltd

Bartilow H A (1997) *The Debt Dilemma: IMF Negotiations in Jamaica, Grenada and Guyana*. London: Macmillan

Becken S (2002) Analyzing international tourist flows to estimate energy use associated with air travel. *Journal of Sustainable Tourism* 10:114–131

British Airways (no date) Carbon footprint—carbon offsetting. https://www.britishairways.com/travel/csr-carbon-offsetting/public/en_gb (last accessed 30 April 2008)

Brown S L (1997) The free market as salvation from government: the anarcho-capitalist. In Carrier J G (ed) *Meanings of the Market: The Free Market in Western Culture* (pp 99–128). Oxford: Berg

Burke R I (2005) *Environment and Tourism: Examining the Relationship between Tourism and the Environment in Barbados and St. Lucia*. (Sustainability impact assessment of the new Economic Partnership Agreements between the ACP & the GCC States and the EU). Neuilly-sur-Seine: PriceWaterhouseCoopers

Carrier J G (1995a) *Gifts and Commodities: Exchange and Western Capitalism since 1700*. London: Routledge

Carrier J G (1995b) Maussian occidentalism: gift and commodity systems. In J G Carrier (ed) *Occidentalism: Images of the West* (pp 85–108). Oxford: Oxford University Press

Carrier J G (2001) Social aspects of abstraction. *Social Anthropology* 9:243–256

Carrier J G (2003) Biography, ecology, political economy: Seascape and conflict in Jamaica. In A Strathern and P J Stewart (eds) *Landscape, Memory and History* (pp 210–228). London: Pluto Press

Carrier J G (2006) The limits of culture: Political economy and the anthropology of consumption. In F Trentmann (ed) *The Making of the Consumer: Knowledge, Power and Identity in the Modern World* (pp 271–289). Oxford: Berg

Cohen E (1979) A phenomenology of tourist experiences. *Sociology* 13:179–201

Duffy R (2002) *Trip too Far: Ecotourism, Politics and Exploitation*. London: Earthscan

Errington F and Gewertz D (2001) On the generification of culture: from blow fish to Melanesian. *Journal of the Royal Anthropological Institute* (NS) 7:509–525

EU (2002) The European Commission's Delegation to Jamaica, Belize, The Bahamas, Turks and Caicos Is. and the Cayman Is. *The EU and Jamaica*, http://www.deljam.cec. eu.int/en/jamaica/projects/environment/negril.htm (last accessed 27 March 2006)

Fahrenthold D A (2008) Value of U.S. House's carbon offsets is murky. *The Washington Post* (28 January), http://www.washingtonpost.com/wp-dyn/content/article/2008/01/27/AR2008012702400.html?wpisrc=newsletter (last accessed 10 November 2008)

Fairtrade Foundation (2008) Images, http://www.fairtrade.org.uk/resources/photo_library/images.aspx (last accessed 29 April 2008)

Fairtrade Labelling Organizations International (2006) Generic standards, http://www.fairtrade.net/generic_standards.html (last accessed 29 April 2008)

Ferren A (2008) It's Zaragoza's turn to make a splash. *The New York Times* 15 June, http://travel.nytimes.com/2008/06/15/travel/15journeys.html?th&emc=th (last accessed 10 November 2008)

Garner A (2009) Uncivil society: local stakeholders and environmental protection in Jamaica. In J G Carrier and P West (eds) *Virtualism, Governance and Practice: Vision and Execution in Environmental Conservation* (pp 134–154). Oxford: Berghahn

Geoghegan T, Smith A H and Thacker K (2001) *Characterization of Caribbean Marine Protected Areas: An Analysis of Ecological, Organizational, and Socio-Economic Factors*. Technical report 287. Laventille, Trinidad and Tobago: Caribbean Natural Resources Institute

Gössling S (1999) Ecotourism: a means to safeguard biodiversity and ecosystem function? *Ecological Economics* 29:303–320

Graburn N H H (1989) Tourism: the sacred journey. In V Smith (ed) *Hosts and Guests: The Anthropology of Tourism* (pp 21–36). Philadelphia: University of Pennsylvania Press

Hickman M and Attwood K (2008) Fairtrade sales double to £500m as supermarkets join trend. *The Independent* 25 February, http://www.independent.co.uk/news/business/news/fairtrade-sales-double-to-163500m-as-supermarkets-join-trend-786931.html (last accessed 29 April 2008)

Hochschild A R (1983) *The Managed Heart: The Commercialization of Human Feeling*. Berkeley: University of California Press

Honey M (2003) A letter from TIES new executive director, Martha Honey, 19 January, http://www.ecotourism.org/letters.html (last accessed 28 April 2005)

Igoe J, Neves K and Brockington D (2010) A spectacular eco-tour around the historic bloc: Theorising the convergence of biodiversity conservation and capitalist expansion. *Antipode* 42(3):486–512

Khan M (2003) ECOSERV: Ecotourists' quality expectations. *Annals of Tourism Research* 30:109–124

Luetchford P (2008) The hands that pick Fairtrade coffee: Beyond the charms of the family farm. In G De Neve, P Luetchford, J Pratt and D Wood (eds) *Hidden Hands in the Market: Ethnographies of Fair Trade, Ethical Consumption and Corporate Social Responsibility. Research in Economic Anthropology* 28:143–170

Lyon S (2009) "What good will two more trees do?" The political economy of sustainable coffee certification, local livelihoods and identities. In P West and J G Carrier (eds) *Surroundings, Selves and Others: The Political Economy of Environment and Identity. Landscape Research* 34(special issue):223–240

Martinson J (2007) The ethical coffee chief turning a fair profit. *The Guardian* 9 March, http://business.guardian.co.uk/story/0,,2029607,00.html (last accessed 9 March 2007)

Marx K (1867) The fetishism of commodities and the secret thereof. *Capital*, Vol 1 (Part 1, Chap 1, Sect 4). Numerous editions

Montego Bay Marine Park (1998) *Montego Bay Marine Park Management Plan*. Computer file

Mowforth M and Munt I (1998) *Tourism and Sustainability: New Tourism in the Third World*. New York: Routledge

Neves K (2010) Cashing in on cetoursim: A critical ecological engagement with dominant E-NGO discourses on whaling, cetacean conservation and whale watching. *Antipode* 42(3):719–741

O'Callaghan P A, Woodley J and Aiken K (1988) *Montego Bay Marine Park: Project Proposal for the Development of Montego Bay Marine Park, Jamaica*. Report for the Organization of American States. TS

Payne A (1994) *Politics in Jamaica* (rev edn) Kingston: Ian Randle

Polanyi K (1957 [1944]) *The Great Transformation: The Political and Economic Origins of our Time*. Boston: Beacon Press

Rudd M and Tupper M H (2002) The impact of Nassau Grouper size and abundance on scuba diver site selection and MPA economics. *Coastal Management* 30:133–151

Scott J C (1998) *Seeing Like a State*. New Haven: Yale University Press

Smith A (1976 [1776]) *An Inquiry into the Nature and Causes of the Wealth of Nations*. Chicago: University of Chicago Press

Story L (2008) F.T.C. asks if carbon-offset money is well spent. *The New York Times* 9 January, http://www.nytimes.com/2008/01/09/business/09offsets.html?_r=1&ref=business&oref=slogin (last accessed 8 November 2008)

Strathern M (1988) *The Gender of the Gift: Problems with Women and Problems with Society in Melanesia*. Berkeley: University of California Press

Thomas-Hope E and Jardine-Comrie A (2005) Valuation of environmental resources for tourism: The case of Jamaica. IRFD World Forum on Small Island Developing States, http://irfd.org/events/wfsids/virtual/papers/sids_ethomashope.pdf (last accessed 15 May 2006)

Trentmann F (2007) Citizenship and consumption. *Journal of Consumer Culture* 7:147–158

West P (2010) Making the market: Speciality coffee, generational pitches and Papua New Guinea. *Antipode* 42(3):690–718

West P and Carrier J G (2004) Ecotourism and authenticity: Getting away from it all? *Current Anthropology* 45:483–498

Chapter 9
Making the Market: Specialty Coffee, Generational Pitches, and Papua New Guinea

Paige West

Neoliberal Coffee

In the late 1980s the popular media in the United States began to carry stories about the relationship between coffee production and environmental sustainability, and by the mid 1990s "sustainable" coffee production was being directly linked to "saving" tropical rainforests (Hull 1993; *The Economist* 1993). Throughout the late 1990s and during the early 2000s, this trend continued with an almost exponential growth in the number of stories linking coffee to the environment.[1] Today the coffee-related popular narrative encompasses not only an environmental message but also a message about how growing particular kinds of coffee can help rural peoples around the world pursue small-scale economic development in ways that allow them to access their fair share of the global circulation of cash (Alsever 2006; Pascual 2006). In addition, the purchasing of coffee and other commodities that have been cast as embedding "ecological, social, and/or place-based values" into market transactions has come to be thought of as a potential "form of resistance to neoliberalization" (Guthman 2007:456). The kinds of coffee that are linked to environmental and social sustainability, economic justice, and resistance to neoliberalization are known as "specialty coffees". These coffees, which include "single-origin" coffees like Papua New Guinean coffee that are marketed as such, organic coffees, Fair Trade coffees, and the other seemingly socially responsible coffees, are usually brought to market by small coffee companies—roasters, distributors, and coffee shops—that gained entry into the global coffee market when it was deregulated in the 1980s.

From the late 1940s to the early 1980s, gigantic corporate roasters like Maxwell House dominated the international market for coffee

(Roseberry 1996).[2] These companies produced what we might think of as "Fordist" coffee: coffee that was standardized and mass-produced through a process that was similar to all other Fordist industrial innovation (Harvey 1989). During this time period the International Coffee Agreement (ICA), a set of international treaties and agreements in place between 1962 and 1989 that set production and consumption quotas, regulated the market and governed coffee industry standards. But the coffee industry, like many other industries, was deregulated in the 1980s during the first global stage of neoliberalization. During this neoliberalization, the ICA was cast as a set of regulations that were unfriendly to business, and as it was phased out the coffee industry was affected by privatization, corporate attempts to minimize labor costs, state attempts to reduce public spending on social welfare, and the receding of the state from the support and regulation of public life in general. The neoliberalization of coffee production and consumption through the retreat of the state in terms of industry regulation also opened spaces for other industry actors in the market and drastic changes in global production (Bacon 2005:499).[3] Transnational trade and roasting companies quickly filled these spaces, and countries like Brazil and Vietnam radically increased their production in the wake of this market liberalization.

Another aspect of this phase of neoliberalization in general was that governance that had once been the purview of the state became the purview of non-governmental organizations and so called "civil society" groups like churches, development agencies, and other international bodies and organizations. These actors moved into various structural positions concerned with the environment, economic development, and human rights—areas that that we think of today as directly connected to coffee production—that had been previously filled by state agencies. Concurrently, within the coffee industry, the loss of the ICA allowed small companies into the global coffee market. Small-scale coffee roasters, traders, and the sellers of coffee-related objects flourished in this new de-regulated market where they could have direct access to producers and consumers. But, as the structure of the global market for coffee has changed, the price of coffee has fallen drastically and the prices paid to coffee producers have declined rapidly (Bacon 2004; Ponte 2002).

Some scholars and activists argue that the sharp decline in prices paid to farmers between 1999 and 2004 (with a global 30-year low in 2001) galvanized NGOs, development organizations, well-meaning companies, and well-meaning consumers to expand the market for socially responsible coffees that bring more money to producers and that contribute to environmental sustainability (Bacon 2004; Bacon et al 2008). They also argue that the trend towards socially responsible coffees is tied to consumer knowledge about the plight of poor farmers and consumer knowledge of "quality, taste, health, and environment" (Bacon 2004:497). According to these scholars, this consumer pressure resulted in the growth of the specialty market and the development of the Specialty Coffee Association of America (SCAA), an industry group—made up of roasters, traders, and sellers—that promotes specialty coffees in North America. Activists and scholars who argue that these coffees can redress inequality do so because they think that production and distribution monitoring through certification processes and labeling can work to protect conditions of production, land–labor relations, and the environment and thus they argue that these coffees counter some of the ravages of neoliberalization (Guthman 2007). But certification practices and labeling have been debated and critiqued in the literature (Guthman 2007; Mutersbaugh 2005), and some scholars have shown them to be typical forms of neoliberal regulation (Guthman 2007).

In contrast to those who see specialty coffee as a corrective to neoliberal changes, others see these coffees, the conditions of their production, and their market as brought about by neoliberalization above and beyond critiques of certification. Roseberry argues that as the market structure changed with the demise of the ICO, small-scale coffee-producing companies, distributors, and roasters began to "envision a segmented market rather than a mass market", and that as they imagined this market, the public relations companies working for them began to attempt to create new consumers through advertising (Roseberry 1996:765). People who had not been coffee drinkers in the past were targeted through the creation of particular stories and images that were designed to appeal to them along generational, political and class lines (Roseberry 1996:765). Certain types of specialty coffee were marketed to appeal to people's ideas about the refined nature of their own tastes and the uniqueness of their position as a certain type of consumer in the marketplace, while others were marketed to appeal to people with particular political beliefs. Marketers wanted coffee consumption to be seen as a way to distinguish oneself in terms of class and to express one's political ideas and they worked to create the consumer desire right along with the growth of the new specialty industry (Roseberry 1996). As with other commodities, marketers worked to produce desire first and then present products that would fill people's sense of need like a "key in a lock" (Haug 1987:93). As these specialty markets developed they came to resemble what has been called a "Post-Fordist" regime (Harvey 1989); they were flexible and supposedly consumer oriented and consumer driven, two of the hallmarks of the neoliberalization of markets.

Whether one thinks that the market for specialty coffees emerged because of consumers' desires to redress global inequities generated by neoliberalization through consumption choices or because of the desires of businesses to find new consumers for their products, we can see that the market for specialty coffee attempts to merge seemingly disparate strands of consumer life: economic choice, political action, and identity production. Economic choice (what people buy, when they buy it, where they buy it, and why they buy it) is wedded to political action (both the ideas and practices people associate with their politics and the sense of communities-of-sentiment that goes along with politics and political affinity) and they then contribute to identity production (how people come to be in the world and see themselves in the world and in relation to others especially in terms of social and environmental equity based identities). At first glance then the contemporary specialty coffee market seems to be countering the process that Karl Polanyi called "disembedding" (Guthman 2007).

In writing about the economy as a "process" in general and the emergence of the modern market as a system of organization specifically, Polanyi argues that the economy:

> is embedded and enmeshed in institutions, economic and noneconomic. The inclusion of the noneconomic is vital. For religion or government may be as important for the structure and functioning of the economy as monetary institutions or the availability of tools and machines themselves that lighten the toil of labour (Polanyi 1968 [1958]:127).

As Polanyi traces how and when markets gain radical importance over other aspects of social life he shows how the process of "disembedding" has taken place. Disembedding, as Polanyi saw it, was when economic activities, like buying coffee, became increasingly removed from the social relationships in which they had historically occurred and when the objects circulating in the economy came to be seen as fetishes in that they were seen as emerging in and of themselves and not from labor. In his discussion of disembedding, Polanyi shows how economic transactions increasingly became abstracted from social relations. Disembedding allowed for the disarticulation of economic choice, political action, social relationships, and identity production through abstraction, and with this it made the market less a web of social bonds and relations and more a clean and sharp set of economic transactions.

Specialty coffees, again whether we imagine them as resulting from consumer desire or market manipulation, seem at first glance to re-embed coffee into a web of social relationships. Well-meaning consumers are connected to well-meaning producers by well-meaning coffee industry businesses. A veritable love fest of learning about each other, supporting each other, and contributing to environmental sustainability while still taking part in the global market seems to have emerged. But these specialty coffees are first and foremost neoliberal coffees; coffees that have a place in the market because of structural changes in the global economy that, in addition to many other things, target individuals as the seat of economic rights and responsibilities. In what follows I show how individual coffee consumers and producers are portrayed and produced through and by people associated with the coffee industry. I then call into question the validity of these portrayals by presenting ethnographic data from the Papua New Guinea, where people grow coffee, and from New York City, where people drink coffee. I conclude by returning to the idea of re-embedding and argue that the seemingly re-embedded coffee market is an eco-neoliberal fiction; a fiction that is meant to divert our attention away from the structural causes of environmental degradation and social injustice.

Consumer Production

The tall blond man from Nebraska wears the clip-on microphone like a professional. He towers above us, the participants in his seminar on marketing at the Specialty Coffee Association of America's Annual Meeting, and smiles a radiant row of perfect white teeth. He breaks the ice by revving up the fairly caffeinated crowd when he says, "Okay. OKAY. We are here to sell coffee! YEAH." People in the audience cheer enthusiastically.

We are all (coffee shop owners and an anthropologist who studies coffee consumption) here in this conference room in our attempt to understand why people buy specialty coffees. Our first task, before we begin any discussion of coffee, consumption, or anything really, is to break into groups and come up with a list of the "essential qualities" (I know, too perfect for an anthropologist, right?) of our own "generation". I am put into the Generation X group. We were all born between 1964 and 1982 and although I momentarily hope that we will bond over our great love for the music of The Replacements, powerful memories of anti-apartheid protests and divestment campaigns, and our ability to quote long bits of the movie *Point Break*, we don't, as a group, seem to have much in common. So we get to our task and try to make a list of essentialisms. We have trouble because we don't seem to agree on any of them.

After a break, Mr Nebraska smiles us back to our seats and we get started. People yell out the answers to his questions.

> Mr Nebraska (MN): "Okay, so you Silent Generation folks [those born between 1927 and 1944], give us your qualities."
> "We are loyal and dependable," says one man in the front.
> "We built and defended this country," says another who seems to be wearing a hat with a battleship's name on it.

> MN: "Okay, now for the Baby Boomers [those born between 1945 and 1963], what do you have to contribute as a generation?"
> "We are tenacious and idealistic," says a woman wearing a perky little red suit.
> "We are free thinkers!" shout several people at the same time.
> MN: "What about the Xers?"
> Several people from my seemingly stoic group now perk up and yell, "We are individuals," "We question authority," and "We are fast technology!"
> MN: "Now, what about you Millenials? Hello, Millenials? Where are my Millenials?"
> Two young guys shyly raise their hands. They appear to be just-out-of-college and sort of out of place in this older business-suited crowd. One of them says, "We are much faster technology."

Everyone laughs. Then Mr Nebraska begins his lecture.

For the next hour he talks about different American generations and how they hold the key to marketing. He begins with his analysis of the essential characteristics of each generation. "The Silent Generation" is "defined by World War II and the Korean War". They are hardworking-loyal-sacrificing-dedicated-conformist-never questioning authority-respectful-patient-delayed gratification-duty before pleasure kinds of folks. Mr Nebraska smiles broadly when he talks about these people, calling them "folks" at several points and mentioning his grandparents. Then he tells us that we won't talk about them anymore because as a generation they don't have any purchasing power in the retail world so they are a waste of time for the seminar.

He then moves on to the Baby Boomers. They are "All about civil rights, Vietnam, and Woodstock" and they can be summed up as essentially full of "optimism", "team-oriented", dedicated to "personal-growth" and "personal-gratification". They work long hours and have a "hard core" work ethic but a "youthful mindset" which they keep up with "health and fitness".

He says, "GUYS, come ON. There are some values going on here, right? VALUES." He says this meaningfully, pacing the stage and smiling at his own insight.

It turns out that my Generation, Generation X, is defined by "Three-Mile Island", "the fall of the Berlin wall", and "Rodney King", and that we are "liquid". We have "liquid value" and "a liquid mindset". We can "adjust to anything" because we are independent-individualist-selfish latchkey kids who are "all about experience" and who have "no loyalty to anyone or anything". We are hard to work with because we have a "totally flat view of organization", which means we have "no respect for authority".

Finally, he moves on to the Millenials. Premising his discussion of them by saying, with no hint of insight into his role in the creation of this marketing fetish, "Isn't it just weird? It is just weird that generations are getting shorter. Isn't it?"

For Mr Nebraska, Millenials are defined by the Oklahoma City bombing, the Clinton–Lewinsky affair, 9/11, and the Columbine shootings. They want "achievement" but are "not driven". They value "globalism" but are "community focused" and think that by "looking inward" they can "change the world". They are also apparently "t-totalers" who "don't want drugs or alcohol".

When he is finished with his description of the Millenials, he looks at us thoughtfully, pauses, and says meaningfully, "This, THIS, is at the very core of people, it is who they are".

Next we move onto how to market to the different generations. Mr Nebraska says, "The logo, product, service and atmosphere, or CULTURE of a business", is "key" to making your "generational pitch". And he cautions the audience, "You want to listen to this, the cultures I'm talking about, they are in people's DNA".

Baby Boomer DNA is apparently encoded with the deep and abiding desire for iconic logos that symbolize gratification, indulgence, and the "unyielding" defiance of age and ageing. Their DNA forces them to desire lots of choices among products, quick and thoughtful professional services, and "up-scale" consumer-comes-first type "retail culture".

My "Generation X Cultural-DNA" makes me skeptical of logos, desirous of multiple similar products with a unique story behind each of them, wanting service that is "authentic" and during which I can "make a connection" and "share a story". Culturally, I desire casual, flexible, liquid space where I can read the paper, check my e-mail, and chat with friends. I "can't abide" images of control.

Millineals are "encoded" with the desire for brands and logos. They "value the symbols of products" more than anything else "about the retail world". They want "global products" that are "political" and "environmentally friendly", things that allow them to "express" their "self knowledge" and "politics". Service-wise they want to "be coddled" and "made to feel important". They want to "see people who know, really know, how to work the equipment". And culturally, they desire and can find "a meaning filled experience" during "retail time".

After the description of the generations, what is in their "DNA", and the sort of "retail culture" that appeals to them, Mr Nebraska begins to talk about specialty coffee and its emerging market. He focuses in particular on the "stories" behind the coffee and the ways in which they can be made to appeal to the different generations. The stories exist on two scales: the first is that of the coffee shop and the second is that of the coffee producer.

Mr Nebraska's Baby Boomers, constructed against a social mirror of 1960s activism (the civil rights movement and protests against the war in Vietnam) and the constrained/restrained rebellion of going to a musical show (the concert in Bethel, NY on 15–17 August 1969 which became known as Woodstock), and who are produced as deeply desirous of validation of their continued youth even in the face of their 60th birthdays and deeply connected to the idea that they have spent their working lives working harder than others, can easily be sold specialty coffees and specialty coffee venues that appeal to their ideas of work and activism. He discusses their work ethic, how they "worked long hours themselves" when they were young and how they "understand" labor. Because of this, stories about coffee shops will appeal to the Boomers. He says that they "love Starbucks" because it started out as one shop and is now, "the biggest and the best". They like a story of success that somewhere along the way meanders through a 1960s sense of helping "the down trodden". If small coffee shops and roasters can tell a story that shows now they have "fought hard" for their market share and that they have "made hard choices" along the way, the Boomers will flock to their shops. If people selling coffee can write stories about producers that appeal to that 1960s sense of rights and war protests, they can win consumers. He suggests that Boomers are more likely to buy coffee that is grown by people who live in a war torn country ("Guatemala really appeals to their sense of post war hardships"). Since they are health conscious and since they "really wrote the first book on organics" they are particularly interested in organic certified coffee in that its story is one of a "more healthy" drink than regular coffee. He also argues that Boomers want the standardization of a chain retail outlet but the "feel" of an "up-scale" personalized experience. This is why chains that are meant to feel like local coffee shops appeal especially to them (eg Peets, Caribou, and Starbucks).

Since Generation X is defined against depressing Regan-era events, and since we are "liquid", we are hard to sell to. We are "cynical" when it comes to retail and we are the reason for "diverse venues" for standard consumer products. We don't want the same experience over and over again (the aforementioned chain coffee shops). We want a coffee shop that has an authentic story that we can connect with. We like alternative venues that might have been begun as anti-establishment shops. We like the "Seattle connection" to be articulated in the shop stories. We want to know the story of the shop and the stories of the people who work there. We also want stories behind each of the products that are all similar but that are marketed to us as "unique". We like the idea of authenticity when it comes to the people who grow the coffee. And we like the idea of experiencing some aspect of their lives through drinking the coffee. We want to connect to the authenticity of others in some way, and that way can easily be through buying product. We also like the idea of supporting people whose story shows that they are "bucking" the establishment in some way.

The people that Mr Nebraska called Millenials are, for him, the "driving force" behind the "globalism" that is emerging in the specialty coffee market. While Boomers and Xers appreciate certain aspects of the stories behind origin-marketed coffees and Fair Trade and organics, it is the Millenials who "thrive" on these stories. "They want to change the world and they know that they can do it through coffee." They also, "know that the politics of their parents are not their politics" and that their politics "can change the world one village at a time". They are much less concerned with the shop and its story and much more concerned with the ways in which coffee can connect them with "people all over the world" and the ways that it can allow them to "participate in (the grower's) struggles". They define self through their consumption in that they see themselves as politically active through their connection with "these stories about growers and the environment".

The marketing seminar is wrapped up by Mr Nebraska with a long discussion about how each of the generations wants a particular story about the products that they buy. He talks about the process of creating a story for a business and the ways in which coffee works to "sell itself in today's market universe" because of the stories of growers that can be associated with it. He is passionate about the reality that he has just laid out for us—he repeatedly talks about how the "DNA" of the consumer is set along generational lines and how these generations want to "know and experience" stories about their coffee.

Mr Nebraska works to both create "virtual consumers" for his audience and to imbue them with particular sets of values. James Carrier and Daniel Miller have built on Polyani's previously discussed notion of abstraction in order to propose a new set of theories about how we might think about contemporary social relationships in general and people's roles as consumers more specifically (Carrier and Miller 1998). They call this set of theories "virtualism". Carrier defines virtualism as the attempt to make the world around us look like and conform to an abstract model of it (Carrier 1998:2). He uses the concept to criticize thinking and policies in economics where there is a common tendency to abstract human decision making from its complex social context, and build models of the world and its workings that cannot take the full range or complexity of people's daily social activities, practices, and lives into account. This much is normal for modelers. But these abstractions become virtualism when the real world is expected to transform itself in accordance with the models:

Perceiving a virtual reality becomes virtualism when people take this virtual reality to be not just a parsimonious description of what is really happening, but prescriptive of what the world ought to be; when, that is, they seek to make the world conform to their virtual vision. Virtualism, thus, operates at both the conceptual and practical levels, for it is a practical effort to make the world conform to the structures of the conceptual. (Carrier 1998: 2)

Miller argues that we can see this virtualism at work in the production of the contemporary consumer in economic discourses (Miller 1998:200). He sees the creation of a "virtual consumer" or the image of a person who wants and desires and buys certain things according to models of consumer behavior based on aggregate figures used in economic modeling. The world of retail then comes to resemble the world in this economic modeled world—storeowners and shopkeepers alter the physical space of their businesses and the social actions of their business in attempts to catch part of the market of this virtual consumer. Eventually, if we follow Miller's argument, because the physical world has come to look like the virtual world, the consumer's actual behavior comes to mirror the virtual behavior and the virtual consumers become real (Miller 1998).

Mr Nebraska creates virtual consumers and then lays out the process by which the physical world should and will come to reflect the wants, desires, and values of this consumer. He takes generational stereotypes and casts them as immutable biophysical characteristics that work to guide tastes, desires, politics and economic choices. He also argues that the evaluative processes that people make with regard to consumer choice have to do with these immutable generational characteristics. But "Generations" as objects and entities are a production of contemporary public culture; they are social artifacts. As artifacts they give demographers, journalists, and marketers a way of describing social and economic trends by age group without attending to race or class (Ortner 1998). In her analysis of the social creation of "Generation X", Sherry B. Ortner shows that the idea of "generations" began as a way of describing people in terms of economic living standards like buying homes, competing for jobs and promotions, and competing for places in universities and colleges (Ortner 1998). In the past, generations were used to predict and talk about "identity through work: jobs, money, and careers" (Ortner 1998:421). Today Mr Nebraska relies on "generational marketing" to construct his virtual consumers and to set the stage for story telling about producers and production that draw on the assumed values of these virtual consumers. It is therefore a fictional, virtual consumer and her values that guide how coffee production is to be told as a marketing narrative.

How is coffee production in Papua New Guinea, for example, where I have conducted anthropological research for the last 12 years, turned into a story that consumers can "thrive" on? How does it become a story that makes people "want to change the world" or feel like they learn something special about the world and themselves through its consumption? And how are stories told that will work to re-connect economic choice, about what coffee one buys, to politics in a way fosters both social relationships between producers and consumers and allows consumers to feel good about themselves?

Producer Production

The following Blog entry, entitled "Papua New Guinea—Back to the Future", is one example of how coffees from Papua New Guinea are given a story by marketers and roasters and how that story is conveyed to consumers. It was written by an employee of Dean's Beans and placed on the company's website. The company is a small, extremely successful specialty coffee roasting company in Massachusetts that specializes in organic and Fair Trade certified coffee. They sell only certified specialty organic and Fair Trade coffees and they associate each of their coffees with certain origins. They focus not only on commerce but also on "people-centered development", which they define as, "An approach to international development that focuses on the real needs of local communities for the necessities of life (clean water, health care, income generation) that are often disrupted by conventional development assistance".[4] The website juxtaposes this form of development with "conventional development", which includes "military aid, large dams, free trade zones and export economies that bring lots of money to the contractors and aid organizations, but often result in massive deforestation, resettlement of communities, introduction of pollutants and diseases". And the website states that the company is "committed to small, meaningful projects that the community actually wants, and that are sustainable over time without our continued involvement". They specifically link organic and Fair Trade certified coffee with their critique of "conventional development" and state repeatedly that these specialty coffees help growers get their fair share of profit and that they contribute to the ecological health of the planet. Dean's Beans is therefore specifically positioning itself as a company that attempts to redress the social and environmental inequity generated by neoliberalization.

This commitment to countering the evils of neoliberalization is noble but the way that the company presents images of and ideas about Papua New Guinea reveal a global vision that has been produced by an antiquated sense of unilinear evolution and people without history and that feeds directly into eco-neoliberalism. The Dean's Beans employee who visited the Eastern and Western Highlands of Papua New Guinea in 2005 writes:

> Chiseled warriors in Bird of Paradise headdresses and spears, impassable mountain roads, stunning vistas, abundant gardens of coffee and vegetables. Papua-New Guinea is the final frontier of dreams, of images from the pre-colonial past. Yet here I am, the first American anyone can remember coming into these Highlands, many say the first white guy. I have dreamed of this land since I was a child, looking at National Geographic (yeah, those photos!), reading about its wildness in my Goldenbook Encyclopedia.
>
> There are no roads connecting the capital, Port Moresby, with the rest of this island, which is the size of New England. We have to fly to the interior, and I am glued to the window of the small plane, knowing that below me are anacondas and pythons, tree kangaroos and Birds of Paradise, wild rivers and still uncontacted tribes.
>
> There is also coffee, introduced to the Highlands only in the 1950's from rootstock taken from the famed Jamaican Blue Mountains. Coffee is the only cash crop in the Highlands. The people grow all of their own food, using the coffee money to buy cooking oil, sugar, used clothes and other necessaries. They depulp the cherries by hand using round rocks. This is the only place in the world where coffee is depulped this way. It is a family affair, and I visit with several families singing and depulping by the river. After sun drying the beans, the villagers have to carry the sixty pound sacks on their backs for up to twenty miles, over mountains, through rivers via rocky paths.
>
> Historically, they would sell their beans to a number of middlemen who wait by the only road, giving the farmers pennies for their labor. But we are here to change that. We are here to work with several farmer associations to create legally recognized cooperatives, and to create more direct trade relationships that should increase the farmer's income fourfold, as well as increase sales.
>
> As I am the first coffee buyer to come into this area, the farmers organize a Coffee Cultural Show. I thought that meant a few dancing and singing groups, a feast and a gift exchange. Wrong! As we rolled into a distant village after three hours over rivers, boulders, mudpits and bridges that shook beneath the land rover, we were greeted by ten thousand people! It was the largest gathering ever seen in these parts. Traditional warrior societies, women's clans, singing groups, hunters and every possible combination of feathers, noses pierced with tusks, and painted bodies festooned with coffee branches and berries greeted us riotously. I was hoisted into the air and carried almost a mile by joyful men, while the women called a welcoming chant. There were

speeches by every village's elders, by coffee farmers and of course by me.

For two days the festivities roared on, segued together by an all-night discussion around a fire about coffee techniques, trade justice, the role of women and every imaginable subject for people who have never met an American or a Fair Trader. Wild pigs were cooked on hot stones in pits, covered with banana leaves. Huge plates of yams (they laughed when I told them about research which links yam consumption to twin births—and they have a lot of twins there!). Of course, we brewed up lots of Dean's Beans Papuan coffee (Ring of Fire). It was the first time these farmers had ever had their own coffee, and they loved the taste almost as much as they loved seeing their own tribal names on the coffee bags, tee-shirts and hats I had made for the visit. As we passed through the Highlands, we had to stop at each tribal boundary for permission to enter the territory.

Considering that there are over eight hundred tribes in PNG, we were crossing boundaries every ten miles or so. At each boundary we were greeted by warriors in full dress, with welcoming chants and speeches, and invited to feast and speak. Needless to say, it took a long time to get a short distance, but we were well fed and made hundreds of new friends every day.

Back in the capital, we went on the radio (four million listeners nightly, as there is no electricity in the villages, only battery powered radios) and talked about making strong cooperatives and quality coffee to insure vibrant communities. Our meeting with the Prime Minister didn't happen, so we spent a day on an island of fisherman and their families, cooking the bounty of the sea and playing with the kids. My kinda day. Papua-New Guinea. A lifelong dream come true. It was a profound honor to be able to go as an emissary of peace and positive social change. If you ever get to go, DO IT! You can be assured of a warm welcome and a great cup of coffee. Just tell them you're a friend of mine.[5]

This blog entry is a good example of what Mr Nebraska suggested that roasters, importers, and marketers do with regard to creating a story for specialty coffees. This is an unsettling example to be sure, one that locates Papua New Guinea in a morass of primitivist imagery, colonial nostalgia, self aggrandizing travel narrative bravado of white exploration, outright falsehoods (eg anacondas and un-contacted tribes do not exist there and the claim that the writer is the "first coffee buyer" to enter places in the Western and Eastern Highlands is absurd), and inaccurate information (eg the description of middlemen and the claim that cooperatives will "increase the farmer's income fourfold"), but it is representative of many of the narratives one finds in blogs about Papua New Guinea in general and with regard to coffee specifically.

The coffee growers that this narrative constructs seem to have the same values as the virtual consumers for which it was constructed. These growers seem to want to maintain tradition, maintain benign ecological relations, based on market economics, with their forests, they want to see and know and understand modernity but not to lose their souls to it. They are produced as both the ecologically noble savage and the fallen from grace but deeply wanting to maintain ecological stability primitive. These images penetrate deeply into the Euro-American psyche and are a reflection of Euro-American fantasies about indigenous peoples.

When businesses tell stories like this one about virtual producers, they want consumers to see natives who are poor Third World agricultural laborers who value and contribute to the ecological sustainability of the earth, while at the same time make just enough money to maintain their coffee-producing ways of life, without wanting to gain access to all the things that consumers have including the feeling of right to over consume the world's resources. They also want to provide consumers with an aura of social responsibility, political action, exotic locality, environmental sustainability, and social status through a capitalist marketing version of Geertzian "being there" narrative. The idea is to market meaningfulness without actually going all the way down the road of consumer education. Coffee companies like Dean's Beans add value to their products by going half way, by creating virtual producers and hoping that these narratives appeal to the virtual consumers that have been made for them by the likes of Mr Nebraska.

This narrative also attempts to repackage poverty as uniqueness and to make primitivism as a form of scarcity. Scarce things have value and by producing a fantasy of Papua New Guinea's coffee industry as primitive and of primitivism as scarce, this narrative adds value to the coffee at the expense of people from Papua New Guinea by turning them into virtual producers. The virtual producers created by Dean's Beans are poor farmers yearning for a benevolent and right-thinking American businessman to come in and create economic equality through the softer side of capitalism.

But what does coffee production look like on the ground in Papua New Guinea and how do people there think about and understand the industry? How does that story above relate to the lived experiences of people in Papua New Guinea who grow coffee? Does the Dean's Beans story achieve the goal of creating a narrative that people can connect to and thrive on?

Producting Lives

Although coffee had been grown in what is now Papua New Guinea since the late 1800s, the post war period in the 1950s and early 1960s gave rise to the indigenous coffee production industry that exists today. In the 1950s, the Agricultural Experimental Station at Aiyura near Kainantu, in what is now the Eastern Highlands, began to experiment with coffee. In 1951 Bimai Noimbano, a Papua New Guinea national from Watabung, began working at the station, and in 1953 he developed the first nursery for coffee seedlings near Goroka—the city that is now the capital of the Eastern Highlands Province (Finney 1987:5). Upon seeing the success at Aiyura, George Greathead and Ian Downs, two important colonial officials in the Highlands who lived in Goroka, advocated coffee as a means for encouraging both settler colonialism and the development of a cash-crop industry that Highlands residents could take part in.[6] Greathead was particularly vocal in his insistence that coffee provided a crop that would foster a strong agricultural backbone for the territory's emerging economy. The national workers at Aiyura had social networks across the Highlands and they used these networks and the established relations and routes between colonial patrol officers and local peoples to move seedlings in to remote areas very early on.

The industry developed rapidly in the 1960s and 1970s with a series of minor setbacks during the pre and post independence years (roughly 1972 to 1977).[7] This development of the industry brought many of the major infrastructural developments that came about directly prior to independence. Colonial patrol reports from the Lufa District of the Eastern Highlands Province provide census related data about coffee production in the early 1960s and by the late 1960s they provide detailed demographic and production information (who has how many trees and where). By the early 1970s the patrol officers are advocating road and walking track construction projects so that rural growers can get their crops to market. During the 1980s and 1990s the industry became the backbone of the Highlands regional economy.

Today in Papua New Guinea one out of every three people in the country is connected to the coffee industry. Either as a coffee grower, a worker in the processing and transporting industries, a businessperson in the processing, transporting and distributing industries, or as one of the thousands of people who support the industry as security guards, cleaning women, clerical and accounting staff, and truck drivers. In the Highlands region most other regional industry depends on coffee to keep the cash flowing: the second-hand clothing industry sees an increase in their business during coffee season, the craft and fresh vegetable markets attract more sellers and buyers when people are flush with coffee cash, the trade stores see their profits increase during the season, and the restaurants, shops, car dealerships and other small businesses depend upon the coffee season to make their yearly profit margins. Many people call it "the people's industry" because it directly links people throughout the country with each other socially and economically, and it links Papua New Guineans with other people across the world.

Between 86% and 89% of the coffee grown in Papua New Guinea is "smallholder" coffee. This means that it is grown by landowners who live in relatively rural settings with small family-owned and operated coffee gardens with little to no support from private or government agricultural extension. Among most of these smallholders, families run these small coffee businesses and they produce the only cash income that people have. People may own as many as 2000 trees or as few as 200. During the coffee season, men, women and children work on the coffee plots and other social life (weddings, compensation cases, school fees, head payments) revolves around the coffee season.

We can take the history of coffee production by Gimi speakers in the Lufa district as an example of how coffee spread across the Highlands.[8] By the mid 1960s coffee has been introduced as a cash crop to people in all of the Lufa district Gimi-speaking villages, and by the mid 1970s it has taken off as a system of production. Gimi men had taken part in the Highlands Labor Scheme and left their villages during the late 1960s and early 1970s to work on coffee plantations around the Highlands as well as plantations in coastal areas (Lindenbaum 2002:67, fn 5). They came home from their travels with agricultural knowledge about coffee production and the social knowledge of what cash income from coffee could provide. Coffee itself had come to Gimi through rural agricultural extension and through traditional networks of exchange.

In Gimi territory the coffee cherry begins to ripen around the beginning of June and men begin to take interest in their coffee groves again after leaving the work associated with clearing brush and weeding up to their wives throughout the rest of the year. When a man thinks that he has enough cherry on the branch to begin the harvest he and his entire family go to the lowest altitude grove of trees that he possesses and begin to harvest the coffee. People hand-harvest the coffee and then process it using the wet method of coffee preparation. Once the coffee has been harvested, the cherries are washed and pulped. Most people gain access to one of the coffee-pulping machines in the village so that they do not have to undertake this hand pulping, and indeed hand pulping is extremely rare. With a machine, the woman pours cherries into the top of the machine and then as the man turns the crank she pours water in with the cherries. This turning forces the cherry off the beans and the slimy beans drop out of the machine. Once the pulp has been removed, the berries are placed into clean bags and allowed to ferment. They must be fermented so that the sticky enzyme-rich substance on the beans can be washed off. The fermentation process should take between 12 and 24 hours. Once fermented, the coffee is washed over and over again until the water runs perfectly clear. It is then placed in the sun to dry and when this is complete people are left with dry silver-skin covered beans called "parchment". This parchment is then placed in bags and left to wait until a coffee buyer comes to the village.

Gimi sell their coffee to intermediary buyers who visit their villages and then take the coffee to the capital of the Eastern Highlands Province. These buyers are often the only link that rural people have to the national coffee market and in rural villages these buyers are seen as providing a necessary service. The network of rural coffee buyers in the Eastern Highlands alone employs over 700 people. Buyers take the coffee and sell it to factories where the coffee is processed and packaged for export. The factories employ thousands of people; there are factory machinists, accountants, drivers, secretaries, coffee sorters, executives, drivers and security guards, among many other professional types of labor. Across Papua New Guinea, coffee production has helped to create a middle class with these industry jobs.

From these factories, the coffee goes to the international port where it is then shipped all over the world to major ports in Australia, Germany, the UK, and the USA. For the past 20 years the total production of bags in Papua New Guinea has hovered around 1 million bags per year. In terms of the world market, Papua New Guinea produces only about 1% of the total amount of coffee worldwide. But the industry brings about US$1.9 billion into the Papua New Guinea economy each year, making it about 4% of the total export revenue for the country. Given that the three main export commodities for Papua New Guinea are gold, crude oil and copper and that the other main agricultural commodities are timber and palm oil (which have slightly higher revenues than coffee), and that these aforementioned commodities are owned and operated by the state and joint ventures between large multinational companies and the state (with the minor revenue streams flowing down to landowners not being widely distributed), coffee is the only export commodity that is owned and operated by the people for the people. In addition, across the entire Highlands region of the country coffee is the only product that people produce for the global market and that they gain income from. In many places it is the only economic development that people have and people see it as just that, as economic development and as a marker of their place in the modern global economy.

In Papua New Guinea the remoteness of many of the smallholder growers today is a consequence of rollback neoliberalization. From the 1950s to the 1980s colonial officials and then post independence government officials worked to build roads and walking tracks and maintain them so that coffee growers could get their coffee to market. Since structural adjustments in the 1980s these networks for movement have declined as have most other services provided by the government. Across the Highlands people have decreasing access to schools, health care, agricultural extension, and other services that foster a healthy population in general and a healthy coffee industry in particular. Remoteness and inaccessibility, two of the themes of Dean's Beans blog entry, are actually the affects of neoliberalization not some condition that can be corrected by it.

How do consumers hear the stories about people who grow coffee; what do they take away from them? How do they understand places in the world like Papua New Guinea as they are related to coffee? And how do they define themselves through consumption? If specialty coffee is indeed re-embedding political action and social relationships into economic choice, what does this re embedding look like in terms of consumers?

Consuming Lives

The marketing of specialty coffee has taken place in numerous ways. One of the most powerful market forces today, according to the marketing seminar discussed above, is the Internet. Mr Nebraska told us that people "research products on the Internet" and that this is the perfect place to "tell your story". Dean's Beans took that advice and used the form of a blog to craft a particular fantasy of Papua New Guinea through story telling. This story, if we take seriously what Mr Nebraska told us, should appeal to and work to craft the consumer consciousness of people born 1983 and 2000, the "Millenials".[9] Over the course of one semester my student research assistants and I conducted 100 interviews with people born between 1983 and 1989.[10] All of the people interviewed were undergraduate students attending one of the colleges associated with a large private university in New York City. The interviews were concerned with people's knowledge about coffee production, distribution, and consumption and in particular with their ideas about the stories associated with certain kinds of specialty coffees.[11] Below I summarize the data collected and show, quite clearly, that the students we interviewed do not fit the consumer image produced by Mr Nebraska.

First, what did these Millenials know about Papua New Guinea? Ten people knew exactly where it is and gave us specifics like, "sixty kilometers north of Australia" and "the eastern half of the island of New Guinea". Thirty-five interviewees knew that it is in the southern Pacific. They gave us answers like, "near Australia in the Pacific ocean", "sort of near the Philippians and Indonesia but closest to Oceania" and "near Malaysia and Indonesia". Ten people told us that Papua New Guinea is in Africa. Eleven people told us that it is somewhere else incorrect like "South America", "near Brazil", "close to India but not near China" and "in the Middle East but not near Iraq". Six interviewees freely admitted that they had no idea where it is while six refused to answer the question instead saying things like "the fuck I know", "fuck off" and "I refuse to answer that on the grounds that I incriminate my shitty knowledge of geography". So about 45% of our interviewees had an idea of where the country is but only about 10% had a good grasp on the geography of the region.

Next, what did Millenials know about Papua New Guinea coffees? Forty-eight people interviewed had no idea that Papua New Guinea produces coffee. Thirty people knew that Papua New Guinea produces coffee and out of those 30, 18 could tell us something about what they think coffee production is like there. They said things like "people live in villages and grow coffee", "I think indigenous people grow it there" and "yeah, its huge there, small farms all over the country". The remaining 22 people had a vague idea about Papua New Guinea producing coffee saying things like "yeah, I kind of know it does" and "I've seen it in Starbucks". Of these 22, nine people mentioned Indonesian coffee saying things like "well, I know it is near Indonesia and they grow coffee so I assume they grow it there also". So 48% of the people interviewed had no idea that Papua New Guineans produce coffee at all.

In terms of consumption, 19 of the people interviewed always drink Fair Trade coffee, 51 people never drink it, and 30 people said that they drink it sometimes. Eighteen people always drink organic coffee, 52 people never drink it, 20 people sometimes drink it, and 10 people said that they did not know how often they drink it. Of the 20 people who always drink Fair Trade coffee all of them mentioned "producers" and "labor" and about half of them mentioned either "ethical" or "fair" uses of their money as consumers. Of the 18 people who always drink organic coffee not one of them mentioned the health of workers or the earth's ecosystems but all of them mentioned their own health (but see below, some people do understand that organic is about ecosystem health). All of the people who say that they sometimes drink Fair Trade and organic said that they do not seek them out but if given a choice they prefer to drink them. Of the 50 people who never drink Fair Trade, 20 said it was because they did not know what it is, 18 said it was because it is more "expensive" or "costs more", eight gave no specific reason, and four gave other responses ("is brewed in self-righteousness", "reminds me too much of poor people", "tastes like peasants" and that it is "left wing"). Of the 52 people who never drink organic coffee 35 of them said it is because it is more "expensive" or "costs more", 15 said that it simply never occurred to them to look for it in stores of shops, and two people said "too fucking hippy" and "too liberal". So less than 20% of the people interviewed actively seek out and consume Fair Trade or organic coffees and over 50% of the interviewees never drink it. The rest of them drink it if it is convenient.

We also interviewed people about their level of knowledge concerning Fair Trade certified and organic certified coffee and then broke people's responses into categories of "understands", "has a good general idea", "has a vague idea", "has no idea" and "has the wrong idea". In order for people to be coded as "understands" they had to mention certification, but this does not necessarily mean that they understand the process of certification. Out of the 100 people interviewed eight fully understood what "certification" means and entails, eight vaguely understood it, 19 wrongly thought that it was a process of ensuring quality by the FDA and the USDA, and 65 had no idea what it was. Of the 16 people who either understood or vaguely understood the process, not one of them knew that growers must pay to have their coffee certified.

Fourteen people understand what Fair Trade coffee is, 23 have a good general idea about what it is, 25 have a vague idea, 27 have no idea, and 11 have the wrong idea. For Fair Trade we coded their answers as "understanding" if they could convey to us that Fair Trade is a set of relationships through which producers are meant to earn a better wage for their labor and through which labor is regulated by external bodies of some sort.[12]

For answers that we coded as "a good general idea" interviewees had to be able to link the term "Fair Trade" with at least five of the following ideas: living wage, minimum price, long-term relationships, monitoring, certification, small farms, cooperatives, credit to farmers, cutting the middle men, sustainable development, and fair labor practices. They did not have to use the specific terms but they had to allude to the ideas behind them. For "vague idea" they had to mention three of them.

For organic certified coffee we coded the answers as "understanding" if they could convey to us the process by which coffee is grown and that organic coffee uses no pesticides or fertilizers and is connected to the health of the planet, workers and consumers. Only six people fell into this category while 10 fell into it if we take away comments about planet and worker health as criteria. Seventeen people, all of whom mentioned that organic means there are no pesticides or fertilizers used, have a good general understanding of what organic means. Thirty-five people had a vague idea of what organic means, mentioning that it has something to do with how and where it is grown, 23 people had no idea, and nine had the wrong idea.

The first question during our interviews was "what are the different kinds of coffees?" and we were all surprised by the answers we got to it. Overwhelmingly people mixed their kinds of specialty coffees. For example, one 19-year-old man said "dark roast, light roast, Fair Trade, Hazelnut, Decaf, French Vanilla, Mocha, Columbian, and Guatemalan", a 20-year-old woman said "Origin, I mean the place they were grown, organic, and flavored". Another 19-year-old woman said "Columbian, chocolate-hazelnut, Ecuadorian, American, Maxwell House". These answers were typical. Fifty-eight percent of the informants mentioned a type of flavored coffee like "hazelnut" and 45% mentioned a process like "dark roast" or "decaffeinated". Forty percent mentioned a form of coffee shop production like "espresso", "latte" or "cappuccino". The majority of people, 78%, mentioned a Latin American country when answering this question (Brazil, Columbia and Guatemala were mentioned most often). Fifteen percent mentioned "Africa" and 6% mentioned countries within Africa (Ethiopia, Kenya, Tanzania, Uganda and Rwanda). Four percent mentioned Indonesia.

The one question we asked that had a uniform answer (85% of the interviews) was "who grows coffee?" The answer was: poor farmers. Thirty-five percent of these farmers are "impoverished". Twenty-two percent of them are "minorities" and 18% of them are laboring under "slave" labor conditions. One respondent says:

> The stereotypical image is of a South American working the fields—
> these aren't rich people. Depends on what conditions they're in—some
> of the farmers are abused, but others have gotten deals with artisan
> coffee makers who want to show off how much they've made from
> coffee.

Another reports that "people with mustaches" and "people who tried to grow cocaine and got in trouble" grow coffee. While 38% of the answers included both "farmers" and some permutation of "businesses" or "corporations" the answers overwhelming reflected an image of poor people living in the tropics living on the edge of the modern. Other descriptors used when students talked about these farmers were "men", "brown people", "poor fuckers", "downtrodden men and women", "tropical people", "people in Africa and places like that" and "donkey riding farmers".

The deregulated coffee market was supposed to be fairer and more flexible but it has resulted in the lowest coffee prices in the history of the market. Growers in places like Papua New Guinea are cast into fantasy images of ecologically noble savages and pure guileless economic primitives so that people will see their coffee as a scarce resource and thereby pay higher prices for it. Coffee consumers, and potential coffee consumers, born between 1983 and 2000, a market segment with the largest ratio of disposable income relative to their total income, are cast into fantasy images of wide-eyed well-meaning ecologically and socially noble actors who wish to change the world through their consumption practices.

But, the Internet-based marketing of specialty coffees through particular narratives has not created consumers who can recall the stories. Nor has it created consumers, at least the ones we interviewed, who wish to express their politics through consumption. The students we interviewed know almost nothing about specialty coffee, Papua New Guinea, or the ways in which the millions of poor people who grow coffee around the world have been adversely affected by neoliberalization. They most certainly do not seem to view consuming coffee as a political strategy meant to counter the ravages of neoliberal economic change. But they do, overwhelmingly, have the image of a poor farmer in their heads when they think about coffee. Their fantasy farmers are poor, brown, downtrodden, rural and underdeveloped. These students had absolutely no sense of the sort of multi-classed coffee industry that exists in Papua New Guinea or in any of the other nations where coffee is grown. These people have internalized the virtual producers that they have been given and the structural poverty that many farmers live in becomes something to make a joke about (cf Carrier this volume).

Re-embedding Social Relations and Political Action into Economic Choice?

Neoliberalism fetishizes the market by turning it into something, conceptually, that seems to work on its own apart from human social practices. Since the decline of the ICO the argument, at least with regard to coffee, has been that if left alone and unregulated the market will produce consumer and producer behaviors that bring about environmental stability and social equity. It will do this by re-imbedding the economic choice of what coffee one buys into social relations in which consumers and producers understand each other and have a connection across vast physical distances and vast economic disparities. Consumers are meant to come to understand producers in ways that result in the consumers making ecologically and socially progressive political choices about their buying habits. But all of this is built on a set of fictions. The consumers and producers in this story are all made up— they are produced, crafted, and constructed by the likes of Mr Nebraska and Dean's Beans. But these fantasy figures that populate the specialty coffee media and marketing world are not simply benign images used to sell coffee; they are careful productions that have material consequences.

The consumers fashioned by Mr Nebraska's seminar, the producers fashioned by Dean's Beans blog, the Papua New Guineans who make up the vast coffee industry in their country, and the students we interviewed in New York City seem to have little in common. But they are all brought together as individuals by the contemporary neoliberal coffee market. They are targeted as individuals because that is what neoliberalization does; it focuses on individual producers and consumers as the locus and scale of intervention and disallows for regulation and intervention at other scales. All marketed forms of "ethical consumption" make individuals seem and feel responsible for both the conditions of production and the ecological and social justice issues that stem from these conditions of production. Based on this logic governments, regulatory bodies, and the organizations that forced the structural adjustment programs in the 1980s that resulted in drastic economic changes that disadvantaged coffee farmers worldwide are let off the hook for the problems, contradictions, and negative effects inherent in the global capitalist system and particularly apparent today in the wake of neoliberalization.

In an essay that clearly shows the contradictory nature of labels and certifications as both a form of neoliberalization and corrective to neoliberlization, Julie Guthman argues that "neoliberal valuation rests on the presumption that the market will assign high prices to scarce resources" (Guthman 2007:470). Organic, Fair Trade, and origin labeling and certification supposedly add value to coffee because they guarantee that in an unfair world the product in question was produced in socially and ecologically sustainable ways. Value seemingly accrues to the coffee because the monitoring systems in place guarantee sustainable production and sustainable production is scarce. The entire system of labeling is built on the assumption that most coffee production is unethical and unsustainable but that specialty coffees, because of certification and labeling, are sustainably produced. Here sustainable production, and by that I mean socially and environmentally equitable conditions of production, is the scarce resource.

The stories told about coffee production by the likes of Dean's Beans bring that scarce resource to the consumers. They package and market fair labor as something unique and scarce. These stories also do something else, they bundle together a set of images of Papua New Guinea as remote, biodiverse, primitive and impoverished and present and market them as scarce. So in a sense they use the poverty, which was itself in part created by the processes of neoliberalism, to create a scarce resource that is also meant to add value to the coffee. Poverty, as a necessary prerequisite to specialty coffee, itself becomes unique and valued in this skewed system of value creation. Poverty also becomes a condition that is disembedded from its structural causes when it is conflated with and linked to primitivism.

Socially and environmentally equitable conditions of production and poverty are then the scarce resources that add value to specialty coffees. Three issues emerge here. First, there is an inherent tension between equitable conditions of production and poverty insofar as if we increase one the other should decrease. With truly equitable production, poverty will be ameliorated and the value that poverty adds to the coffees will cease to exist. It therefore behooves companies like Dean's Beans to articulate poverty even when that poverty may not be the entire story, as is the case with the coffee industry in Papua New Guinea. Certainly, people who grow coffee in Papua New Guinea are poor by global economic measures but they own their own land and thereby the means of production. Because of this they are in a radically different structural position than people in parts of the world where this is not the case. Although they have been pushed into a kind of new remoteness by neoliberalization they have not suffered the worst of these reforms in terms of alienation from their land. In addition, the narratives that cast the entire coffee industry in Papua New Guinea in terms of poverty erase the lives of the thousands of people who work in middle class jobs in the industry. By articulating coffee production through a fantasy vision of Papua New Guinea, Dean's Beans adds value to the coffee by misrepresenting those they purport to help. Second, a condition in which fair labor and good environmental conditions have become scarce and thereby valued for their scarcity should be unacceptable to people with truly progressive politics. This state of affairs means that everyone else, people not advantaged by access to sustainable systems of production, is written off in terms of deserving our political attention. This allows us to forget that all production should be made to be equitable, not just that production that can add exchange value when marketed to a particular set of consumers. Third, consumers are asked to address the issue of poverty by "making regulatory decisions about ecological and public health risk, working conditions, and remuneration, and even what sort of producers of what commodities should be favored in the world market", through their economic choices (Guthman 2007:472). They are made to feel like once they have done this they have done their best to redress inequality and they are allowed to forget about all of the people out there who do not have access to certification and labeling. The majority of the impoverished coffee farmers on the planet become someone else's problem and they are disconnected from the lives of consumers. Consumers are meant to feel connected to the poor people who produce coffee sustainably and are allowed to turn their backs on the much larger world of people who don't.

Mass consumer consumption is predicated on the creation of the desire to want and choose and the creation of exchange value is fueled by the creation of unrealistic hope that a given good will fill some need. This perceived value does not stem from anything inherent in the sensual nature of the commodity itself but from projections created by advertising, marketing and consumer culture. Thus, consumers are used to purchasing goods whose qualities, features or usefulness do not meet expectations. What is innovative about Fair Trade marketing is that, by and large, this promised value (ie fair production practices and good environmental standards) is overtly external to the commodity and the consumer truly has no way of knowing whether this value has been received or not—even after the product has been purchased and consumed—other than by reference to the very marketing narratives that promised satisfaction in the first place.

During his presentation Mr Nebraska expressed the belief that a generation's consumer choices are formed by its history and its social and political background. The far more dangerous process, however, is the way that consumer culture shapes political and social choices and actions. In her book, *A Consumer's Republic* (2003), Lizabeth Cohen showed that the atomized nature of consumer culture and the creation of consumer identities ultimately comes to supplant peoples' identities as political actors. In the case of Fair Trade marketing this is the very point. Thus, the hazard created by Fair Trade marketing is that people's desire for greater social justice is co-opted and satisfied by buying into a narrative such as the one presented by Dean's Beans. Because the professed value of Fair Trade products is by definition outside of the commodity itself, the consumer is in no position to learn to what extent these sorts of narratives depart from the real lives of the coffee producers.

The stories encouraged by Mr Nebraska and told by Dean's Beans are a new form of product endorsement. Poverty adds value to specialty coffee as does sustainable production. Poor primitive people, instead of famous wealthy people, increase this value again with their endorsement. In the past, people bought a box of cereal with Michael Jordan on the front because they wanted to "be like Mike"—talented, successful and rich. The "endorsement" from the primitive is about endorsing a set of relationships as being good, ethical and fair. If the Michael Jordan endorsement sells by creating desire to be better than one is, the endorsement from the primitive sells by telling the consumer that they are a good person just as they are and that through their consumption choices they have an ethical and meaningful relationship with the economic have-nots of the world.

The great trick of the specialty coffee marketing by distributors is that it highlights the global economic inequality created by the capitalist system from which they and their customers benefit while at the same time cabining the extent to which consumers should be concerned about such injustices (eg "bad exploitative capitalism only gave people pennies for their backbreaking labor, now thanks to certification those people get four times as many pennies for their backbreaking labor"). The production of the primitive as so backward and so impoverished that even a pittance would be a vast improvement allows for the marketers and consumers to let themselves (and capitalism as a whole) off the hook for no significant cost to themselves. If you present narratives about how awful things were before you got there, then even the insignificant bonus of Fair Trade is seen as a good deal.

The fact that the "Millenial" students don't really think about the plight of the have-nots and that they take for granted that coffee is produced by people at the very bottom end of the economic scale shows on the one hand that certification is not having a huge substantive impact on people's consciousness about the exploitative nature of coffee production (in the sense that producers are alienated from almost all of the value they produce). On the other hand, to the extent that people are concerned about this inequality they are told that certification helps to alleviate some of the worst effects of this exploitation even if consumers are not quite sure how. This supposed embedding of the political and social into capitalist consumption may make for a tasty cup of coffee but it makes lukewarm political action. Ironically, the message of the endorsement of the primitive is not to create desire among the nascent socially conscious consumer to demand greater social justice but to be satisfied in the knowledge that they have done their part to improve the world. This only works because the image of the primitive does what it always does, it tells us that the other is so different from us that it might make sense that receiving a quadrupling of your pennies and a t-shirt is "fair" for hundreds of hours of back-breaking labor. Obviously, consumers wouldn't do that kind of work for that amount of money but the producers are different and they, according to this logic, should be happy with what they get because the market has done the best that it can.

Acknowledgements

I would like to thank the Wenner-Gren Foundation and Barnard College for funding the research on which this chapter is based. I would also like to acknowledge helpful comments on previous drafts from James G. Carrier, Molly Doane, J.C. Salyer, three anonymous reviewers, and the members of my 2006 and 2008 Anthropology of Consumption seminars at Barnard College and Columbia University. Thanks also to Diane Shir-Rae Chang, Sarah Federman, David Hylden, Leila Orchin, Christopher Shay, Julia Turshen, Marissa Sien-Mun Van Epp, and Jesse Waldman who worked on the interview project discussed in this chapter.

Endnotes

[1] A LexisNexis search shows that between 1970 and 1979 there are no articles in popular magazines or major newspapers linking coffee and environmental sustainability. Between 1980 and 1989 there are 77 in newspapers and 12 in magazines. Between 1990 and 1999 there are 802 in newspapers and 187 in magazines. Between 2000 and 2001 there are 422 in newspapers and 154 in magazines and the growth continues such that between 2005 and 2006 there are over 1000 in newspapers and 402 in magazines.

[2] In the USA coffee began as an elite drink but by the first part of the 1900s it was a drink accessible to all, it was consumed in both working-class homes and elite homes (Jimenez 1995). In 1864 Jabez Burns invented an inexpensive roasting machine and small roasting companies began to emerge in the northeastern United States (Pendergrast 1999:55–57). These small companies grew and by the 1890s there was a strong coffee industry in the northeast. During the first three decades of the 1900s a true national market for coffee was created in the United States and the process of standardization in terms

of quality, taste, and production began (Jimenez 1995; Roseberry 1996). The Second World War was a "boon for the coffee industry" worldwide (Pendergrast 1999:222). This was in part because the US army began to requisition about 140,000 bags of coffee a month and serve it to the troops, and in part because of changes in the supply chain for coffee (Pendergrast 1999:222). US troops were being supplied with vast quantities of coffee, and Maxwell House and other large factories began to manufacture coffee specifically for the military (Pendergrast 1999:224). In 1942 the War Production Board in the USA took over all control of the coffee entering the US market and began to regulate and ration coffee (Pendergrast 1999:222). This regulation meant that coffee was rationed for civilians and that both civilians and coffee industry people panicked. Although the rationing was ended in July 1943 the idea of coffee being a limited and luxury good had been planted in consumers' minds (Pendergrast 1999:223). The war created enhanced desire for coffee among civilians and soldiers and it pumped money into the major coffee manufacturers who then, after the war, created expensive and expansive advertising campaigns to keep coffee in people's heads as an item that was an important part of their daily life. This influx of cash into the industry, in the pockets of big companies like Maxwell House, allowed for continued standardization and set the stage for a "trend toward coffee of the lowest common denominator" (Roseberry 1996:765). In the late 1940s several international coffee agreements were signed and the International Coffee Organization, a body to oversee global trade in the commodity, was formed (Roseberry 1996).

Through the late 1940s and 1950s coffee consumption in the USA was "flat" with little fluctuation in levels of coffee bought and sold (Roseberry 1996:765). But between 1962 and 1980 coffee consumption declined radically (Roseberry 1996:765; see also Pendergrast 1999:ch 16). Fewer people were becoming coffee drinkers and people who were already coffee drinkers were cutting back. Even more troubling for the coffee marketing industry was the fact that coffee drinking was "skewed toward an older set" (Roseberry 1996:765).

[3] Another aspect of this phase of neoliberalization was that governance that had once been the purview of the state became the purview of non-governmental organizations, so-called "civil society" groups like churches, development agencies, and other international bodies and organizations.

[4] http://www.deansbeans.com/coffee/people_centered.html, accessed for this chapter on 13 November 2006.

[5] http://www.deansbeans.com/coffee/deans_zine.html?blogid=829, accessed for this chapter on 13 November 2006.

[6] Data presented about Mr Greathead are derived from colonial reports, colonial era newspapers, and interviews with one of his surviving sons.

[7] See Finney (1968, 1973, 1987) and Sinclair (1996) for details on this history.

[8] I have worked with Gimi speakers since 1997 and have conducted over 40 months of field-based research in Papua New Guinea. All names in what follows are pseudonyms and all interviews were conducted in the Gimi language.

[9] In terms of how respondents defined themselves generationally, only three people we interviewed defined themselves as "Millenials", but 52% mentioned "Generation X" in that they were either the "tail end" of it or that they were "the generation behind" it, with 40% saying "Generation Y" at some point during the interview. Thirty-four percent of them mentioned that they were the children of "Baby boomers". Ten percent of the interviewees mentioned 9/11 but no one mentioned the Oklahoma City bombing, the Clinton–Lewinsky affair, or the Columbine shootings (the events that Mr Nebraska used to mark the generation) while only 4% mentioned "global", "globalized" or "globalization" (the main focus of the generation according to Mr Nebraska). Only 3% said the word "community" and only 10% talked about social and ecological change

or justice. Twenty-seven percent mentioned that they use alcohol, not as Mr Nebraska predicted that they are "teetotalers" and 12% mentioned openness towards sexuality (eg "we are not shocked by homosexuality or not clearly defined sexuality"). Thirty percent of our interviewees mentioned how they feel that the media specially targets their generation and that they are under more pressure to consume than other generations. Fourteen percent mentioned that their generation has a sense of "entitlement" or a "spoiled" nature.

Interestingly, 73% of the interviewees mentioned technology when asked to describe their generation, 66% mentioned the Internet, and 35% mentioned speed of communication (eg "people over the age of twenty-five just don't understand the instantaneous nature of communication"). Twenty percent of the interviewees mentioned Facebook or My Space, 25% mentioned television, and 10% mentioned cell phones. Over half of them mentioned the Internet as a source of information for consumer goods.

[10] The campus interview sample presented here is not characteristic of most of the consumers of Dean's Beans coffees. Dean's Beans is part of a small group within the specialty coffee industry called "Cooperative Coffees" and their markets are in unusually progressive communities (Madison, Ann Arbor, Berkeley, Amherst) where there is high general awareness of social issues around Fair Trade and Organics. But this is a very small niche market.

[11] We limited the age range to people who would be classified by marketers as "Millenials". We further limited the age range to people between 24 and 18 because we did not want to engage with interviewing a "special class" of people and we would have had to if we interviewed people below 18 years of age. We did not interview a random sample of students rather the student interviewers and I determined interviewees during observation periods at eight coffee-related establishments on or around the Morningside Heights neighborhood. These establishments included a Starbucks, the coffee shop area of a student dining hall, the College café, an old elite coffee house near campus, a new non-chain coffee shop on Broadway, and a new "funky" one near campus. We wanted to include a range of types of shops and price ranges. We only interviewed people who identified themselves as coffee drinkers.

Seventeen people initially interviewed had taken Interpretation of Culture, our version of introduction to cultural anthropology, with me. We threw out those interviews and replaced them with 17 additional interviews because all of these students would have heard "endlessly" (interview, 10 December 2006) about Papua New Guinea during that course.

[12] During analysis we used the following definitions as guides, only counting people as understanding if they gave us these answers in, of course, their own words:

"Fair Trade is a trading partnership, based on dialogue, transparency and respect, that seeks greater equity in international trade. It contributes to sustainable development by offering better trading conditions to, and securing the rights of, marginalized producers and workers—especially in the South. Fair Trade organisations (backed by consumers) are engaged actively in supporting producers, awareness raising and in campaigning for changes in the rules and practice of conventional international trade. Fair Trade's strategic intent is to deliberately to work with marginalised producers and workers in order to help them move from a position of vulnerability to security and economic self-sufficiency, to empower producers and workers as stakeholders in their own organizations, to actively to play a wider role in the global arena to achieve greater equity in international trade" (http://www.fairtrade.net/faq_links.html?&no_cache=1 accessed on 1 August 2006).

And "Fair trade coffee is coffee that is traded by bypassing the coffee trader and therefore giving the producer (and buyer) higher profits. Fair Trade does not necessarily mean that the extra money trickles down to the people who harvest the coffee. TransFair USA is an independent 3rd party certification that ensures that: Coffee importers agree to purchase from the small farmers included in the International Fair Trade Coffee Register. Farmers are guaranteed a minimum 'fair trade price' of $1.26/pound FOB for their coffee. If world price rises above this floor price, farmers will be paid a small ($0.05/pound) premium above market price. Coffee importers provide a certain amount of credit to farmers against future sales, helping farmers stay out of debt to local coffee 'coyotes' or middlemen. Importers and roasters agree to develop direct, long-term trade relationships with producer groups, thereby cutting out middlemen and bringing greater commercial stability to an extremely unstable market" (http://www.coffeeresearch.org/politics/fairtrade.htm accessed on 1 August 2006).

References

Alsever J (2006) Fair prices for farmers: simple idea, complex reality. *New York Times* 5 May

Bacon C (2005) Confronting the coffee crisis: Can fair trade, organic, and specialty coffees reduce small-scale farmer vulnerability in northern Nicaragua? *World Development* 33(3): 497–511

Bacon C M, Mendez V E, Flores Gomez M E, Stuart D and Flores S R D (2008) Are sustainable coffee certifications enough to secure farmer livelihoods? The millennium development goals and Nicaragua's Fair Trade cooperatives. *Globalizations* 5(2):259–274

Carrier J (1998) Introduction. In J Carrier and D Miller (eds) *Virtualism: A New Political Economy* (pp). New York: Berg Publishers

Carrier J (2010) Protecting the environment the natural way: Ethical consumption and commodity fetishism. *Antipode* 42(3):672–689

Carrier J and Miller D (eds) (1998) *Virtualism: A New Political Economy*. New York: Berg Publishers

Cohen L (2003) *A Consumer's Republic*. New York: Alfred A. Knopf

Finney B R (1968) Bigfellow man belong business in New Guinea. *Ethnology* 7:394–410

Finney B R (1973) *Big Men and Business: Entrepreneurship and Economic Growth in the New Guinea Highlands*. Honolulu: University Press of Hawaii

Finney B R (1987) *Business Development in the Highlands of Papua New Guinea*. Honolulu: Pacific Islands Development Program

Guthman J (2007) The Polanyian way? Voluntary food labels as neoliberal governance. *Antipode* 39:456–478

Harvey D (1989) *The Condition of Postmodernity*. London: Blackwell

Haug W F (1987) *Critique of the Commodity Aesthetic*. Minneapolis: University of Minnesota Press

Hull J B (1993) Can coffee drinkers save the rain forest? *The Atlantic Monthly* August

Jimenez M (1995) "From plantation to cup": Coffee and capitalism in the United States, 1830–1930. In W Roseberry, L Gudmundson and M S Kutschbach (eds) *Coffee, Society, and Power in Latin America*. Baltimore: The Johns Hopkins University Press

Lindenbaum S (2002) Fore narratives through time. *Current Anthropology* 43:S63–S73

Miller D (1998) Conclusion. In J Carrier and D Miller (eds) *Virtualism: A New Political Economy* (pp). New York: Berg Publishers

Mutersbaugh T (2005) Just-in-space: Certified rural products, labor quality, and regulatory spaces. *Journal of Rural Studies* 21(4):389–402

Ortner S (1998) Generation X: Anthropology in a media-saturated world. *Cultural Anthropology* 13(3):414–440

Pascual A M (2006) Peace, love and coffee; Woodstock Coffeehouse opens third store. *The Atlanta Journal-Constitution* 17 December

Pendergrast M (1999) *Uncommon Grounds: The History of Coffee and How It Transformed Our World.* New York: Basic Books

Polanyi K (1968) [1958] *The Great Transformation.* Boston: Beacon Press

Ponte S (2002) The "Latte Revolution"? Regulation, markets and consumption in the global coffee chain. *World Development* 30(7):1099–1122

Roseberry W (1996) The rise of yuppie coffees and the reimagination of class in the United States. *American Anthropologist* 98(4):762–775

Sinclair J (1996) *The Money Tree: Coffee in Papua New Guinea.* Bathurst, NSW: Crawford House Publishing

The Economist (1993) The greening of giving. *The Economist* 25 December

Chapter 10

Cashing in on Cetourism: A Critical Ecological Engagement with Dominant E-NGO Discourses on Whaling, Cetacean Conservation, and Whale Watching[1]

Katja Neves

I will always remember the sight and smell of sperm whales being processed at the whaling factories of Sao Roque do Pico and Horta Faial in the Azores, which I first visited as a child in the mid 1970s. To see such a magnificent animal cut into pieces and melted for blubber was a gruesome experience. But it was also marked by an atmosphere of joy as the men who hunted the whales relied on this activity for income in an otherwise extremely poor community. Later, in my teens, I witnessed the sad image afforded by the decaying whale boats and whaling canoes that signaled the end of a prosperous whaling epoch in the islands of Pico and Faial when Portugal signed the Bern Convention. Thus the century old economic history of human–cetacean relations ended in less than a decade (Neves-Graca 2004, 2006).

Shortly thereafter, however, cetaceans began to regain their economic importance with the establishment of the first whale watching company in Pico, which quickly became a major source of revenue for investors in many other Azorean locations (Neves-Graca 2004, 2006). Tourists visiting the Azores now find the legacy of whale hunting (in museums and/or recovered whaling canoes currently used for nautical sports) existing side by side with many whale-watching businesses. What was once a deadly economic activity appears to have been transformed into an ecologically and economically sustainable source of revenue.

These transformations in the Azores reflect a global trend. Whale watching has now become an important economic earner for maritime communities around the world, though many of these never practiced whale hunting (Neves-Graca 2002). The growth of whale watching since the 1960s can be attributed to several historical trends, including international treaties forbidding continued whale hunting, a drastic decline in market demand for whale-derived commodities, and major campaigns by environmental non-governmental organizations (E-NGOs) to support whale watching as a sustainable alternative to whale hunting (Epstein 2008).

This promotion of whale watching by E-NGOs and tourist companies has contributed to a predicament in which whale watching is widely equated with cetacean conservation. Taking this assumption for granted, however, undermines the possibility of distinguishing between different types of whale watching and the degree to which they effectively live up to conservationist goals (Neves-Graca 2002, 2004). Consequently, E-NGOs have diminished their own capacity to identify cases where whale watching may actually have damaging effects on cetaceans. While for the most part whale-watching businesses do indeed reflect pro-environmental and pro-conservation principles, the activity itself does not constitute a set of homogeneous practices (Neves-Graca 2002, 2004, 2006). Given that cetourism continues to grow exponentially,[2] its purported ecological soundness and ability to generate socio-economic gains for communities certainly merits closer scrutiny.[3]

This chapter tackles uncritical assumptions about whale watching from a historical perspective. It compares classic whaling with contemporary whale watching in order to reveal the lines of continuity—as well as gaps—that exist between them. Relying on Castree's (2000, 2007) arguments concerning the "production of nature", and the Marxist concept of "metabolic rift" (Foster 1999, 2000), I will show that the two types of commercial activity share core features of a capitalist mode of production, although they each produce different natures at distinct junctures in the history of capitalism. While whale hunting was essentially tied to a mode of capitalist production and growth that relied on the extraction of natural resources from the ocean to distant localities, whale watching produces nature as a provider of services to be consumed and enjoyed in situ. Whale hunting instated a metabolic rift by removing nutrients from the ocean and by decimating entire whale populations. Whale watching can potentially create a metabolic rift by disrupting whale feeding patterns and animal sociability communication. Because each of these two capitalist activities produces a different kind of metabolic rift, it is necessary to rethink this concept towards a better understanding of its relevance in a contemporary setting of human–cetacean relationships. Especially when aspects of dominant environmental discourse have the unexpected effect of eliding potential harms posed by cetourism.

This chapter contextualizes the production of cetacean natures vis-à-vis socio-culturally, politically, and historically situated understandings and uses of cetaceans as they intersect with global economic processes as well as with major concerns in late modern societies. I will demonstrate that the promotion and practice of whale watching tends to obscure what O'Connor (1988) has labeled "the classic contradictions of nature and capitalism". As I have argued elsewhere (Igoe, Neves and Brockington this volume; Neves 2008, 2009), the equation of marine ecotourism with ecologically sound conservation reflects a much wider late capitalist trend in which conservation is increasingly conflated with consumption. I will develop these arguments as follows.

First, I will argue that the efforts of some of the world's most prominent E-NGOs[4] to save whales from being hunted to extinction have produced and propagated whale watching as a quintessentially and uniformly benign activity. This homogenized promotion of whale watching includes the forging of explicit linkages between whale watching and an endless variety of commodities and consumptive experiences, while simultaneously portraying cetourism as an enlightened form of biodiversity conservation.

Second, I will demonstrate that there is greater continuity between whale hunting and whale watching than the above presentations would suggest.[5] The transition from one to the other is more closely related to transformations in the global capitalist economy than to enlightened progress in human–cetacean relations.

Finally I will reveal that whale-watching business models are a prime factor in shaping how whale watching is designed and practiced in specific contexts. Some business models contribute to significant ecological harm while failing to make significant contributions to community economic sustainability. I illustrate this argument through a comparative analysis of two different whale-watching business models: Azores and the Canary Islands.[6] I thereby contest the reductionism that is entailed in taking for granted that the relation between cetourism, economic development/growth, and conservation is essentially and universally benign.

Whale Watching and E-NGO Discourse: Conservation-*Cum*-Ecobusiness

E-NGOs occupy pole position in promoting whale-watching businesses as the means to achieve marine cetacean conservation, not only as more ecologically enlightened than whale hunting but also in yielding unprecedented returns to investment on cetaceans[7]. In this context, consumption is often presented as the road to conservation. In keeping from view the contradictions that may stem from the concurrent pursuit of conservation and capitalist growth, the relationship between conservation and consumption is presented in overly simplified terms as essentially benign. Accordingly, whale-watching businesses are homogenized as being necessarily "good for conservation".

This conceals the fact that different whale-watching businesses have distinct ecological and social impacts—some carrying negative dimensions (Cawardine 2007; Duffus 1996; Hoyt 1998; Hoyt and Hvenegaard 2002; Orams 2000). E-NGOs thus miss out on their potential to take a more effective role securing the implementation of sound ecological practices and in discouraging bad whale watching conduct. It is therefore important to briefly describe the main features of this increasingly dominant discourse and discuss the ways in which it can at times have the unintended effect of hiding linkages between the commoditization of cetaceans and the potential ecological harms they may incur as a result. While I only mention key examples of this type of E-NGO discourse (cf Einarsson 1993; Epstein 2003, 2006; Freeman 1994, 1996, 2001; Kalland 1994a, 1994b; Pearce 1991), these examples are derived from a systematic survey of E-NGO websites promoting whale watching

On its UK anti-whaling website Greenpeace informs us that "whale watching has shown the potential to become far more profitable than whaling ever was. It is already generating a staggering $1.25 billion per year".[8] The site further informs us that in addition to these major economic advantages, whale watching provides unique opportunities for cetacean research and that it stimulates increased public appreciation of marine environments and greater public sensitivity in relation to conservation issues.[9] Another Greenpeace website evokes social benefits:

> Its benefits are spread out over a larger portion of the local population. It is not just the (by now) ex-whalers who would benefit from charging for the actual boat safari. Other locals running shops, hotels, and restaurants would also benefit. This in areas where employment opportunities are desperately needed.[10]

The research I have conducted on whale watching does indeed support this positive view of the activity. Nevertheless, the discourse leaves unmentioned the many cases around the world where whale watching has failed to live up to such promises and even caused major environmental problems, as discussed in the final section of this chapter.

Greenpeace's discussion of the potential dangers that cetaceans still face after the cessation of whale hunting is illustrative of dominant representations of whale watching that emphasize its positive outcomes while leaving out its damaging effects. Instead of addressing the problem of underwater noise pollution or the major stress that an excessive number of whale-watching boats causes to cetaceans, or even the disruptions that are caused when humans insist on swimming with cetaceans, the site mentions only that "whales face an increasing number of environmental threats from contaminants such as PCBs, and other forms of pollution" (Greenpeace nd). Similar patterns for discussing the risks and dangers faced by cetaceans can be found in most of the E-NGO sites I have researched. Given that many of these E-NGOs have actually participated in critical studies of whale watching, the omission of potential negative impacts of whale watching can only be understood as a strategy that is meant to avoid sending mixed messages to the public about the good and bad of this activity.

A similarly uncritical view of the exponential growth and benefits of whale-watching businesses is observable in a joint press release by the Whale and Dolphin Conservation Society (WDCS), the International Fund for Animal Welfare (IFAW), and Global Ocean.[11] Reporting a "massive growth" for whale watching in Latin America, the press release states that it "highlights the economic value of the industry as an alternative to whaling", in spite of the fact that whaling has not been a viable economic activity in Latin America in recent times. While it makes reference to internationally renowned whale watching expert Erich Hoyt, stating that "responsible whale watching offers substantial, diverse, community benefits compared to narrowly focused, out-of-touch whaling industry", it contains no warnings about the effects of irresponsible whale watching, which Hoyt has warned against in many other contexts (eg Hoyt 2001).

Of all the major E-NGO websites that I have analyzed, the World Wildlife Fund (WWF) draws the clearest connections between whale watching and conservation when it states that: *"It's not just good fun; whale watching is good for conservation too"*.[12] (emphasis added) The site goes on to add informative numbers concerning the economics of whale watching. It states that 10 million people go on whale-watching trips every year around the world (one wonders what the ecological footprint of 10 million ecotourists might be) spending more than US$1.25 billion a year. To really drive home the point that whale watching is a serious player in the realm of international economics, the site adds that: "The number of whale watchers is increasing at 12 percent a year, three times faster than overall tourism numbers."

Since there is no doubt that the WWF has played a major role in saving whales from being hunted to extinction, its intervention would surely make a big difference in places where bad whale-watching practices are having negative effects on cetaceans. Unfortunately, as is the pattern with other NGOs, when it comes to a discussion of the negative impacts of whale watching, WWF's site is mute. It lists the following as the major risks still faced by whales: by-catch, climate change, collisions, seismic and sonar activity, loss of habitat, and over-fishing.[13] None of these are related to whale watching excepting potential collisions, but the information that is available on the site refers only to cetacean collisions with cargo ships. With the exception of over-fishing, none is related to whale hunting either.

Still, insofar as this chapter is concerned, what is most surprising about WWF's website is that it actually provides a link to purchase wilderness trips through a partnership it established with "Natural Habitat Adventures". The ways in which conservation and consumerism intersect in this context are so striking that it is hard to determine to what extent Natural Habitat Adventures is—or is not—a subsidiary of WWF.[14]

Of the many trips it offers, one is to Spitsbergen, Norway to see beluga whales. The rates for such a trip are US$9880 per person based on double occupancy, with solo rates starting at US$16,580. Note that the top banner of the page advertising reads: "WWF works globally to protect endangered whales" thus establishing a connection between conservation efforts and the consumption of nature services.

Finally, the website raises the issue that traveling to such a remote location produces a quite large ecological footprint since most ecotourists need to travel by airplane in order to reach that destination. The website readily offers a solution to this problem by providing a link to a page that offers climate-friendly travel.[15] At the click of a few electronic buttons, one can thus not only calculate the amount of carbon emission such a trip would accumulate, one can also—with a few more clicks and a credit card—offset carbon emissions by spending a respective dollar amount.

The WWF website reveals a close relationship between contemporary mainstream conservation, whale watching and consumption. Based on the data I have just shown, it is indeed easily argued that conservation and consumerism have become closely intertwined. Collaborative research with Jim Igoe (Dartmouth College) and Dan Brockington (Manchester University) shows that this tendency is not peculiar to whale watching but is rather a general tendency in conservation discourse and practice (Igoe, Neves and Brockington this volume). This is an issue of concern and one that calls for a more careful analysis of core assumptions in conservation discourse.

The inception of a more critical approach to cetourism can be found within the international regulatory context of the IWC, which has made a firm commitment towards a more objective and scientific-based assessment of whale watching. A 2008 IWC report reads: "the sub-committee agreed that a review of whale watching in Portugal (Azores,

Madeira), the Canary Islands and the Strait of Gibraltar would be of interest for next year's meeting" (IWC 2008). I contend that efforts in this direction must clearly engage the capitalist nature of cetourism, which lies hidden behind a dualistic understanding of whale watching vis-à-vis whale hunting, and include a more reflexive understanding of the current ambiguity that marks the production of nature (Castree 2000) at the intersection of capitalism and conservation (Igoe, Neves and Brockington this volume).

Are Whale Hunting and Whale Watching Really Polar Opposites? A Theoretical Framework

In modern western societies interactions with cetaceans have been at the forefront of major socio-economic historical transformations that characterize different socio-economic epochs. Commercial whale hunting, for example, was a definitive element of the industrial revolution. When the finiteness of natural resources became inescapably evident in the 1960s and 1970s, images of industrially hunted whales became themselves a new type of commodity. This commodity was essential to the ascendancy of E-NGOs like Greenpeace (Blok 2007; Mowat 2005 [1972]) who relied on images of bleeding whales to run funding and sensitization campaigns, and to the creation of the pro-conservation context out of which whale watching grew into a premier mass scale ecotourism enterprise.

An especially significant outcome of these transformations was the institutionalization of flagship species in the branding of NGOs and in raising awareness about environmental causes. The danger in these otherwise fortuitous changes, however, is that they can obscure the embeddedness of species in larger ecosystems as well as the nature of their relationships to human beings. Likewise, iconized presentations of whale watching as universally beneficial to cetaceans and community sustainability can at times conceal the potentially negative aspects of some types of whale-watching practice. I am particularly concerned with the ways in which dominant conservation discourse contrasting whale hunting with whale watching at times obscures the capitalist nature of cetourism therefore inhibiting the emergence of a more critical view of this activity in places where practices are more closely dictated by short-term profit goals than by environmental concerns. Indeed, the major aspect of concern in this discourse is the presentation of whale watching as quintessentially ecological as if it wasn't first and foremost a capitalist enterprise.

In order to compare the capitalist nature of whale hunting and whale watching I deploy a Marxist approach to tackle four capitalist pillars I identify as pertinent to the present context. First, a relationship between human and non-human worlds that—given the logic of capital investment—tends to disrupt the dynamic processes by means of which ecosystems endure. These disruptions fall under a term which Marx theorized as a "metabolic rift", and which I employ and expand upon with reference to the practices of cetourism. Second, a process that renders invisible the material and social activities by means of which goods are produced and subsequently exchanged for money such that they appear in the market as if by magic; following Marxist terminology, I call this process the fetishization of commodities. Third, a system that is based on exploitative class relationships whereby those who are in a position to make capital investments are enabled to extract labour value from those who have no other means to subsist but to offer their labour in exchange for money; this process leads to the alienation of workers, including alienation from the ecological conditions of production (see Burkett 1999). Fourth, the conceptualization of "nature" as a bountiful pool of resources that exist either in the form of material resources or, more recently, in the form of services that are meant to satisfy human needs.

The conceptualization of nature and the material transformation of nature are so interlinked in the context of capitalism that it is misleading to talk of nature and society as autonomous systems (Castree 2000). To be sure, within the currently dominant mode of production, Capital and Nature are always implicated in one another such as to suggest "a perspective ... in which causality and agency is complex, relative, and contingent: in short, difficult to generalize about" (Castree 2000:18). In the context of whale hunting, nature was produced conceptually and materially from a Promethean standpoint whereby whales were seen as endlessly plentiful resources that existed to be harvested in order to serve human material needs. The whale-hunting industry acted accordingly, creating a material reality for whales that is well known. In the context of whale watching, nature is produced from a conservationist standpoint whereby cetaceans are produced conceptually and materially as unique and precious species to be protected from the damages that they incurred historically to serve human material needs. This produces cetaceans materially as service providers meant to satisfy a new array of non-material (ie psychological, emotional, ludic) human needs.

When one considers the devastating impacts that hunting had on whale populations all over the world, the differences between whale hunting and whale watching appear abysmal. From the perspective of the processes outlined in the previous paragraph, however, the two are more similar than they first appear. Both have reduced cetaceans to commodities, while fetishizing the social and ecological costs of this commoditization. While the potential damages of whale watching may seem trivial in comparison to the damages of whale hunting, the important issue it raises concerns the fetishization of ecotourism *as* conservation practice which, as others have argued (Brockington, Duffy and Igoe 2008; Igoe, Neves and Brockington this volume), is an increasingly common occurrence in the context of neoliberal conservation.

Marx's notion of metabolic rift, as taken up in recent years by John Bellamy Foster, is core to the formulation of these arguments. Foster has demonstrated that Marx's insights are key to a critical understanding of twentieth and twenty-first century ecological predicaments. Drawing mostly from *Grundrisse*, Foster (1999, 2000) argues that Marx conceptualized the human relationship with nature as a dynamic activity whereby humans and nature exchange nutrients that are essential to living processes both in humans and in nature. Marx called this process *Stoffwechsel*, which is commonly translated as "metabolism".[16] Humans intervene and affect nature's metabolism as they procure their own metabolic sustenance and in the process the "two" (human and non-human) metabolisms become intertwined. While in certain modes of production this relationship may entail higher degrees of reciprocity, in a capitalist context it entails a destructive schism marked by a unidirectional flow of nutrients from the non-human world to the human sphere. This unidirectionality compromises the viability of nature's metabolism in the long term. According to Marx this is further accentuated in the context of an extractive relation between rural areas and cities (Foster 2000).

While Marx was particularly concerned with the effects of industrial agriculture on the long-term fertility of the soil, the concept of the metabolic rift is more generally applicable to the fetishization of human–environmental relationships, such that the environmental costs of capitalist modes of production are hidden from view (see Burkett 1999).[17] From this more generalized perspective, Marx's concept of the metabolic rift remains a highly relevant contribution for theorizing and analyzing human–nature relations in a capitalist context. First, the concept draws attention to the serious disruptions that capitalist modes of production introduce to ecosystems due to profit accumulation imperatives. Perhaps even more importantly, Marx's articulation of the metabolic rift provides a theoretical framework for explaining the social and ecological conditions under which nature is transformed into consumable products and the ways in which these processes are rendered invisible to consumers. It helps explain how commodities appear on the market as if by magic, that is, in fetishized form, since most consumers never have a chance to know the social and ecological costs/effects that are associated with the production of commodities.

Whale hunting emerged in global economic conditions similar to those from which industrial agriculture emerged. As such it entailed crucially similar types of relationships to which the metabolic rift concept is easily applicable. First, there were significant geographical distances between production and consumption. Whales were extracted from the ocean to be processed and consumed at distant locations. This included the decimation of whales to produce oil and other industrial products. Significantly, this included the production of fertilizers for the emerging forms of industrial agriculture with which Marx was so concerned (see endnote 15). These relationships clearly constituted unilinear flows of nutrients from the non-human world to the human sphere. They also entailed brutally exploitative class relationships. Because of the distances they entailed, however, the ecological and social costs of these relationships were effectively concealed from the average consumer.

It was not until the 1960s that the destructive ecological and social impacts of these relationships were widely exposed through the anti-whaling campaigns of E-NGOs. However, these campaigns involved new types of commoditization and fetishization, as images of slaughtered whales became important commodities in the highly competitive world of NGO fundraising. The promotion of whale watching as an alternative to whale hunting created new possibilities for metabolic rifts that reflect emerging types of social and economic relationships in the context of late market capitalism. In stark contrast to whale hunting, the production and consumption of whale-watching experiences are not characterized by tremendous distance. In fact, they tend to occur simultaneously at precisely the same location. The metabolic rift that occurs in whale watching is thus not rendered invisible by distance, but because the average whale watching consumer does not possess the knowledge (Beson 1998) to identify the patterns that signal stress and behavioral disruptions amongst cetaceans (Neves-Graca 2002, 2004, 2006). The metabolic rifts that can occur in such contexts are related to the disruption of a whale's ability to locate and consume essential nutrients. The invisibility of these disruptions in turn conceals the potential ecological contradictions of market-driven whale-watching business models. Thus they also conceal important continuities between whale hunting and whale watching as historically specific forms of capitalist production. Sadly, this invisibility is most likely reinforced by E-NGO discourses that promote whale watching as synonymous with conservation.

Continuities and Gaps in the Transition from Whale Hunting to Whale Watching

My substantive analysis begins with the historical period of industrial whale hunting in the USA. The core traits of this enterprise stayed constant throughout its 400-year history (eg Bullen 1902; Dolin 2007; Francis 1990; Mawer 1999; Robertson 1954; Tripp 1938). Since these traits are where one can see the most obvious continuity to whale watching, they are worth describing in some detail. First, whale hunting revolved around the extraction of material goods from whales and transforming them into highly valuable commodities. Two of these commodities, oil and spermaceti candles, were essential to the industrial revolution.[18] In Marxist terms the relationship between humans and whales was a classic example of the metabolic rift that capitalist systems tend to establish with nature: one where "nutrients" are extracted from ecosystems into capitalist markets to the point of compromising the metabolic health of ecosystems—that is, their ecological sustainability (Foster 2000; Marx 1952). When whales became scarce in one place, whalers simply expanded into another.

Class relationships in the context of whale hunting were exploitative to the point of atrocity (Dana 2001; Hohman 1928). A captain might receive 1/6 of the profit of a whaling trip while a whaler might receive as little as 1/250. To make matters worse, every piece of clothing, tobacco, or paper that whalers took from the ship's store was taken into account once they finally received their share of the profits. Prices for these goods were highly inflated too. Many accumulated such large debts that they hardly saw any pay for their labor.

This reality is directly related to the fetishization of whale commodities. Since the processing of whales occurred on ship, consumers interacted with whale-derived products that were completely removed from the ecological, political and social processes that produced them. With the exception of Melville's descriptions in Moby Dick, the relation between these goods and the ecology of the animals they were extracted from, or between these goods and the class relations that brought them into existence, remained mostly invisible from the public eye.

Because whales were valued exclusively for their potential transformation into goods with high market value during this period, their aesthetic value was almost never considered. Within this early capitalist logic cetaceans were divided into a hierarchy with sperm whales, the most economically profitable species, at the top. Some captains did not even slow their ships for species that weren't sperm whales. For everyone who depended on this extractive industry, the only good cetacean was a hunted and processed whale of a particular species.

The emergence of the global fossil fuel economy in the late nineteenth century was the beginning of the end of global whale hunting. Compared to a global economy run almost exclusively on whale-derived products, currently only a fraction of whales are hunted globally every year. Nevertheless, major E-NGOs continue to imply that whales are facing a near extinction crisis due to *contemporary* hunting. This is in no way to deny that even in small numbers the hunting of whales today may have devastating effects for the survivability of whales. However, it is important to distinguish between different species of whales because not all are threatened with equally devastating consequences. Of greater concern to me here is that the current commoditization of whales and cetaceans in general is founded on the presentation of whale watching as key to replacing whale hunting as an economic activity.

In reality most whale watching occurs in places where whale hunting was either never practiced or died out long ago. People thus often make fictional historical claims to whale hunting in order to make whale watching viable for their communities. In cases where whale hunting previously did exist it requires converting skills and technology to whale-watching activities, something that local people often recognized and undertook without any external prompting. These transformations are indicative of larger transformations away from material commodities to service-based commodities in the context of global neoliberalism.

As a service-based enterprise, whale watching is intimately associated with the rise in popularity of the environmental movement, and the exponential growth in the demand for eco-friendly commodities that such movements often promote. Most of the world's largest corporations are stepping up the production of presumed green commodities to satisfy these growing demands. New realms for expansion include ecotourism, spectacle, services and entertainment. I therefore contend that it is equally accurate to view whale watching as a product of shifting market conditions and as the result of more enlightened understandings of non-human natures. Moreover, as previously mentioned, whale watching in some contexts has resulted in new kinds of metabolic rifts.

Understanding metabolic rifts produced by whale watching requires a closer look at cetacean ecology, and especially echolocation. Echolocation consists of emitting sound waves that bounce back to the point of emission once they reach an object. This is how cetaceans find food and navigate within their ecosystems. Sound is also essential to cetacean communication and sociability. Outboard boat engines, which produce high-pitched underwater sounds, interfere with cetacean echolocation and communication. While the extent of these consequences is not fully known, Monteiro (1998) argues that extreme interference, such as that caused by many boats concentrated in a small area, could seriously inhibit cetaceans from finding their food. In any case, underwater noise pollution causes major stress on cetaceans that may have devastating long-term effects.

When this sort of disruption impedes cetaceans from using echolocation to find food over extended periods of time, it might introduce metabolic disruptions that are serious enough to be called a metabolic rift since whales may as a consequence be deprived of crucial nutrients (this would be, as my friend and colleague Luis Monteiro argued in 1998, particularly acute in the case of the resident sperm whales of Lajes do Pico in the Azores, Portugal—see Neves-Graca 2002, 2004; see also, for example, Hildebrand 2005; Higham and Lusseau 2007; Hoyt and Hvenegaard 2002; Lusseau, Slooten and Currey 2006; Richter, Dawson and Slooten 2006). Over time, this inability to locate food, combined with the stress that noise causes, may even have gravely debilitating, or even deadly, effects.

The notion of a possible metabolic rift in the context of whale watching calls for adjustments to Marx's understanding and use of the term. Here the rift is not instituted by the extraction of nutrients from an ecosystem but rather from disruptions to the cetacean ability to locate and consume nutrients. Also, the fetishization of the ecological costs of whale watching occurs at the very moment that the consumption of cetaceans takes place, right in front of consumers. It remains, nevertheless, invisible to most consumers. Having conducted extensive participation onboard whale-watching boats in the Azores since 1998, I have learned that the majority of cetourism consumers that travel to the Azores cannot recognize and identify the signs of a metabolic rift in human–cetacean relations. This is not to say that they are ecologically illiterate or insensitive but rather that such signs are indeed hard to identify for the untrained person (Neves 2004, 2006). This situation is likely exacerbated by E-NGO discourses that present whale watching as the diametrical opposite of whale hunting, thus transmitting a false sense of assurance to the average tourist that all whale watching is sound (Benton 1998).

In my field research I have found that the idea of positive synergies between whale watching and cetacean ecology has become commonsensical for whale-watching tourists. Whenever I bring up the notion that whale watching may have negative impacts, most lay people initially react with disbelief and shock to my arguments. Often while interviewing members of E-NGOs I have been told that rectifying this situation is not of their concern, since their primary mission is to end whale hunting through the promotion of alternative economically viable uses of cetaceans. I am quickly informed that sending out mixed messages that whale watching may be good and bad at the same time confuses people and takes away the strength of much simpler anti-whaling slogans.

It is not so much that these organizations are not able to access and understand the kinds of data that challenge reductionist views of the world. In fact, all the E-NGOs I mention in this chapter have developed exhaustive and well-conceived codes of conduct for whale watching. Many have also participated in critical assessments of this activity.[19] However, E-NGOs keep these complex and critical views separate from what the average person sees on their website, which amounts to a simple message that whale hunting = bad and whale watching = good. This discrepancy is rendered poignant by the fact that "save the whale" campaigns, combined with the marketing of associated high-end whale watching, has become an important source of revenue for these organizations.

This discrepancy is indicative of the types of fetishization that pervades whale watching as an economic activity. Here again, hasty analysis might lead one to think that in the context of whale watching cetaceans are neither fetishized nor commoditized. After all, they are used "just as they exist naturally and in wilderness".[20] This gives the appearance that in appreciating cetaceans during a whale-watching trip one is also appreciating the whole of their ecosystem.

In reality, however, whale-watching trips are constructed in ways that make it impossible to see cetaceans in their full ecological context. Much to the contrary, one sees only part of their bodies and only the surface of the environment in which they live. Moreover, like whale hunters before them, whale watchers construct hierarchies of cetaceans based on their commercial worth. Humpback whales are most popular because of their interactivity with whale-watching boats. Dolphins are also highly valued for their playful and interactive demeanor, especially when children are onboard. Less entertaining species are for the most part left alone.

One outcome of this fetishization is a conflation of the symbolic value of whales with their ecologic value (see Paine 2006). This conflation is especially visible when the ecological value of cetaceans is inflated by E-NGOs for their use in conservation campaigns. A more scientific grounded perspective suggests that whales play a relatively small role in the health and maintenance of ecosystems. From this perspective the value that cetaceans have for humans is their high intelligence and quasi-human capabilities for communication and socialization. Hence, the value of cetaceans seems to be determined by ethical judgments rather than by ecological reality (Heller 2007; Watson 1996).

This conflation of the symbolic and ecological values of cetaceans also contributes to the illusion that little or no human ingenuity has gone into the "transformation" of cetaceans as resources. Because cetaceans are not sold as material resources in the context of whale watching there is a common misconception that they are not commodities in the capitalist sense. This is the basis of the spurious claim that whale watching is a non-consumptive enterprise. I suggest instead that cetaceans are being sold in this context as a new type of commodity. While no longer consumed directly as a material commodity, cetaceans are now providers of ecological services. These include entertainment, amusement, catharsis and even therapeutic healing.

Additionally, whale watching entails a high degree of human technological intervention which amounts to extractive relations with nature even though these too tend to remain invisible in the romanticized and naturalized presentation of this business. First, whale watching is more often than not conducted with engine boats that rely on fossil fuels and create various forms of pollution, including noise, fumes and oil by-products. Second, ecotourists need to travel long distances to reach whale-watching sites, which again leaves a major ecological footprint since it normally requires airplane travel. Third, most whale-watching companies have developed highly profitable franchise goods such as plush renderings of whale and dolphins, posters, t-shirts, sweatshirts, caps, postcards and an endless array of similar types of branded products that invite ecotourists—most successfully—to continue to engage in consumptive practices that ultimately require an extractive relation with non-human natures.

Finally, whale watching creates new forms of exploitative class relations and uneven distribution of profits, which brings into question the belief promoted by E-NGOs that whale watching automatically promotes *community* sustainability. For example, because it is a seasonal activity in most places of the world, it creates sub-optimal labor conditions where laborers often find themselves having to live without waged employment for most of the year.

Not all whale-watching businesses are equal: some have greater ecological and social impacts than others, while some of these may even be quite detrimental for both cetaceans and humans. However, this is rarely visible within the context of dominant E-NGO conservation discourses that celebrate whale watching as a fit for all solutions to both ecological and economic problems all over the world.

The Ecology of Business and the Business of Ecology: Two Models of Whale-Watching Enterprise

While conducting research on whale watching I have observed different business models for implementing and conducting this activity, two of which I analyze here.[21] Each business model reflects basic understandings of cetaceans and their ecosystems, as well as premises about the extent to which businesses should aim to secure the economic sustainability of maritime communities (see Neves-Graca 2004, 2006). What these models share at their core is the use of cetaceans to satisfy market demands for entertainment, amusement, learning and therapeutic catharsis. An implicit assumption is that whale-watching companies have the right to profit from putting humans in contact with the animals that provide these services. In short, all whale-watching businesses that I know of are capitalist enterprises, regardless of how much energy they put into justifying their *raison d'etre* as a means to achieve marine conservation.

Nevertheless, there are also big differences between distinct whale-watching business models and their social and ecological implications. In order to be able to compare these three different models I consider the following business aspects: market conditions; the type of tourists attracted; existing legislation and/or other factors that affect the behaviors and decisions of whale-watching operators. From the intersection of these aspects emerge different types of whale-watching business, each with very distinct ecological and social implications.

The first of the two models, based on high volume and mass consumption, is characteristic of whale-watching businesses in the Canary Islands. Because they offer access to large numbers of cetacean species, the Canaries were one of the first places where whale watching boomed. The large number of tourists that was already present on the islands, coupled with the absence of strict regulations and enforcement, created conditions under which the number of whale-watching operators boomed exponentially during the 1980s and 1990s. In other areas of the world highly informed ecotourists put pressure on whale-watching businesses to abide by ecologically informed handling of cetaceans. Mass tourism, by contrast, tends to seek ecologically uninformed tourists, looking for a quick thrill ride out to see cetaceans. Hence, zodiacs (with large outboard engines that produce high levels of underwater noise) are the preferred choice of business operators. To make matters worse, the large number of operators puts enormous competitive strains on these businesses to offer low-price tours to large numbers of tourists. This in turn requires them to increase the speed at which they conduct each tour.

The consequences of all this are most serious and for sure not what one would expect from a practice popularly seen as being essentially conservationist. Indeed, the Dolphin Fund warns ecotourists against going to the Canary Islands; even though "whale watching is regulated, harmful impacts occur due to the sheer number of operators. It is recommended to carefully select your operator as the majority of operators is to be considered unqualified".[22] Eric Hoyt says of the Canary Islands that:

> we are operating in the dark, with large numbers of boats in some areas spending large amounts of time with the same whales day in and day out. In some cases, the whales may be spending large parts of their day on a year-round or nearly year-round basis with boats of whale watchers.[23]

He adds that:

the whale watch participant may be spending most of the time watching other boats, rather than actually seeing the whales. It is not unusual to have 40 or 50 boats in the general vicinity of a group of whales, all vying for a look.

These numbers are striking in and of themselves, but they assume alarming relevance when one considers scientific findings concerning the effects that whale-watching boats have on pilot whales—one of the species most popularly observed in the Canary Islands (Glen 2003): "In the presence of one or two vessels, 28% of sightings involved avoidance behaviors, rising to 62% of sightings in the presence of three or more vessels".[24] One is left to guess the effect that 40 or 50 boats mentioned by Hoyt may have. Long-term consequences are particularly worrying, given that these are a resident species in the archipelago.

By the time that whale watching was introduced in the Azores, the Canary case was well known, and to a great extent it served as an example of the mistakes that should be avoided. This was in fact a core discussion during the first Biannual Conference on Whales and Dolphins held in Lajes do Pico in October 1998, which brought together business operators, E-NGOS, government authorities, and scientists (Neves-Graca 2002, 2004). As a consequence, one of the main concerns in the Azores at the inception of whale watching was to carefully regulate this activity in order to avoid the presence of excessive numbers of boats in the vicinity of cetaceans. These concerns are reflected in the legislation that was eventually approved by the Azorean parliament.

While it has been difficult for the Azorean authorities to invigilate and enforce the implementation of whale watching, several factors have contributed to prevent the Canary Islands scenario. First, there is an inherent bottleneck in relation to the numbers of tourists that reach the Azores every year. Flight availability is limited as are the number of hotels where tourists can stay. In addition, tourism in the Azores is highly seasonal and restricted mostly between May and October. For these reasons, mass tourism never developed. Moreover, Azorean legislation limits the number of whale-watching licenses that can be issued. Most tourists who come to the Azores for whale watching are attracted precisely because of the absence of mass tourism. Many are highly informed about the potential harm of cetourism to cetaceans, and put pressure on operators to abide by regional law and international whale-watching guidelines. Still, not all aspects of cetourism practice in the Azores are above reproach. Between 1998 and 2000 (ie the amount of time that elapsed between the presentation of the first draft of whale-watching legislation and its approval in parliament) there was a big debate as to whether some areas of the Azorean waters were populated by resident whales and whether these constituted nursery areas. If so, then special precautionary measures would have to be introduced

(Neves-Graca 2004, 2006). This would include keeping greater distances from cetaceans, and using inboard engines to mitigate underwater noise pollution. The debate was never resolved, though there was sound evidence in support of the nursery area argument (Monteiro 1998). One of the main unstated worries at the time was that this would result in lower profits. First, in spite of high levels of ecological knowledge, many tourists desire closer encounters with whales than a precautionary approach would allow. Second, the return on investment on zodiacs is faster than on inboard boats.[25] The debate was put aside without relevant research ever being conducted, and part of the Azorean marine ecosystem thus became invisible due to profit concerns.

The Azorean case also raises some doubt about presumed increased community-level benefits. This is evident in Lajes do Pico, where I conducted most of my research, and which is also the most important cetourism destination in the archipelago. While cetourism has attracted large numbers of tourists to the island, the first few people who invested in whale watching were able to benefit from an advantageous position where they quickly made profits that allowed them to invest in parallel economic activities (horizontal integration in business parlance).

The same company that started whale watching in Pico has bought the main hotel, opened a restaurant, and created a brand that now includes all sorts of whale watching memorabilia ranging from plush renderings of cetaceans, to clothing, videos, books and art. As a result, much of the money that cetourists spend stays mostly in the hands of this company and a rival company that was founded at around the same time. Finally, because of its seasonality, most people who work in whale watching are forced to live on welfare during the off season. This includes not only those who work directly for whale-watching operations but also for the hotels and restaurants. Thus, while whale watching has brought increased flows of capital to the Azores, it is unclear the extent to which the entire community has benefited, especially in a sustainable way.

Concluding Thoughts

The emergence and rapid proliferation of whale watching offers a prime example of one of the roles that conservation has in the production of nature in late capitalism. Many economic investors, politicians and conservation organizations now celebrate the economic use of nature services as *the* solution to the world's most pressing environmental problems. This undeniably attractive idea is aggressively promoted by E-NGOs concerned with the protection of cetaceans and their ecosystems. While there is in principle nothing wrong with this notion, it does become a problem when it inhibits a more critical assessment of distinct practices pursuant of ecological goals.

In the case of cetourism, diminished ability to assess the ecological soundness of different practices stems from establishing a clear-cut dichotomy between whale hunting and whale watching. As I have shown, this discursive dichotomy creates potential for blind spots whereby E-NGOs fail to see the continuity of the capitalist logic that links the two activities historically. In failing to see this continuity, they risk being unable to acknowledge the extent to which the goals of most cetourism businesses are in essence capitalist goals which at times may collide with conservation objectives.

The reality is that in the context of a late capitalist society cetaceans have undergone a very peculiar process of commoditization where they are no longer the source of material commodities but rather the providers of services. Cetacean conservation is increasingly intermingled with the highly profitable business of cetourism. The power of these interests silences alternative voices and makes it all the more important for research on the ecological effects of whale watching to be conducted in different contexts and conditions (as called for by Higham and Lusseau 2007), and for E-NGOs to intervene in cases of poor ecological conduct. Not all cetourism is alike, but distinctions will remain invisible so long as whale watching is presented and approached in homogenized form. The IWC has called for further research in this direction. In the spirit of this initiative, I have sought to show why such research is so fundamentally important and to invite E-NGOs to consider developing more complex presentations of cetourism for the average consumer: acknowledging that the masses are capable of understanding less simplistic messages would do great service to the conservation cause.

Acknowledgements

I would like to thank Jim Igoe for encouraging me to continue to develop my earlier work on human–cetacean relations in the context of biodiversity conservation; it has opened new and fascinating doors in the context of my intellectual landscape. I am also extremely grateful to Rosaleen Duffy and Dan Brockington for having invited me to the conference where this chapter was originally presented. Finally, I am enormously grateful to the anonymous reviewers of this chapter and the generous manner in which they helped fine-tune and improve its arguments.

Endnotes

[1] I call whale watching "cetourism" in this chapter to emphasize the use of cetaceans as economic resources where the logic of neo-liberal markets and contemporary conservationism intersect.

[2] In the Azores, for example, the number of tourists going out on whale-watching trips has grown 400% since 1999 (Incentivo 2 August 2009).

[3] For other examples of instances where whale hunting and whale watching go far beyond the more narrow terms of E-NGO discourse, see Evans (2005) http://www.sicri. org/ISIC1/h.%20ISIC1P%20Evans.pdf; Moyle and Evans (2008) http://www. shimajournal.org/issues/v2n1/f.%20Moyle%20&%20Evans%20Shima%20v2n1.pdf

[4] The IWC is now calling into question these assumptions and is planning to produce a more rigorous assessment of the diverse ecological impacts of different types of whale watching around the world.

[5] In no way do I wish to diminish the devastating effects of whale hunting that brought most whales to a near extinction status—this particular aspect bears no comparison to the impacts of whale watching. The point rather is to show that the uncritical acceptance of fundamental differences between these two activities is at the root of the conditions that render the negative aspects of whale watching invisible.

[6] My knowledge of these models is based on my first-hand fieldwork experience in the Azores, interviews with experts in the field, and interactions with marine biologists at the Biannual Conference on Whales and Dolphins in the Azores in 1998.

[7] I will not be able to provide a critique of assumptions regarding the higher profitability of whale watching in this chapter, but I would like to point out that it is highly problematic to make such claims without engaging in a comparative economic analysis of these two commercial activities—including with due inflationary corrections.

[8] http://www.greenpeace.org.uk/oceans/solutions/whale-watching-overview (accessed 16 June 2008).

[9] Again it must be pointed out that this understanding of the use of cetaceans for research purposes is part of wider historical processes in the context of paradigm shifts in biology. During the whale hunting period, and even well into mid twentieth century, marine biologists—as biologists in general—relied heavily on the dissection of whales as their main source of data on these animals. Thus, marine biologists were often found conducting research in places where whaling was still practiced (see Neves-Graca 2004, 2006). The discourse that promotes research among living cetaceans is therefore a reflection of recent historical transformation whereby the interests of whale-watching enterprises are as aligned with those of scientists as were the interests of whale hunters and those of classic modern marine biologists (Neves-Graca 2004, 2006).

[10] http://www.greenpeace.se/norway/english/9camp/4whales/index944.htm (accessed 13 July 2008).

[11] http://www.ifaw.org/ifaw/general/default.aspx?oid=230318 (accessed June 2008).

[12] http://www.panda.org/about_wwf/where_we_work/europe/what_we_do/arctic/what_we_do/species/whales/watching/index.cfm (accessed 11July 2008).

[13] http://www.panda.org/about_wwf/where_we_work/europe/what_we_do/arctic/what_we_do/species/whales/index.cfm (accessed 11 July 2008). The use of the term overfishing is ambiguous in this context because it is unclear whether it refers to the overfishing of whales or of the fish that cetaceans prey on. This is because for many years whaling was called "whale fishery" due to a dispute as to whether cetaceans are fish or mammals.

[14] http://www.worldwildlife.org/travel/item7716.html (accessed 3 July 2008).

[15] http://www.worldwildlife.org/travel/item7716.html (accessed 3 July 2008).

[16] I find it important to note that translated literally from German, the word is actually composed of two words as "material (nutrient) and swapping" which places great emphasis on the dynamic aspects of the process.

[17] For further critiques of Foster and for a rebuttal by Burkett, please see June 2000 edition of *Capitalism, Nature and Socialism.*

[18] There was also a temporary resurgence in the global demand for spermaceti during the two world wars and during the Korean War.

[19] See, for example, IFAW's review of whale watching in Vava'u Tonga produced in conjunction with Tonga's SPREP PROE.

[20] http://www.ngo.grida.no/wwfap/whalewatching/about_wwf.shtml (accessed 16 June 2008).

[21] I have also identified a third whale-watching model which is best illustrated by the Hawaiian case. Indeed, Hawaii has created a whale sanctuary where whale-watching regulations stipulate that this activity must be first and foremost concerned with the welfare of whales, and only then with profit.

[22] http://www.dolphinfund.eu/en/whalewatching/spain-canary_islands.htm (accessed 12 June 2008).

[23] http://whale.wheelock.edu/archives//ask03/0131.html (accessed 23 June 2008). It is important to note that at the time Eric Hoyt wrote this reply to a query, he also called for a precautionary approach to whale watching given that studies on the long-term effects of this activity on cetaceans were not yet known.

[24] See http://www.cms.int/reports/small_cetaceans/data/G_macrorhynchus/g_macrorhynchus.htm (accessed 14 July 2008).

[25] This argument is based on a business model analysis that was done by Dr Guerra in 1996 for a local whale-watching company (*Baleias a Vista*) and which was shared with me by two of the company's owners.

References

Benson A J and Trites A W (2002) Ecological effects of regime shifts in the Bering Sea and eastern North Pacific Ocean. *Fish and Fisheries* 3:95–113

Benton T (1989) Marxism and natural limits: An ecological critique and reconstruction. *New Left Review* 178:51–86

Blok A (2007) Actor-networking ceta-sociality, or, what is sociological about contemporary whales? *Distinktion: Scandanvian Journal of Social of Social Theory* 15:65–89

Brockington D, Duffy R and Igoe J (2008) *Nature Unbound: Capitalism and the Future of Protected Areas*. London: Earthscan Publishers

Bullen F T (1902) *A Whaleman's Wlfe*. London: Hodder and Stoughton

Burkett P (1999) *Marx and Nature*. New York: St Martin's Press

Carwardine M (2007) *Animal Records*. London: Natural History Museum

Castree N (2000) Marxism and the production of nature. *Capital & Class* 72:5–33

Castree N (2007) Neoliberalizing nature: The logics of de- and re-regulation. *Environment and Planning A* 40:131–152

Cawardine M, Hoyt E, Fordyce R and Gill P (1998) *Whales and Dolphins: the Ultimate Guide to Marine Mammals*. London: Harper Collins

Dana R H Jr (2001 [1840]) *Two Years Before the Mast: a Personal Narrative of Life at Sea*. New York: The Modern Library

Dolin E J (2007) *Leviathan: the History of Whaling in America*. New York: W W Norton and Company

Duffus D (1996) The recreational use of grey whales in southern Clayoquot Sound, Canada. *Applied Geography* 16:179–190

Einarsson N (1993) All animals are equal but some are cetacean. In K Milton (ed) *Environmentalism: The View from Anthropology* (pp 73–84). London: Routledge

Epstein C (2003) World wide whale: Globalisation and a dialogue of cultures? *Cambridge Review of International Affairs* 16(2):309–322

Epstein C (2006) The making of global environmental norms: Endangered species protection. *Global Environmental Politics* 6(1):47–67

Epstein C (2008) *The Power of Words in International Relations: Birth of an Anti-Whaling Discourse*. Cambridge, MA: The MIT Press

Evans M (ed) (2005) *Refereed Papers from the 1st International Small Island Cultures Conference*. Sydney: Small Island Culture Research Initiative

Foster J B (1999) Marx's theory of metabolic rift. *Annual Journal of Sociology* 105(2):366–405

Foster J B (2000) *Marx's Ecology: Materialism and Nature.* New York: Monthly Review Press

Francis D (1990) *A History of World Whaling.* Markham, Ontario: Viking

Freeman M (1994) *Gallup on Public Attitudes to Whales and Whaling.* In High North Alliance (ed) *11 Essays on Whales and Man.* Reine. Norway: High North Alliance http://www.highnorth.no/Library/Opinion/ga-on-pu.htm (accessed 10 May 2010)

Freeman M (1996) Polar bears and whales: Contrasts in international wildlife regimes. *Issues in the North* 1(40):174–181

Freeman M (2001) Is money the root of the problem? Cultural conflict in the IWC. In R Friedheim (ed) *Towards a Sustainable Whaling Regime* (pp 123–146). Seattle: University of Washington Press and Edmonton: Canadian Circumpolar Institute

Glen R (2003) "Behavioral responses of the short-finned pilot whale, Globicephala macrorhynchus, in relation to the number of surrounding whale-watching vessels." Paper presented at the Annual Meeting of the European Cetacean Society, Tenerife, Spain

Greenpeace (nd) *Iceland Whaling.* http://archive.greenpeace.org/whales/iceland/Whale Watching.htm

Hohman E P (1928) *The American Whaleman: A Study of Life and Labor in the Whaling Industry.* New York: Logmans, Green and Co

Heller P (2007) *The Whale Warrior: The Battle of the World to Save the Planet's Largest Mammals.* New York: Free Press

Higham J E S and Lusseau D (2007) Urgent need for empirical research into whaling and whale watching. *Conservation Biology* 21(2):554–558

Hildebrand J (2005) Impacts of anthropogenic sound. In P Reynolds, M Reeves and T J Ragen (eds) *Marine Mammal Research: Conservation Beyond Crisis* (pp 125–136). Baltimore: The John Hopkins University Press

Hoyt E (1998) Watch a whale: Learn from a whale. Enhancing the educational value of whale watching. *Proceedings, Swan Festa, Third International Whale Watch Forum* (pp 5–19), July 1998. Muroran, Japan

Hoyt E (2001) *Whale Watching 2001: Worldwide Tourism Numbers, Expenditure, and Expanding Socioeconomic Benefits.* Yarmouth: International Fund for Animal Welfare

Hoyt E and Hvenegaard G (2002) A review of whale watching with implications for the Caribbean. *Coastal Management* 30:381–399

Igoe J, Neves K and Brockington D (2010) A spectacular eco-tour around the historic bloc: Theorizing the convergence of capitalist expansion and biodiversity conservation. *Antipodei* 42(3):486–512

Incentivo (2009) A Evolucao do Whale Watching nos Acores [The evolution of whale watching in the Azores]. *Incentivo* August:1–4

IWC (2008) Scientific Committee Report on Whale Watching. *Journal of Cetacean Research and Management*, http://www.iwcoffice.org/sci_com/screport.htm (last accessed 30 June 2008)

Kalland A (1994a) Whose whale is that? Diverting the commodity path. In M Freeman and U Kreuter (eds) *Elephants and Whales: Resources for whom?* (pp 159–186). London: Gordon and Breach

Kalland A (1994b) Super whale: The use of Myths and symbols in environmentalism. http://www.highnorth.no/library/myths/su-wh-th.htm (last accessed 4 May 2010)

Lusseau D, Slooten L and Currey R (2006) Unsustainable dolphin watching tourism in Fiordland, New Zealand. *Tourism and Marine Environment* 3(2):146–163

Marx K (1952) *Capital: A Critique of Political Economy in Three Volumes.* Chicago: Encyclopedia Britannica

Mawer G A (1999) *Ahab's Trade: The Saga of South Seas Whaling.* New York: St Martin's Press

Monteiro L (1998) Open letter on the harmful effects of whale watching. Sent to all the local mass media, the Azorean government and the University of the Azores. Personal copy obtained by the author in 1998

Mowat F (2005 [1972]) *A Whale for the Killing.* Mechanicsburg, PA: Stackpole Books

Moyle B and Evans M (2008) Economic development options for island states: The case of whale watching. *International Journal of Research into Island Cultures* 2(1): 41–58

Neves K (2007) Elementary methodological tools for a recursive approach to human–environmental relations. In J Wassmann and K Stockhaus (eds) *Experience New Worlds* (pp 146–164). Oxford: Berghan.

Neves K (2009) The great Wollemi saga: Betwixt genomic preservation and consumerist conservation. In C Casanova (ed) *Contemporary Issues in Environmental Anthropology* (pp 213–241). Lisbon: ISCSP University Press

Neves-Graca K (2002) *A Whale of a Thing: Transformations from Whale Hunting to Whale Watching in Lajes do Pico.* Doctoral Dissertation, York University, Canada

Neves-Graca K (2004) Revisiting the tragedy of the commons. *Human Organization* 63(2):289–300

Neves-Graca K (2006) Politics of environmentalism and ecological knowledge at the intersection of local and global processes. *Journal of Ecological Anthropology* 10:19–32

O'Connor J (1988) Capitalism, nature, socialism: A theoretical introduction. *Capitalism, Nature, Socialism* 1:11–38

Orams M (2000) Tourists getting close to whales, is it what whale-watching is all about? *Tourism Management* 22:281–293

Paine R T (2006) Whales, interaction webs, and zero-sum ecology. In J Estes, D DeMaster and T Williams (eds) *Whales, Whaling, and Ocean Ecosystems* (pp 7–13). Berkeley: University of California Press

Pearce F (1991) *Green Warriors—The People and Politics behind the Environmental Revolution.* London: Bodley Head

Richter C, Dawson S and Slooten E (2006) Impacts of commercial whale watching on male sperm whales in Kaikoura, New Zealand. *Marine Mammal Science* 22(1):46–63

Robertson R B (1954) *Of Whales and Men.* New York: Alfred A Knopf

Tripp W H (1938) *There She Flukes.* New Bedford, Massachusetts: Reynolds Printing

Watson P (1996) *Ocean Warrior: My Battle to End Illegal Slaughters on the High Seas.* Toronto: Key Porter Books

Chapter 11
Neoliberalising Nature? Elephant-Back Tourism in Thailand and Botswana

Rosaleen Duffy and Lorraine Moore

Introduction

In this chapter we explore the neoliberalisation of nature debate by examining the elephant-riding tourism industry in Botswana and Thailand. The spectacular growth in the global tourism industry has been one of the core drivers of neoliberalism in the last 20 years. It constitutes one of a number of global processes that allows neoliberal norms and values to travel over time and space (Castree 2009). Despite claims that alternative tourisms such as ecotourism, responsible tourism and nature-based tourism offer a challenge, our analysis of elephant-back safaris demonstrates that they have been central to the expansion and deepening of neoliberalism at a global scale.

Within the debate about the neoliberalisation of nature there have been calls for more research on "actually existing neoliberalisms" (Brenner and Theodore 2002; see also Castree 2008b) and the purpose of this chapter is to do just that. Brenner and Theodore (2002) note that there is a tendency in the literature on neoliberal nature/neoliberal environments to assume that neoliberalism is hegemonic, and therefore it is invested with greater powers than it really has (see also Peck and Tickell 2002). In this chapter we interrogate the elephant-riding industry across two different contexts to determine what is precisely neoliberal about the industry, and are the dynamics of neoliberalism hegemonic or challenged, reshaped or resisted by existing context-specific processes? We do this by asking four questions applied to two different contexts (Botswana and Thailand). Where does the industry obtain elephants from and what are the patterns of ownership? How are the elephants used as a workforce in the tourism industry? Are the captive elephants valued and used in other ways apart from their role as a workforce for global tourism?

And finally, what are the welfare implications of using elephants as a workforce?

In answering these questions, we make four arguments about the role of the tourism industry in processes of neoliberalising nature. Firstly, that the global tourism industry is not simply reflective of global neoliberalisation, but is in fact an important constitutive element which expands and deepens processes of neoliberalisation, especially in the South. Secondly, that the tourism industry is one means by which nature is neoliberalised, since it allows neoliberalism to target and open up new frontiers in nature (Castree 2008a:141; Castree 2009). Neoliberalism, through tourism, reconfigures and redesigns nature for global consumption (West and Carrier 2004). Thirdly, examining the ways trained elephants are used in the tourism industry in two very different contexts allows us to analyse the variations in "actually existing neoliberalisms" (Brenner and Theodore 2002) and thereby demonstrate that the effects are not unremittingly negative (Castree 2008b:166). Fourthly, such a cross comparison highlights the ways that neoliberalisation is challenged, resisted and changed by context-specific processes, values, ideas and institutions. In essence, the elephant-back tourism industry demonstrates that neoliberalisation of nature through the global tourism industry produces a kind of "palimpsest" effect; it inscribes new values and uses for nature so that it can be opened to international markets, but does not completely obscure or obliterate existing ways of valuing, using, owning and approaching nature.

Neoliberalisations, Nature and the Global Tourism Industry

We analyse the growth of elephant-back tourism in two different contexts (Thailand and Botswana) in order to critically examine the neoliberalisation of nature debate. Defining neoliberalism itself and identifying what is especially "neoliberal" about elephant-back tourism is no easy task. Debates about the precise nature of neoliberalism are already well covered in the literature. Castree (2008a, 2008b) points out there is a large and growing body of critical scholarship on the neoliberalisation of nature, which aims to understand why the natural world has become such a vital target (see Castree 2009). For example, McCarthy and Prudham argue the connections between neoliberalism, environmental change and environmental politics are deeply, if not inextricably interwoven. In this way neoliberalism can be regarded as an inherently and necessarily environmental project because it changes the relationships between human communities and biophysical nature; neoliberalism and environmentalism have emerged as powerful ideological foundations for social regulation; and finally because environmental concerns are the most powerful source of opposition to neoliberalism (McCarthy and Prudham 2004:275–277; see also Heynen et al 2007).

First it is important to define what neoliberalism is in order to understand what neoliberalisation of nature means. Neoliberalism can be briefly defined as a specific form of capitalism which is privatisation, marketisation, deregulation and various forms of re-regulation. Critical scholars defined it as a hegemonic project that produces a "nebuleuse" of ideas, institutions and organisations which create conditions favourable to neoliberalism so that it appears as natural, neutral and as if there were no alternative (see Cox 1996). During the past 20 years we have witnessed the global expansion of neoliberalism, including the roll back of states coupled with a roll forward of new forms of regulation to facilitate private interests, the expansion of market-based mechanisms to new natural resources such as water and genetic material, as well as the privatisation of public services (Castree 2008a; see also Harvey 2005; Heynen et al 2007; Heynen and Robbins 2005; Liverman 2004; McCarthy and Prudham 2004:275–277; O'Neill 2007; Peck and Tickell 2002). Neoliberal approaches rose to international prominence during the 1970s and by the mid 1990s they seemed globally dominant, pushed forward by a range of global actors including states, corporations and International Financial Institutions (IFIs) (Brenner and Theodore 2002; Cox 1996).

However, it is critically important not to reify neoliberalism and ascribe it a greater level of coherence and dominance than it really deserves (Bakker 2005; Castree 2008a; Brenner and Theodore 2002; Mansfield 2004; McCarthy and Prudham 2004). Instead it is important to interrogate how neoliberalism plays out "on the ground", to probe its complexities, unevenness and messiness (see Peck and Tickell 2002). In this chapter we concentrate on comparing the practices of neoliberalism in order to draw out these messy entanglements; this demonstrates how neoliberalism can be challenged, resisted and changed by its encounter with nature (Bakker 2009; Castree 2008b:161). Therefore, we do not rehearse the well worn debates on definitions of neoliberalism, but rather take up the challenge of comparative research on "actually existing neoliberalisms", which involves engaging with contextual embeddedness in order to complicate neat theoretical debates. As Brenner and Theodore (2002:356–358) suggest, to understand actually existing neoliberalism we must explore the path-dependent, contextually specific interactions between inherited regulatory landscapes and emergent forms of neoliberalism. As such, the neat lines and models generated via theoretical debates can be traced, refined, critiqued and challenged through engagement with specific case studies (Bakker 2009; Castree 2008b). We examine how nature-based tourism relies on the neoliberalisation of nature through the transformation of experiences with elephants.

Tourism, as a global industry, is not just reflective of neoliberalism, it drives, expands and deepens it. Tourism has experienced a sustained period of growth; this occurred from the 1970s onwards against a backdrop of global shifts and especially benefited from the expansion of neoliberalism across the world. Tourism development fitted very well with the new faith in markets, decentralisation and roll-back of the state. Since the late 1970s global tourism flows have rapidly increased in response to greater prosperity and social and economic shifts in the industrialised world, which allowed larger numbers of people to engage in overseas travel. This has further developed into markets for ethical/responsible/green travel that reflect and draw on the changing holidaying tastes of societies in the North (for further discussion, see Butcher 2003). Statistics on international tourism clearly demonstrate how much it has grown, and despite warnings of the impact of a global recession, it is still a healthy industry. The UN World Tourism Organisation (UNWTO) figures show that international tourist arrivals grew at 3.7% between January and August 2008, compared with the same period in 2007; international tourism receipts grew to US$856 billion, an increase in real terms of 5.6% on 2006; and international tourist arrivals were 903 million in 2007, up 6.6% on 2006. However, the first quarter of 2009 indicated a sharp drop in international tourist arrivals, down by 22% on the same period in 2008. The UNWTO states that the reduction in numbers is due to fears about an international influenza pandemic coupled with the global financial crisis. Nevertheless, the UNWTO still believes the industry is robust and estimates there will be 1.6 billion visitor arrivals in 2020.[1] Tourism has remained a highly attractive option for governments, the private sector and international organisations as a potential means of delivering economic growth, and even "development".

The wider context of growing international tourism has led to the promotion of nature-based tourism as a key policy agenda for IFIs, national governments, private sector and international environmental non-governmental organisations (NGOs). As a result, tourism has been identified as a strategy by which many states in the South can diversify their economies and produce environmentally sustainable development (Bramwell and Lane 2005:53). Furthermore, the UN declared 2002 the International Year of Ecotourism, a clear reflection of the global expansion of tourism and the expectations placed on it. The World Tourism Organisation also claims that ecotourism can contribute to conservation of natural and cultural heritage in natural and rural areas, as well as improving living standards in those areas (UNWTO 2003:2). Elsewhere Duffy has argued that there is little difference between various forms of "alternative tourism" (such as ecotourism) and mass tourism: they are both part of the same continuum and heavily interlinked with global capitalism through their reliance on international markets (Duffy 2008). One of the core justifications for nature-based tourism is that nature can be conserved or saved precisely because of its "market value" to tourists willing to pay to see and experience it. While supporters of tourism development argue that natural resources, landscapes and wildlife have intrinsic, cultural and ecological values, they also point to their economic value, which can be harnessed through the application of market-based mechanisms (see McAfee 1999). In effect, wildlife and landscapes are produced, reproduced and redesigned as tourist attractions. In the process they are commodified and drawn in to the global tourism marketplace as products to be consumed (see West and Carrier 2004). The tourism industry is particularly adept at designing and creating new commodities that clients will pay to see or experience. It relies on the transformation of places into desirable "must see" locations, and the development of new "must do" activities that people will be willing to pay to experience. This includes the production of new sensory experiences centred on close encounters with animals (for further discussion, see Bulbeck 2004).

Comparing Neoliberal Nature in Thailand and Botswana
This chapter is based on fieldwork conducted during 2008 with a total of 3 months' fieldwork in Botswana and 3 months in Thailand. The research primarily used qualitative methods as the most appropriate to uncover the complexities of the elephant-riding industry in both countries. We conducted 75 interviews in Kasane, Maun and Gaborone (Botswana) plus Bangkok and Chiang Mai (Thailand). Kasane, Maun and Chiang Mai are the main areas for elephant-related activities, whereas undertaking research in the capital cities of Gaborone and Bangkok allowed us to access the relevant wildlife, conservation and tourism authorities associated with each country. Interviewees were selected from a diverse range of regulatory authorities, wildlife and human rights NGOs, tour operators, lodge/camp owners, tourist guides and the tourists themselves. These interviews were supplemented by participant observation at tourist sites, such as elephant shows and elephant riding tours which allowed us to observe how tourists interacted with elephants (eg riding, washing, feeding, playing with them and watching them). The interviews and observations are also supported by analysis of documentary evidence from NGOs, national tourism authorities and tour operators.

Bakker (2009) argues it is important for research on neoliberal nature to take up the challenge of comparative research; in line with Bakker's argument we examine particular liberal environmentalist strategies (nature-based tourism) as applied to a specific resource (captive/trained elephants). This kind of comparative research allows us to examine the contours, boundaries, challenges and limitations placed on neoliberalism by its encounter with "nature". Thailand and Botswana provide interesting but contrasting contexts in which to research the character of neoliberal nature. Thailand has a long history of training and using elephants as a workforce: elephants have an important cultural-heritage value and the tourism industry has expanded to become a major income earner for the country. In contrast, Botswana has no history of using trained elephants as a workforce. As training elephants is an emerging activity in Botswana, there is no cultural heritage associated with the practice. The tourism industry in Botswana is based around the wildlife safari industry and has been expanded by the government so that it is now a critical economic sector (Mbaiwa 2004). The different trajectories of development of elephant riding in Thailand and Botswana provide us with an interesting set of questions around the role of the global tourism industry as a driver of the expansion of neoliberalised nature.

Neoliberalising Nature: Elephant-riding Tourism

Elephant Trekking in Botswana

Botswana, formerly the British Protectorate Bechuanaland, gained independence in 1966. It is often presented as one of Africa's success stories, demonstrating the benefits of following a neoliberal development model centred on diamond mining and luxury safari tourism. During the 1980s sub-Saharan Africa was the site of a series of external interventions inspired by donors and IFIs which developed and promoted economic liberalisation packages. This was followed in the 1990s by a decade of political conditionalities centred around ideas of good governance (Clapham 1996). Botswana was not immune from these global shifts and Melber (2007:5) points out that Botswana is "currently widely considered as a relative success story in terms of democratisation and 'good governance' ...". Botswana is often identified as a model African democracy, and a beacon of economic development in a region with high rates of poverty and underdevelopment. However, Good (2008) argues that Botswana is dominated by a ruling party which revolves around an elite clique. He further states that the immense wealth produced by diamond mining (and by the tourism sector) has not led to greater prosperity for all, but instead the country is marked by serious disparities of wealth (also see Taylor and Mokawa 2003; and Poteete 2009a, 2009b).

Ideas of democratisation and decentralisation of natural resource management were also taken up by the Ministry of Environment and Tourism, and implemented via community-based approaches to conservation (see Twyman 2001). The Community Based Natural Resource Management (CBNRM) approach developed during the 1990s as a critically important, more effective and more community-oriented means of managing natural resources (see Hutton, Adams and Murombedzi 2005). The notion that communities can manage natural resources and develop ecotourism fits well with neoliberal approaches to regulating, organising and implementing conservation that include extension of the market as the most efficient manager of natural resources. In particular, it intersects with the argument that decentralised networks of "stakeholders" can govern resources rather than leaving them in state hands (Hulme and Murphree 1999; Ribot 2004). This resonates with the agendas of IFIs and NGOs that claim to engage in participatory methods of development and conservation with local communities.

Botswana also promoted CBNRM as a potential solution to conservation. The decentralisation of natural resource management to local communities became a central pillar of environmental policies. This intersected with the rise in faith in markets as a means of "producing" development in Botswana and in sub-Saharan Africa more generally. Under the Natural Resources Management Project, rural communities were encouraged to enter into partnerships with private safari operators, with promises that such links would result in better wildlife management and would bring material benefits (Hoon 2004; Mbaiwa 2004). However, as Hoon (2004) points out, these new market-oriented, decentralised and community-oriented policies did not obliterate existing social dynamics; instead they opened up new kinds of interactions. The results of CBNRM were mixed in Botswana. While some communities derived benefits from their partnerships with safari operators, critics pointed to the lack of genuine participation, the limited economic benefits and the conflicts it produced within communities (Hoon 2004; Mbaiwa 2004, 2008; Twyman 2001).

Tourism is one of the largest income earners for Botswana,[2] and it has been identified by the government as a critical sector for sustained economic development in the future (Keitumetse 2009; Mbaiwa 2004). Wildlife-based tourism has been promoted by a range of actors, including the government, private sector, international conservation NGOs and IFIs. It has been developed as a potential driver for economic development, environmental sustainability, wildlife conservation and community empowerment, among others. The National Ecotourism Strategy for Botswana (drawn up in 1990) promotes wildlife/wilderness as the central "brand" and attraction for the country. However, Mbaiwa notes that the upscale luxury safari tourism industry in Botswana produces similar problems to enclave tourism in the Caribbean and elsewhere; namely that the main profits go to foreign companies, there is little benefit for local communities and the lodges/safari concessions are highly exclusive and potentially exclusionary (Mbaiwa 2004, 2008).

One interesting twist in the wildlife/wilderness safari industry is the development of "elephant-back safaris". It is one small, but growing, part of the southern African tourism industry product and relies on the development of close interactions with elephants, including elephant riding and walking with elephants. The development of elephant riding is the combined result of entrepreneurs such as Randall Moore (owner of Elephant Back Safaris) and Uttum Corea (the owner of the Mokolodi Nature Park), seeking to create niche forms of tourism at a time when Botswana wanted to diversify its tourism package. The way in which elephant-back riding has developed in Botswana (as well as the wider southern African region) is strongly related to the wider preference for market-based approaches to natural resource management.

Elephant trekking in Botswana accounts for less than 1% of the total elephant population in Botswana, which is one of the largest in sub-Saharan Africa. In stark contrast to the case of Thailand, discussed later in this chapter, the Botswana Department of Wildlife and National Parks estimates that it has 152,000 wild elephants.[3] Trained elephants are used for transport, essentially as safari vehicles for tourists. But encounters with elephants are also marketed and promoted as a "back to nature" experience, despite the obvious contradiction that tourists are interacting with a trained animal, not one of its "wild" counterparts. Historically, experiences with captive trained elephants are not part of the standard safari package, but have been developed as an additional high-end/luxury tourism product in the last decade, and can now be found in South Africa, Botswana, Zimbabwe and Zambia.

The ways that elephants are obtained for the elephant-riding industry is an important factor in understanding how non-human nature is neoliberalised. The absence of a history or culture of domesticating elephants in Botswana means elephant camps rely on elephants that formerly lived in circuses, zoos and safari parks, or as calves who survived elephant culling operations. For example, one of the reasons for developing an elephant trekking industry was because they had to find a new role for wild-caught calves that had survived culling operations, elephants that were no longer suitable to work in circuses, or were being kept in poor circumstances in zoos. One researcher based in Maun suggested that since they could not be returned to the wild, one way of dealing with them was to train them for work in the tourism industry.[4] The use of trained elephants in Botswana is limited to three privately owned safari camps and one publicly run nature reserve, Mokolodi (near Gaborone), and cannot be done without approval and monitoring by the Department of Wildlife and National Parks. The elephants have been purchased or donated to three camps in the Okavango Delta: Abu Camp and Seba Camp (both owned by Elephant Back Safaris, EBS) and Living With Elephants, which works with Stanleys Camp and Baines Camp. The District Wildlife Co-ordinator in Maun stated that although the Parks Department retains oversight of the captive elephants and had to give their permission for the elephants to be used in the camps, the captive elephants are privately owned.[5]

The elephants funnelled into the tourism industry were trained and developed as a labour force to be deployed for international tourism. By allowing these elephants to be removed from other captive situations and come together in newly formed and privately owned elephant camps, they have, in a sense, been re-packaged and re-regulated to occupy a space within the tourism industry to improve their circumstances, and to diversify the tourism product in Botswana. The status of these captive elephants as privately owned sets them apart from their wild counterparts because they are state owned. The unique nature of this situation does not end here though. Twyman (1998, 2001) argues that while Botswana has been committed to decentralisation of natural resource management, the actual process has been a real challenge. Yet, in the case of captive elephants, this has been achieved because wild elephants are state-owned, but captive elephants have been devolved to the private sector. As such, the trained elephants in Botswana can be characterised as fully neoliberalised because they exist only to serve international safari tourism markets and have little opportunity for employment outside tourism. Even their limited use as educational tools for local communities is inextricably linked to their use as a labour force in international tourism.

The experiences with elephants occupy a high-end niche market, and elephant rides cost at least US$150 per person per hour. In Botswana close interactions with elephants are sold as part of an accommodation package in fly-in luxury camps costing between US$500 and 2000 per night (see Mbaiwa 2004). It is possible for visitors to Botswana, however, to incorporate elephant trekking as a day trip during a mobile safari, or as an independent traveller in Kasane where they can take advantage of shorter elephant rides in Victoria Falls (on the Zambian and Zimbabwean sides).[6] There are key differences between the camps and elephant experiences. The camps are located in the Okavango Delta, which is the centre-piece of Botswana's "wilderness tourism" product.[7] Abu Camp is the most expensive luxury resort in the Okavango Delta— costing approximately US$2000 per night.[8] It is owned by a former animal trainer from America, Randall Moore, who brought the first trained elephants to the camp in 1990. Moore has hired his elephants out for film, television and advertising. For example, the camp is named after Abu, the original male elephant which was brought from the USA to South Africa to star in the film *The Power of One*. Since then the Abu Camp elephants have appeared in diverse formats including adverts for Cote d'Or chocolate, and when the Miss World Contest was held in South Africa the elephants were used as part of the show.[9] Apart from their media appearances, the elephants are used for wildlife safaris, when tourists can ride the elephants or walk with them while viewing wildlife. Abu Camp promotes the activity as a means by which tourists can then feel they are part of the elephant herd and will get closer to the wildlife. Abu Camp also has an active research programme into how domesticated elephants can reintegrate into wild herds, but this too is offered as a spectacle for tourists visiting the camp.[10]

In contrast, Baines Camp and Stanley's Camp offer walking with elephants, but are quite clear that they do not allow or promote elephant riding. EBS and Living with Elephants state that they are committed to returning captive and domesticated elephants to the wild. Baines and Stanley's Camps are exclusively for clients booked through Sanctuary Lodges—a division of the global safari company Abercrombie and Kent.[11] The elephants used at the camps are privately owned by Sandi and Doug Groves who run a company "Grey Matters" and an NGO "Living with Elephants" (LWE).[12] LWE was founded in 1999 with three adopted elephants, Jabu, Thembi and Morula. The elephants are used by the camp for tourist experiences but during the low season, LWE use their elephants to provide free educational tours for school children in the delta area. Sandi Groves of LWE suggested that the trips are to educate children into having a more "positive" view of elephants.[13] LWE's mission statement is that it is "dedicated to creating harmonious relationships between people and elephants. Doug and Sandi have striven to find ways in which their foster elephants can act as ambassadors for their wild counterparts." The key motivation for LWE is to "reduce competition between elephants and human populations in Botswana".[14] The use of captive elephants for educational purposes and school trips to the luxury lodges are one way in which the lodge and tour operators maintain their concession and operator licences. It is a government stipulation that all safari concessions must provide a package of benefits for local communities.[15] Mokolodi Nature Park offers similar educational activities as LWE, but also allows visitors to ride the elephants.

The elephant-back safari industry in Botswana is a good example of how nature-based tourism acts as a driver for global neoliberalism. The animals used in this "alternative" form of tourism are privately owned, and in contrast to their wild counterparts their management has been decentralised away from state authorities. In essence they have been privatised, re-regulated and re-packaged to be consumed by the global tourism industry. Wild elephants remain state owned, and are the core concern of state authorities. They are a central part of the promotion of Botswana as a leading safari tourism destination, a key strategy for economic development which mirrors the state approach to ownership and management of diamonds as a national resource (see Poteete 2009b). The approach to management and use of captive, trained elephants is rather different. While the state retains ultimate oversight of captive trained elephants, they are effectively under private sector management and have been re-configured and re-packaged for tourist use. Essentially, this is an example where resources become de-regulated and re-regulated by the state (see Bakker 2005; Brockington, Duffy and Igoe 2008; Castree 2008a; Heynen et al 2007). Captive elephants in Botswana exist only as a labourforce to work in the tourism industry. Therefore, the development of elephant-back safari tourism in Botswana acts as an important driver of neoliberalisation. The elephant-riding industry has targeted and opened up new frontiers in nature (elephants) for colonisation by neoliberalism.

Neoliberalisation of nature has been rendered more complex by locally specific contexts, histories and social processes. Understanding the precise patterns of ownership (a particular blend of state and private associated with de-regulation and re-regulation) and use of captive elephants in Botswana complicates the neat lines of theoretical frameworks used to understand the nature of neoliberalism. First, its effects are not unremittingly negative (Castree 2008b:166). Second, neoliberalisation of nature is not a neat process, easily implemented in standard ways; instead it plays out in particular ways on the ground. The challenges it faces from context-specific dynamics shape the character of neoliberalism in different places around the world. In the case of Botswana, the specificities of ownership of captive versus wild elephants and the lack of history of training and using elephants as work animals shapes the character of neoliberalised nature. This further develops McCarthy and Prudham's (2004:275–277) argument that environmental concerns are the most powerful source of opposition to neoliberalism (also see Heynen et al 2007). Our analysis of elephant-back tourism reveals the ways that existing approaches to nature encounter and mix with neolibralism to produce new forms on the ground in different contexts. This is made clear in the ways captive elephants are used as a labour force for the tourism industry in Thailand. In that context it is the history of using elephants as domesticated, working animals which shapes the precise character of neoliberalised nature.

Elephant Trekking in Thailand

Thailand's political history is rather complex and differs substantially from Botswana. It was never a colony but was governed by an absolute monarchy until Rama VII, King Prajadhipok was challenged by a military coup in 1932. Prajadhipok was invited to remain a constitutional monarch, he was succeeded by Ananda, and then by his brother, King Bhumibol Adulyadej, the current King of Thailand (Wyatt 2003). Since 1932 Thailand has effectively switched between military dictatorships and civilian governments. Despite this, since the end of the Second World War Thailand has largely followed a development trajectory which fits well with a global context of neoliberalism. Along with the other ASEAN-5,[16] Thailand aligned with the USA during the Cold War and followed a growth model influenced by modernist principles, including IMF reforms and the development of a US style consumer society modelled on the USA. Thailand went on to develop export-orientated industrialisation from the 1980s, and by 1995 74% of Thailand's exports were manufactured commodities. However, Thailand's economic success appeared much more fragile with devaluation of the Thai baht in 1997, which is often credited with triggering the wider Asian Financial Crisis. The crisis led to high levels of unemployment and economic recession, leading to further neoliberal reforms under the IMF (McGregor 2008). In the late 1990s the increasing powers of the IMF and of foreign capital were used as a focal point for anti-government campaigns by the Thai Rak Thai Party, established by the Thai billionaire, Thaksin Shinawatra; local big businesses was keen to protect their own interests and supported the election of Thaksin in 1997 (see McCargo and Pathmanand 2005 for further discussion). Economic growth returned to pre-crisis levels by 2002 and Thailand remains a lower middle-income country, but one characterised by extreme variations of wealth and poverty (McGregor 2008). Neoliberal approaches to the environment have been applied in Thailand since 1980. Government and independent think tanks, such as the Thailand Environment Institute, have focused on the "technocratic task" of developing economic instruments for "sound environmental management" and the right market signals (Hirsch 1997:23). Thai NGOs have embraced the neoliberal ideals of democracy and participation at the grassroots level, particularly in the forestry sector, which translated into community-based projects in rural areas.

The development of the tourism industry pre-dates the global expansion of neoliberalism, and the economic reforms inspired by the IMF in the 1980s and 1990s. The current form of international tourism developed out of the creation of "rest and recreation areas" for the US military during the Vietnam War, which themselves drew on a longer tradition of a sex industry in the country (see Enloe 2001; Peleggi 1996). By the late 1970s, the Thai government had identified tourism as a critically important sector for expansion, and made it a top priority in the 1978–1991 National Economic and Social Development Plan.

Between 1998 and 1999 the government aimed to attract 17 million tourists through the *Amazing Thailand* Campaign to counter the negative impacts of the Asian economic crisis (Kontogeorgopoulos 1999:317). The government was keen to diversify the tourism product and attract clients from the Asian region, especially Japan (see Enloe 2001:36, 43–65) as well as Europe and the USA (see Kontogeorgopoulos 1999).

The longer-term history of tourism development is important for understanding how elephant trekking occupies a particular niche within a wider industry that caters to a much wider range of interests than the Botswana tourism industry. Unlike Botswana's focus on high-cost–low-impact international safari tourism, Thailand is a well-worn tourist destination. Its product is highly varied, and includes standard packages of the three or four "s"s: sun, sea, sand and sex; it also markets itself as a destination for cultural tourism (especially to see Northern Hill communities), adventure tourism (eg scuba diving, sea kayaking), nature-based tourism (visiting national parks, wildlife viewing) (for further discussion of the profile of the Thai tourism industry, see Cohen 2008; Peleggi 1996).

There are 40–50 elephant camps in Thailand, and they form a critical part of the tourism industry in northern Thailand, especially around Chiang Mai (Kontogeorgopoulos 2009:431). Thailand's use of elephants in tourism differs from that in Botswana in a number of ways. For example, unlike Botswana, the longstanding cultural practices of owning and training elephants as work animals in Thailand means that it is much easier to acquire them as private property than in Botswana (see Lair 2004). In contrast to Botswana, Thailand has 3000 elephants, 2000 of which are privately owned trained elephants; only 1000 are "wild" animals living in national parks. The 2000 captive elephants were largely bred within captivity, but there are concerns that a small number of baby elephants have been wild caught and trafficked in from Burma to serve the demand for "cuter" animals in circuses and shows. The government authorities responsible for Thailand's captive elephants are the Department of Livestock, the Department of Transport, and the Forest Industry Organisation, rather than the Department of National Parks or Ministry of Environment. This serves to underline the ways that captive elephants are regarded as working animals rather than as wildlife. Each elephant is owned by someone or something. Lair (2004) identifies two broad categories of ownership: government and private. Lair further divides privately owned elephants into mahout-owned and non-mahout-owned. Non-mahout-owned elephants usually require a hired mahout to look after the elephant (for further discussion of ownership categories, see Lair 2004:15–30; also see Kontogeorgopoulos 2009:440).

These privately owned elephants have been deployed as an important labour force for the tourism industry, but unlike Botswana they have a series of other roles as working animals. Elephant trekking in Thailand is marketed in a variety of ways to attract the vast range of tourists that visit Thailand seeking different kinds of tourist experiences, unlike Botswana where elephant riding is a specific niche market within a dominant safari industry. In a sense, Thailand has reconfigured and mobilised the historical practice of using elephants as a labour force in the logging industry to attract international tourists. Elephant shows vary in form and include circus-style shows in the major package holiday destinations where elephants perform tricks including standing on their heads (such as Fantasea in Phuket) to demonstrate the historical skills used in the logging industry (stacking and pulling logs) and displays of elephants painting pictures, or the Elephant Orchestra where elephants play tunes on a glockenspiel (such as the elephant show at the National Elephant Research Institute in Lampang). Other common activities found in elephant shows (for example Mae Sa and Maetaman elephant camps) include elephants playing football/basketball and re-enactions of the trust between mahout and elephant where the elephant steps over a mahout lying on the ground.[17]

Some of the differences between neoliberalisation of nature in the Thai and Botswana tourism industries can be explained via the differing histories of using captive elephants as a labour force. Prior to 1989 elephants were primarily used in teak logging by the government of Thailand and by private operators. However, following the 1989 ban on logging the elephants in the logging camps faced an uncertain future. While a limited amount of logging still continued in government reserves, many elephants were put out of work by the decision in 1989. Each trained elephant requires a mahout to work with it—and without the mahout the elephant can be extremely dangerous. For the mahout the elephant is a source of income, but it eats between 120 and 200 kg of food per day. As the head of the Mahout Training School in Thailand pointed out, without access to work in the logging camps the elephants and their mahouts had to find alternative employment.[18]

Following the logging ban, some elephants were released, but the amount of land under national parks is not large enough to provide sufficient habitat for all the elephants in Thailand. Since elephants are a key source of income for their mahouts, the majority had to look for alternative forms of employment (for further discussion, see Kontogeorgopoulos 2009). However, this proved to be difficult. In the initial years after the ban some elephants were funnelled into the illegal logging industry. One of the vets working with in the National Elephant Hospital suggested that a major concern at the time was that elephants were often working long hours and during the night to avoid detection, and there were numerous accounts of elephants being given amphetamines to make them work faster.[19] The head of the Training School for Thai Elephants and Mahouts pointed to a second problem that arose after the ban: it created the phenomenon of "street wandering elephants"—unemployed elephants and mahouts who came to the big cities and tourist areas to beg on the streets.[20] In Bangkok the problem was especially acute because of the dangers posed to traffic and pedestrians. At the time, the Food and Agricultural Organisation (FAO) and the Forest Industry Organisation expressed concerns regarding elephant welfare: they were working long hours (including at night) they were being frightened by the noise and lights of city traffic, and walking on tarmac and concrete is painful and leads to potentially crippling disorders in the elephants' feet.[21]

Once the government banned elephants from urban areas, there was a need to find alternative forms of employment for the elephants and mahouts. This is where the tourism industry became the potential solution for the unemployed elephants and their mahouts. In this way the relationship between elephants and global markets switched from their position as a source of labour for the timber industry to a new role as a source of labour and as a commodity in themselves for the global tourism industry. This also meant that, in line with wider processes of neoliberalisation, they shifted from being employed in government-run logging plantations to working in the private sector. The unemployed elephants and their mahouts had the opportunity of employment as tourist attractions, and the industry developed new commodities out of marketing everyday working practices of elephants (such as offering transport and pulling logs) and of mahouts (washing and feeding elephants).

The transfer of working elephants from the public to the private sector came from small beginnings in the government-run Thai Elephant Conservation Centre (TECC). It started out as the "Young Elephant Training Centre" as a place for training mahouts and providing short courses for officials who might encounter elephants as part of their duties (eg police officers who confiscated elephants in urban areas, zookeepers etc). The TECC then began to offer tourist rides for elephants and a short show where elephants displayed the skills they had once used in the logging industry, such as skidding and piling up logs. As the TECC tourist "product" developed the mahouts trained the elephants to play musical instruments. The idea was then taken up by the private sector, and a number of elephant camps have opened in the Chiang Mai area and the product was extended again to include elephant painting (an idea which originally came from the USA),[22] aimed at the international tourist market. Since the TECC is a national government institution it has to be affordable to Thai visitors, and so it has found it increasingly difficult to compete with the privately owned elephant camps that have developed around Chiang Mai, Chiang Rai, and Pai near Mae Hong Son.

These camps offer much the same tourist product as the TECC: elephant rides, elephant shows, the chance to purchase an elephant painting or the opportunity to experience mahout training for one or more days. It is clear that the elephant camps are not a singular phenomenon that offers a homogenous experience which is produced and managed in the same way.

Although international tourism has been Thailand's single biggest foreign exchange earner for the past 10 years, those involved in tourism are painfully aware of how fickle the industry can be. In particular, Thailand has been the site of a major disaster, the tsunami of 2004, and suffered from outbreaks of avian flu and severe acute respiratory syndrome, and each of these has affected the number of visitors to Thailand. Partly as a result of this, the TECC has explored other avenues to demonstrate the importance of conserving wild and domesticated elephants in Thailand. Prasop Tipprasert, chief of the Training School for Thai Elephants and Mahouts (at the TECC), argued that tourist rides and even the mahout training experiences might not be the basis for funding elephant conservation in the longer term. Therefore, the TECC developed new products from their elephants, including elephant dung paper and organic fertiliser. He has also been involved in a medical study with the University of Chiang Mai about the positive influence of experiences with elephants on autistic children. He suggested that through the study on autism he hoped that Thai people would see the utility of elephants and support their conservation.[23] Therefore, their long-term survival might well depend on demonstrating their utility but not necessarily their market value. As such, it is clear that even in an area where the elephant tourism industry is booming, those involved in elephant conservation are wary of relying on a global industry to save the species in the longer term. Indeed it is very interesting to note that Tipprasert's view was that the future lay with demonstrating the wider importance of the elephant to Thai people.

This view of elephants as more than just a commodity taps into the historical relationship with elephants through the practice of mahoutship. For example, mahouts will pray to Ganesha before any journey with an elephant. Ganesha is a very popular idol in Thailand, as well as the wider South and Southeast Asian regions, and it is common for people to make an offering to Ganesha before any undertaking. In addition to this, elephants were once a status symbol in Thailand. The rich could demonstrate their wealth through the number of elephants they owned. As in Africa, elephant symbolism is ever present in Thailand, from the logo of Chang Beer (Chang is Thai for elephant) to the lapel pins worn by women students at Chiang Mai University. However, there are very few people in Thailand who believe that culture alone can save the elephants. Due to the high number of captive elephants, combined with the amount of money required to keep a captive elephant, the survival of captive elephants will be reliant on their ability to pay their way, either in the tourism industry or by providing services for people. This does lend captive elephants to neoliberal principles. Kontogeorgopoulos (2009:443) argues that Thailand has failed to conserve elephants based on their intrinsic worth as living creatures, and so their future depends on demonstrating their economic importance and utility to human beings.

Neoliberal approaches to nature can be used to re-invent traditional practices. For example, an elephant fashion show organised by the owner of the Elephant Life Experience to mark National Elephant Day on 13 March 2008 demonstrates the use of historical practices of training elephants to create a diverse range of products to support conservation. The show featured clothes made from fabrics which used patterns taken from paintings by elephants. The elephants painted pictures onto paper and the patterns were then transferred to fabrics to be used for bespoke designs by a British fashion designer, Lawrence Goldman. Funds generated from the sales of the outfits were to be used for elephant conservation, and it was the first such event anywhere in the world. The Elephant fashion show was intended to draw attention to elephant conservation, the Art by Elephants Foundation and to the artificial insemination programme organised between the National Elephant Hospital at the Thai Elephant Conservation Centre and the privately owned Mae Sa Camp. The event was covered by global television networks, including the BBC, and appeared on youtube.[24] The neoliberalisation of nature in Thailand has become hybridised through its engagement with locally specific contexts, histories and social processes to create new opportunities and ways in which elephants can be conserved. It is clear from these events that the links between the private sector and the national government institutions were reliant on the global tourism industry, but they used the historical practices of training elephants to create a diverse range of products associated with elephants.

The cases of elephant riding in Botswana and Thailand demonstrate neoliberalism can be extended to an increasing range of non-human phenomena through the development of new tourist spectacles. However, the two cases also show that this is an uneven process, and is molded by other values for elephants (especially in Thailand). This reveals that the global tourism industry drives and deepens processes of neoliberalisation, but does so in complex and varying ways. It has not led to an inevitable and all consuming process which displaces all other values and approaches to conservation. The ways in which wildlife can be neoliberalised is dependent on the wider cultural and socio-political contexts in each area. Tourism has developed experiences with elephants as valuable commodities of the market in both Botswana and Thailand, but local cultural traditions associated with elephants and the wider socio-political conditions of each country have produced very different economic possibilities and tourism products.

Neoliberalisation of nature is not a neat process, easily implemented in standardised ways. Instead it plays out in particular ways on the ground, and this allows us to gain greater insight into actually existing neoliberalisms (Brenner and Theodore 2002). In the case of Botswana, the specificities of ownership of captive versus wild elephants and the lack of history of training and using elephants as work animals shapes the character of neoliberalised nature. Thailand's elephants, on the other hand, were re-trained and employed in the tourism industry to manage the problems associated with street-walking elephants since they lost their jobs in the logging industry.

Elephant Welfare in the Work Place

Concerns about elephant welfare are one source of commonality between Thailand and Botswana, and both have much to learn from comparing experiences to ensure that captive elephants are well cared for. The use of elephants in Thailand and Botswana for elephant-back tourism is fraught with a series of ethical issues centred around animal rights and animal welfare, essentially arising from the ways that they are used as a labour force for international tourism. For example, the International Fund for Animal Welfare and the Humane Society of the USA are strongly opposed to any form of elephant training for the tourism industry (or any other industry) on the grounds of cruelty.[25] As Kontogeorgopoulos (2009:430) points out, there are important objections to the ways some elephant training relies on the use of physical force. Critics also object to the ways that male elephants in Asia may be chained for many weeks during musth because they become unpredictable and violent. Furthermore, elephants that live in camps find it difficult to bond with each other as part of a herd, which means they do not reproduce as effectively as in the wild. Finally, captive elephants may not receive a sufficient variety of foodstuffs to provide them with adequate nutrition.

Many of these issues unfold in similar ways in Thailand and Botswana, however both areas also have specific problems they are confronted with. For example, because Botswana does not have a background associated with elephant trekking there is a lack of the knowledge about the long-term management of captive elephants. As one interviewee pointed out, taking on a captive elephant is not a short-term business strategy since the animals can live for 50–60 years and need to be cared for throughout their lives.[26] One of the problems with the expansion of the elephant-riding industry in southern Africa as a whole was that there was little understanding of the lifelong commitment attached to working with elephants. One researcher expressed concerns that lots of elephant-riding experiences were popping up in southern Africa but they were run by people with little or no experience of animal training or working with elephants.[27]

While the stakeholders involved in the elephant-trekking industry see elephant-back safaris as highly successful, some interviewees in Botswana said that they thought African elephants are unsuited to close work with people because of their unpredictable temperament.[28] It is often during the period when mahouts are changed when elephants can become unsettled and dangerous.[29] For example, one interviewee who worked with trained elephants in Botswana stated that they were concerned it would only be a matter of time before a tourist was killed by an African elephant due to the absence of experienced mahouts. One of the criticisms levelled at the elephant-riding industry in Botswana is that it is an industry driven by profit and not by concerns about elephant welfare and conservation. In particular, an interviewee stated there was a perception that you could make "big bucks" from elephant riding in part because of the high prices charged by the lodges in Botswana; and that this has drawn in private operators who have little appreciation of what it takes to work with elephants in the longer term. These concerns about animal welfare led to the creation of the Elephant Management and Owners Association to draw up standards for the elephant-back safari industry. This was regarded as particularly pressing after undercover footage was released which exposed cruelty during training of young elephants.[30]

While elephants have an important cultural status in Thailand, there are concerns that they will not be looked after once they are unable to earn a wage. One interviewee explained that a mahout abandoned his elephant at the elephant hospital in TECC because he believed it would no longer be able to work for a living.[31] However, it is important to point out that mahouts are not intentionally cruel to their elephants, the animals are the main source of income for mahouts, and they may live together for many decades. One mahout commented on the street wandering elephants that it was desperation which drove mahouts to take their elephants to beg in urban areas and tourist resorts; they were aware that the elephants found the environment stressful, but felt they had no choice and did it to survive.[32]

The cross comparison between elephant riding in Thailand and in Botswana highlighted some important issues. On the surface, it may appear that the use of captive elephants in Thailand and Botswana is very similar. The practice was created in both areas to find a use for elephants that were in undesirable situations such as begging on the streets in

Bangkok or performing in a circus. However, there are also significant contrasts between the two cases. Although Botswana (along with the other African range states) has held a few elephants in captivity (eg in zoos) the actual practice of training elephants to carry humans is not a traditional activity in Africa, and it is usually undertaken by foreigners. This is quite unlike Thailand where elephants have been used as working animals for thousands of years. Botswana has seen the introduction of a new practice, whereas Thailand has merely altered an historical practice. As such, the captive elephants in Botswana and people's relationships with them have undergone a far greater transition where they have been re-packaged in order to become a global tourism product. Thailand's elephants, on the other hand, were already employed in work and used as a source of labour: they have merely changed who they and their mahouts work for, and shifted from being public sector (logging) to private sector (tourism) employees. In addition to this, Botswana's captive elephants only exist as safari elephants because there is a market demand for them, whereas Thailand's captive elephants would exist regardless because they had already been taken out of the wild in such large numbers. Elephant trekking in Botswana has created a completely different way in which their elephants can be valued since they can now be valued as a source of labour. In this way, Botswana's elephants are more conducive to the principles of neoliberalism: the development of elephant trekking has allowed the re-regulation of elephants in privately owned elephant camps. Thailand's elephants are more easily acquired as private property than they are in Botswana. The owners of elephants in Thailand have other (albeit limited) options regarding the way they choose to use their elephants. Although the majority of elephants in Thailand are restricted to tourism, there is also demand for elephants to be used in other ways, which include informal forms of transport (particularly in the hills) or in a government-registered logging plantation. Therefore, the application of neoliberalism through the global tourism industry in Thailand has transformed traditional ways in which elephants were used in order to create new products for the tourism industry as well as offering a way in which Thailand's captive elephant population can be conserved. These new ways in which elephants can be used exist alongside traditional values (they do not displace them), but tourism demand has created new practices. The two cases demonstrate that the implementation of neoliberalism adapts to existing circumstances.

Conclusion

To conclude, the case of the elephant-back safari industry in two different contexts demonstrates the ways that the tourism industry acts as a driver of the neoliberalisation of nature, but it also reveals the

contours and boundaries of actually existing neoliberalism (Brenner
and Theodore 2002). First, it shows how the global tourism industry
is not simply reflective of global neoliberalisation, but is a critically
important constitutive element that expands and deepens processes of
neoliberalisation. Second, the elephant-back safari industry in Thailand
and in Botswana revealed that the tourism industry is one means by
which nature is neoliberalised. Through the creation of new work
elephants in Botswana and re-regulation and re-packaging of working
elephants in Thailand as a tourist spectacle, the industry has provided
an opportunity for neoliberalism to target and open up new frontiers
in nature (Castree 2008a:141; 2009). Third, examining the ways
trained elephants are used in the tourism industry in Thailand and
Botswana allowed for comparative research on the same "product"
in two very different contexts; in turn this meant it was possible to
analyse the variations in "actually existing neoliberalisms" (Brenner
and Theodore 2002). The analysis of the commonalities and differences
between elephant riding in Thailand and Botswana revealed that the
effects were not unremittingly negative (Castree 2008b:166). This
is especially important when we consider the case of unemployed
elephants in Thailand following the 1989 logging ban; the tourism
industry provided a very important alternative source of employment
for elephants and their mahouts. Fourth, the comparative analysis of
Thailand and Botswana highlighted the ways that neoliberalisation
adapts to context-specific processes, values, ideas and institutions.
The elephant-back tourism industry demonstrates that processes of
neoliberalisation of nature through the global tourism industry produces
a kind of "palimpsest" effect; it brings new values and uses for nature
that are compatible with its use in international markets, but in Thailand
it does not completely displace existing ways of valuing, using, owning
and approaching nature. It is clear that neoliberalism, through tourism,
reconfigures and redesigns nature for global consumption; the elephant-
riding industry is an excellent example of how tourism remakes the
nature in an acceptable image to appeal to international clients. In sum,
the growth of elephant-back tourism is an example of neoliberalised
nature because the animals are trained, repackaged and developed for
consumption by the global tourism industry.

Acknowledgements

We are grateful to the ESRC for funding this research, grant reference RES-000-22-2599,
*Neoliberalising Nature? A Comparative Analysis of Asian and African Elephant Based
Ecotourism.* We would also like to thank three anonymous referees for their very helpful
comments on an earlier draft of this chapter. This chapter was originally presented to
a Symposium on Conservation and Capitalism, 9–10 September 2008, Manchester
University, and we are grateful to the participants for their comments. Finally, this

research would not have been possible without the support of the International Tourism Research Centre and the Harry Oppenheimer Okavango Research Centre at the University of Botswana, the Thai Elephant Conservation Centre and Maetaman Elephant Camp.

Endnotes

[1] http://www.unwto.org/index.php (accessed 15 August 2009).
[2] Interview with the Principal Tourism Officer, Department of Tourism, Gaborone, 9 October 2008.
[3] Ministry of Environment, Wildlife and Tourism, Draft Elephant Management Plan for Botswana.
[4] Interview with researcher, Maun, 23 October 2008.
[5] Interview with District Wildlife Co-ordinator, Department of Wildlife and National Parks, Maun, 31 October 2008.
[6] African Odyssey Programme of Tourist Activities, Chobe Marina Lodge, Kasane.
[7] Pers. comm., Dr Susan Keitumetse, Okavango Research Centre, University of Botswana/Botswana Tourism Board, Maun, 26 May 2008; also see Mbaiwa (2004).
[8] http://www.abucamp.com/Abu%20Camp.htm (accessed 15 July 20008).
[9] Elephant Back Safaris Media Pack, Elephant Back Safaris, Private Bag 332, Maun, Botswana.
[10] http://www.abucamp.com/Abu%20Camp.htm (accessed 15 July 2008).
[11] http://www.sanctuarylodges.com (accessed 13 July 2008).
[12] http://www.livingwithelephants.org/index.htm (accessed 15 July 2008).
[13] Interview with Sandi Groves, Living with Elephants, Maun, 28 May 2008.
[14] http://www.livingwithelephants.org/project.htm (accessed 15 July 2008).
[15] Pers. comm., tour guide, Kwara Safari Lodge, Okavango Delta, 30 May 2008; also see Mbaiwa (2004) for further discussion.
[16] The original five members of the Association of Southeast Asian Nations (ASEAN), which were Singapore, Malaysia, Thailand, the Philippines and Indonesia.
[17] Observations from Maetaman Elephant Camp, 7 April 2008; Mae Sa Elephant Camp, 10 April 2008; and TECC, 13 March 2008 and 10 April 2008.
[18] Interview with Somchat Changkarn, Mahout Training School, TECC, Lampang, 20 March 2008. Interview with Pat, Theerapat, Patara Elephant Farm, Chiang Mai, 6 April 2008 (see also Scigliano 2004).
[19] Interview with Pornsawan, Pongsopawijit, Faculty of Veterinary Medicine, Chiang Mai University, 31 March 2008; anonymous interview, researcher in Chiang Mai, 18 April 2008.
[20] Interview with Prasop Tipprasert, Elephant Specialist, Forest Industry Organisation/Chief of the Training School for Thai Elephants and Mahouts, TECC, Lampang, 20 March 2008; pers. comm., Pornsawan Pongsopawijit, Faculty of Veterinary Medicine, Chiang Mai University, 17 March 2008.
[21] Interview with Prasop Tipprasert, Elephant Specialist, Forest Industry Organisation/Chief of the Training School for Thai Elephants and Mahouts, TECC, Lampang, 20 March 2008; pers. comm., Pornsawan Pongsopawijit, Faculty of Veterinary Medicine, Chiang Mai University, 17 March 2008; interview with the Manager Thai Permanent Life exhibition, Thailand Cultural Centre, Bangkok, 25 March 2008.
[22] Interview with Anchalee Kalampimjit, manager Maetaman Elephant Camp, owner of the Elephant Life Experience Camp, Maerim, 7–8 April 2008.
[23] Interview with Prasop Tipprasert, elephant specialist, Forest Industry Organisation/chief of the Training School for Thai Elephants and Mahouts, TECC, Lampang, 20 March 2008.

24 Pers. comm., Anchalee Kalmapijit, Maetaman/ Elephant Life Experience, Chiang Mai, 13 March 2008; http://www.elephantart.com (accessed 8 April 2008).
25 "Elephant Back Safaris simply accidents waiting to happen, warns top tourism insurer" on http://www.ifaw.org/ifaw_united_states/media_center/press_releases/5_10_2007_41436.php (accessed 29 January 2009); "Elephants sent into safari slavery from Zimbabwe's world famous Hwange national park" posted 8 November 2007, http://www.ifaw.org/ifaw_southern_africa/media_center/press_releases/11_8_2006_47671.php (accessed 29 January 2009); Humane Society of the US: "Judgement day finally arrives in the Tuli elephant abuse case", posted 17 April 2003, http://www.hsus.org/wildlife/wildlife_news/judgment_day_finally_arrives_in_the_tuli_elephant_abuse_case.html (accessed 29 January 2009).
26 Interview with Sandi Groves, Living with Elephants, Maun, 28 May 2008; also anonymous interviewee, UK, June 2008.
27 Anonymous interviewee, UK, June 2008.
28 Interviews with two guides at the Lyia Guest House, Kasane, 21 October 2008; anonymous guide, Kasane, 11 October 2008; anonymous interviewee, UK, June 2008.
29 Anonymous interviewee, Maun, May 2008.
30 Anonymous interviewee, UK, June 2008; also see http://www.africanconservation.org/component/option,com_mtree/task,viewlink/link_id,89/Itemid,3/ (accessed 19 January 2009).
31 Anonymous interview, Maetaman Elephant Camp, owner, 7–8 April 2008.
32 Interview with mahout 1 from Surin, interviewed in Chiang Saen, 24–25 April 2008; mahout 2 also from Surin, interviewed in Chiang Saen, 24–25 April 2008.

References

Bakker K (2005) Neoliberalising nature? Market environmentalisms in water supply in England and Wales. *Annals of the Association of American Geographers* 95(3):542–565
Bakker K (2009) Neoliberal nature, ecological fixes, and the pitfalls of comparative research. *Environment and Planning A* 41(8):1781–1787
Bramwell B and Lane B (2005) From niche to general relevance? Sustainable tourism, research and the role of tourism journals. *The Journal of Tourism Studies* 16(2):52–62
Brenner N and Theodore N (2002) Cities and the geographies of "actually existing neoliberalism". *Antipode* 34(3):356–386
Brockington D, Duffy R and Igoe J (2008) *Nature Unbound: Conservation, Capitalism and the Future of Protected Areas*. London: Earthscan
Bulbeck C (2004) *Facing the Wild: Ecotourism, Conservation and Animal Encounters*. London: Earthscan
Butcher J (2003) *The Moralisation of Tourism: Sun, Sand . . . and Saving the World?* London: Routledge
Castree N (2008a) Neoliberalising nature I: The logics of de- and re-regulation. *Environment and Planning A* 40(1):131–152
Castree N (2008b) Neoliberalising nature II: Processes, outcomes and effects. *Environment and Planning A* 40(1):153–173
Castree N (2009) Researching neoliberal environmental governance: A reply to Karen Bakker. *Environment and Planning A* 41(8):1788–1794
Clapham C (1996) *Africa and the International System. The Politics of State Survival*. Cambridge: Cambridge University Press
Cohen E (2008) *Explorations in Thai Tourism: Collected Case Studies*. Amsterdam: Elsevier Science

Cox R (1996) *Approaches to World Order*. Cambridge: Cambridge University Press

Duffy R (2008) Neoliberalising nature: Global networks and ecotourism development in Madagascar. *Journal of Sustainable Tourism* 16(3):327–344

Enloe C (2001) *Bananas, Beaches and Bases: Making Feminist Sense of International Politics*. Berkeley: University of California Press

Good K (2008) *Diamonds, Dispossession and Democracy in Botswana*. Oxford: James Currey

Harvey D (2005) *A Brief History of Neoliberalism*. Oxford: Oxford University Press

Heynen N, McCarthy J, Prudham W S and Robbins P (eds) (2007) *Neoliberal Environments: False Promises and Unnatural Consequences*. London: Routledge

Heynen N and Robbins P (2005) The neoliberalisation of nature: Governance, privatization, enclosure and valuation. *Capitalism Nature Socialism* 16(1):5–8

Hirsch P (1997) Environment and environmentalism in Thailand. In P Hirsch (ed) *Seeing Forests for Trees: Environment and Environmentalism in Thailand* (pp 15–36). Chiang Mai: Silkworm Books

Hoon P (2004) Impersonal markets and personal communities? Wildlife, conservation, and development in Botswana. *Journal of International Wildlife Law and Policy* 7:143–160

Hulme D and Murphree M (1999) Communities, wildlife and the "new conservation" in Africa. *Journal of International Development* 11:277–285

Hutton J, Adams W M and Murombedzi J C (2005) Back to the barriers? Changing narratives in biodiversity conservation. *Forum for Development Studies* 2:341–370

Keitumetse S O (2009) The Eco-tourism of cultural heritage management (ECT-CHM): Linking heritage and "environment" in the Okavango Delta regions of Botswana. *International Journal of Heritage Studies* 15(2–3):223–244

Kontogeorgopoulos N (1999) Sustainable tourism or sustainable development? *Current Issues in Tourism* 2(4):316–332

Kontogeorgopoulos N (2009) Wildlife tourism in semi-captive settings. *Current Issues in Tourism* 12(5/6):429–449

Lair R C (2004) *Gone Astray: The Care and Management of The Asian Elephant in Domesticity*. Bangkok: FAO Regional Office for Asia and the Pacific

Liverman D (2004) Who governs, at what scale and at what price? Geography, environmental governance, and the commodification of nature. *Annals of the Association of American Geographers* 94(4):734–738

Mansfield B (2004) Neoliberalism in the oceans: "Rationalization", property rights and the commons question. *Geoforum* 35:313–326

Mbaiwa J (2004) The socio-cultural impacts of tourism development in the Okavango Delta, Botswana. *Journal of Tourism and Cultural Change* 2(2):163–185

Mbaiwa J (2008) The realities of ecotourism development in Botswana. In A Spenceley (ed) *Responsible Tourism: Critical Issues For Conservation and Development* (pp 205–223). London: Earthscan

McAfee K (1999) Selling nature to save it? Biodiversity and the rise of green developmentalism. *Environment and Planning D: Society and Space* 17(2):133–154

McCarthy J and Prudham S (2004) Neoliberal nature and the nature of neoliberalism. *Geoforum* 35(3):275–283

MCargo D and Pathmanand U (2005) *The Thaksinisation of Thailand*. Copenhagen: NIAS Press

McGregor A (2008) *Southeast Asian Development*. London: Routledge

Melber H (2007) Botswana, Namibia, Zimbabwe—anything in common? In H Melber (ed) *Governance and State Delivery in Southern Africa: Examples from Botswana, Namibia and Zimbabwe* (pp 5–8). Nordiska Afrikainstitutet Discussion Paper 38

O'Neill J (2007) *Markets, Deliberation and Environment*. London: Routledge
Peck J and Tickell A (2002) Neoliberalising space. *Antipode* 34(3):380–404
Peleggi M (1996) National heritage and global tourism in Thailand. *Annals of Tourism Research* 23(2):432–448
Poteete A R (2009a) Defining political community and rights to natural resources in Botswana. *Development and Change* 40(2):281–305
Poteete A R (2009b) Is development path dependent or political? *Journal of Development Studies* 45(4):544–571
Ribot J C (2004) *Waiting for Democracy: The Politics of Choice in Natural Resource Decentralizations*. Washington: World Resources Institute
Scigliano E (2004) *Love, War and Circuses*. London: Bloomsbury
Taylor I and Mokawa G (2003) Not forever, conflict diamonds and the bushmen. *African Affairs* 102:261–283
Twyman C (1998) Rethinking community resource management. *Third World Quarterly* 19(4):745–770
Twyman C (2001) Natural resource use and livelihoods in Botswana's wildlife management areas. *Applied Geography* 21:45–68
UNWTO (2003) *UNWTO Assessment of the Results Achieved in Realising Aims and Objectives of the International Year of Ecotourism 2002*. UNWTO: Madrid. http://www.world-tourism.org/sustainable/IYE/IYE-Rep-UN-GA-2003.pdf (last accessed 5 July 2007)
West P and Carrier J G (2004) Ecotourism and authenticity. Getting away from it all? *Current Anthropology* 45(4):483–498
Wyatt D (2003) *Thailand, A Short History*. London: Yale University Press

Chapter 12
The Receiving End of Reform: Everyday Responses to Neoliberalisation in Southeastern Mexico

Peter R. Wilshusen

Introduction

Conservation and capitalism have co-existed prominently in rural Mexico since the 1970s when the federal government began promoting community-based forestry enterprises as a means of encouraging local economic development and combating deforestation. Agrarian communities, federal and state governments, foundations, aid agencies, and non-governmental organizations associated with the forestry sector and rural organizing sought to integrate conservation and development based on the idea that increased local incomes and economic stability would facilitate nature protection. This approach was particularly evident in the southeastern Mexican state of Quintana Roo beginning in 1983, when the national forestry department joined with Germany's development agency, GTZ, to establish a pilot community forestry initiative. The program drew from experiences with rural development projects in other parts of the country but explicitly set out to reverse the decline of the region's tropical forests and to protect habitat linking two large protected areas. Ultimately the community forestry initiative in Quintana Roo became a model for integrated conservation and development projects in other parts of Latin America, such as Guatemala and Ecuador.

Despite significant achievements in establishing permanent forest reserves and locally managed enterprises, community forestry operations in Quintana Roo have faced myriad challenges stemming in large part from their hybrid governance structure. Across Mexico, land grant communities (known as "ejidos") and their support organizations are neither private nor public, and thus face persistent, unresolved tensions between entrepreneurial desires and collective responsibilities.

These tensions have become more pronounced over the last 15 years in the wake of Mexico's embrace of neoliberal economic and political reforms. The shift dramatically altered the country's agrarian sector, bringing an end to agrarian reform and creating the possibility of dismantling the ejido as a collective property regime among other changes. While some initial forecasts predicted that community forestry enterprises, such as those in Quintana Roo, would disappear as a result of the reforms, they continue to play an important role in both rural economic development and regional conservation efforts. Yet, Mexico's neoliberal policy reforms have facilitated several important changes to the ways communities approach both conservation and development. How exactly, then, did the national-level institutional shift play out at the receiving end?

This chapter explores the creative accommodations that rural producers have made in navigating Mexico's neoliberal turn. In contrast to previous work on neoliberalism and environment that emphasizes macro-level processes (eg privatization of public natural resources) and local resistance, I employ Bourdieu's theory of practice to examine the cultural and material dimensions of local responses to neoliberal policy reform. Drawing on research from nine communities in the state of Quintana Roo, I argue that local producers have accommodated neoliberal policies and programs in creative ways by adopting hybrid logics, property regimes, forms of organization, and modes of exchange. Moreover, I contend that these creative responses constitute elements of a longstanding "culture of accommodation" to institutional change that predates Mexico's neoliberal reforms. From this perspective, processes of accommodation at the receiving end are neither purely voluntarist nor simply imposed from without but rather comprise durable practices derived from local interactions with state-sponsored development initiatives over multiple decades. Unlike resistance movements such as the Zapatista uprising in the Mexican state of Chiapas, practices of accommodation emerge subtly and incrementally over time in the course of everyday interactions.

The chapter unfolds in six parts. The first section highlights three dominant themes in the literature on neoliberalism and environment— enclosure/privatization, the state as carrier of neoliberal reform, and contestation/resistance—to explore three points of inquiry that these themes suggest: hybrid logics and governance arrangements, local-level responses to state-sponsored initiatives, and accommodation. The second section introduces Bourdieu's theory of practice as a conceptual frame for examining everyday responses to neoliberal reform. I present Bourdieu's key concepts of field, capital, and habitus to gain a more nuanced view of the nexus between agency and structure, including strategic and routine human actions and the cultural and institutional constraints shaping those actions. The third section discusses the first type of accommodation to neoliberal reform by local communities: discursive. I situate dominant and shifting discourses on neoliberal policy reform in Mexico's agrarian sector in relation to community forestry activities in the state of Quintana Roo to examine how shifting logics of development produced a hybrid institutional and discursive arena rife with contradictions. Despite contradictions, these logics co-exist, inform and challenge activities surrounding community forestry in important ways. The fourth section explores the second type of accommodation: spatial. I detail how community members established a combination of private and communal spaces in response to specific neoliberal policy reforms. The fifth section addresses the third type of accommodation: organizational. I explore how two communities established novel forms of internal organization and commercial exchange that enhanced the flow of capital while maintaining the security of collective resource ownership. In the sixth section, I extend my analysis to assess the extent to which the processes of accommodation to neoliberal policy reform among the community forestry ejidos in Quintana Roo represent longstanding practices constituting a "culture of accommodation" to state programs. I argue that processes of accommodation to neoliberalism are not so much a direct response to specific institutional reforms as a gradual accretion of practices in response to decades of state-sponsored development activities. In this sense, accommodation constitutes creative adaptation to state-led reform rather than passive acceptance.

Contours of "Actually Existing Neoliberalism"

In general, proponents of neoliberalism argue that unfettered markets are the best mechanisms for allocating goods and services within society. Such an approach seeks to minimize state-imposed regulations that might hinder flows of financial capital (Harvey 2005; Heynen et al 2007). In abstract terms compiled by Castree (2008a:142), the strategies associated with neoliberal reform projects typically include all or most of the following: "privatization, marketization, deregulation, reregulation, [the creation of] market proxies in the residual public sector, and the construction of flanking mechanisms in civil society". The state plays a central role in establishing and/or regulating these six conditions and thus granting primacy to the market. Thus, regarding the first strategy—privatization—the state might attach private property rights to natural resources previously considered public or communal property (Castree 2008a, 2008b).

In framing this chapter, I highlight three themes that inspire further points of inquiry relevant to the neoliberal turn in Mexico: enclosure/

privatization; the state as carrier of neoliberal reform; and contestation/ resistance. Regarding enclosure and privatization, much of the literature in critical human geography reflects Harvey's (2003) characterization of capital expansion as "accumulation by dispossession". Studies focus on water (Bakker 2005, 2007; Perreault 2005; Roberts 2008; Swyngedouw 2005), organic food (Guthman 2007), fisheries (Mansfield 2004, 2007; St Martin 2007), life patents (Prudham 2007), land reform (Wolford 2005, 2007), mining (Bury 2005), and conservation (Brockington, Duffy and Igoe 2008), among others. In each case, state-led reform efforts seek to enhance market "efficiency" by redefining and stabilizing property rights via the transfer of resource access and control (land, water, fish, minerals) to capital interests. In the case of rural Mexico, however, policy reforms strongly encouraged but did not mandate resource privatization, producing a mixture of quasi-private and communal spaces within communities. This suggests further inquiry on how hybrid logics and resource governance regimes emerge in practice and what social and environmental impacts such an amalgam might produce (Mansfield 2007; McCarthy 2005).

A related theme contemplates the role of states as carriers of neoliberal reform in conjunction with down-scaling, administrative decentralization, deregulation, and reregulation. Critical geographers tend to emphasize macro-level changes linked to trade agreements or broad policy agendas, such as the Washington Consensus (Essex 2008; Martin 2005; McCarthy 2004; Perreault 2005; Wainwright and Kim 2008). However, since reform processes tend to be fragmented and contested, it raises questions regarding the ways in which agrarian communities receive, challenge and help to shape the application of neoliberal policies and programs in specific contexts. Several recent studies offer detailed ethnographies of such encounters among diverse actors and the tensions, negotiations and adaptations that they produce within the context of conservation/development endeavors (eg Agrawal 2005; Braun 2002; Escobar 2008; Haenn 2005; Kosek 2006; Li 2008; Moore 2005; Tsing 2005; West 2006).

The literature also highlights how marginalized groups such as indigenous peoples have resisted efforts to privatize and commodify natural resources. In Perreault's (2005) work on rural water governance in Bolivia, for example, government attempts to privatize water use rights provoked a national level movement by peasant irrigators to maintain water as a public resource (see also Bakker 2007; Goldman 2005; Sawyer 2004). Mexico is well known for the Zapatista uprising in Chiapas in response to the North American Free Trade Agreement (NAFTA) (Harvey 1998). However, an exclusive focus on resistance removes from view those settings where neoliberal policies and programs are partially or wholly assimilated and the everyday processes by which such accommodations are constructed.

Although the Zapatista rebellion presents an important example of overt resistance to neoliberal reform, the literature on community forestry in Mexico (and agrarian affairs more generally) emphasizes the successes, challenges and adaptations that communities have made in their attempts to maintain locally managed forestry enterprises in the face of institutional change (eg Bray and Merino 2004; Bray, Merino and Barry 2005; Primack et al 1998). Several studies have examined community-based forest management as common property regimes operating within competitive markets (eg Alatorre 2000; Antinori and Bray 2005; Merino 1997). Related work focuses on the impact of community politics on the collective management of natural resources (eg Haenn 2005; Klooster 2000a, 2000b; Nuijten 2003b) as well as the construction of official knowledge by state agencies and forestry communities (Mathews 2005, 2008). Most of the contemporary literature on Mexican community forestry addresses the country's shift toward neoliberalism, including some studies that identify the internal organizational and political adaptations that I highlight below (eg Taylor 2000, 2001, 2003; Taylor and Zabin 2000).

In building on this and related work, I emphasize how specific policies and programs facilitate diverse local-level changes—in both cultural and material terms—that significantly impact power relationships associated with community forestry. The literature emphasizes how neoliberalization at different scales is path dependent (Brenner and Theodore 2002) and embedded within specific institutional contexts (Peck 2004; Peck and Tickell 2002), creating hybrid social constructions comprising "a complex and contested set of processes, comprised of diverse policies, practices, and discourses" (Perreault and Martin 2005:194; see also Bridge and Jonas 2002; Mansfield 2007). Thus the social processes associated with responses to neoliberalization feature frictions, negotiations, conflicts and adaptations where hybrid, often contradictory logics and rule systems—policies, practices and discourses—interface. My analysis centers on a context (Quintana Roo, Mexico) that features not so much protest and resistance as accommodation. Moreover, as has been the case with neoliberal reform in many places, agrarian communities in Mexico, for the most part, currently interact with an absentee rather than a constrictive, authoritarian state.

Practice Theory and Everyday Responses to Neoliberalization

Practice theory as discussed by Bourdieu and others (eg Ortner 1999, 2006; Sewell 2005) provides a useful set of heuristics for analyzing everyday politics. It captures the discursive and institutional constraints but also the longstanding practices associated with everyday life. Moreover, it is both relational and contextual. In this section, I present a conceptual frame for analyzing everyday responses to neoliberalization using the vocabulary of Bourdieu's theory of practice.

The social theoretical weave that I present seeks to intertwine critical political economic and poststructural perspectives on power in ways that maintain the reciprocity or mutually constitutive relationship between agency and structure. This is the core contention and challenge of practice theory—the need to overcome rigid, artificial analytical dichotomies that favor either human action/behavior or social structure/culture (Ortner 1999, 2006; Sewell 2005).

Central to Bourdieu's theory of practice is the interpretation of power relationships as longstanding practices tied to specific institutional and cultural contexts. A dominant thread in Bourdieu's writings compares everyday social interaction to theatrical plays (or games) that reveal underlying power relationships. Thus, in *Distinction*, for example, Bourdieu (1984) examines the formation of social class relationships based in part on observations of how individuals make judgments and choices regarding matters of taste—clothing, food, wine—or enact polite behavior. Thus everyday aesthetic preferences and choices reflect cultural practices charged with meaning in terms of how actors understand their relationships to one another.

Bourdieu refers to these durable dispositions as "habitus" (Bourdieu 1977, 1984, 1990; Bourdieu and Wacquant 1992). Habitus combines elements of both agency and structure—historically and culturally defined tendencies and practices informing a "practical sense" or logic (*le sens pratique*). Actors may be more or less aware of the rules or logics of the game as they pursue longstanding, culturally defined practices or routines. Thus, although Bourdieu's rendering of habitus has been critiqued as overly deterministic and devoid of intentionality (Ortner 1996, 2006; Sewell 2005), it still presents a way to bring the subject back to the center of analysis without relying solely on voluntarist notions of agency.

For Bourdieu, the term capital simultaneously represented both a power relationship and a power resource. In this sense, individuals' dispositions (*habitus*) derive, in part, from relative endowments of different forms of capital, which, in turn, help to define their historically evolving "positions" within social settings. Actors exchange and accumulate capital (material and virtual) in the course of everyday social interaction. Bourdieu (1986) described three forms of capital, including economic, cultural and social. Economic capital constitutes material and financial assets while cultural capital encompasses symbolic goods, skills and titles, such as educational credentials. Social capital comprises a means (set of relationships) by which actors accrue economic and cultural capital as a result of participation in culturally embedded networks. Although Bourdieu's presentation of different forms of capital seems to portray social life in overly mechanistic and economistic terms, his analysis of capital flows emphasizes historically derived differences among actors' material and symbolic "endowments", the relationality of social exchanges, as well as the constant unfolding or processual qualities of power dynamics (Wilshusen 2009a, 2009b).

To capture the structural bounds that shape social life, Bourdieu linked habitus and capital to the concept of "field". Fields are arenas of struggle in which actors attempt to accrue or control economic and cultural capital. The term could include actor networks but also captures formally institutionalized relationships based on explicit codes or rules as well as non-formalized, customary relationships structured by cultural norms, discourses or practices. The dominant or subordinate positions that individual and group actors hold within a field are determined by their relative endowments of economic and cultural capital. As a result, the character and configuration of fields constantly shift as power relationships change.

In addition, as both a structural and cultural heuristic, fields present certain "logics" and thus define the domain of struggle. In other words, even though both dominant and subordinate actors may challenge one another for resource control, they all tacitly accept that the "rules of the game" and that certain forms of contestation are legitimate while others are not (Bourdieu and Wacquant 1992; Swartz 1997). In his empirical studies, Bourdieu used the term field to characterize domains of social interaction—the "artistic field" and the "academic field", for example. Thus as a cultural and institutional form, he referred to the state as "the bureaucratic field", featuring legalistic/technocratic rationalities, formal administrative practices, and control of diverse forms of capital (Bourdieu [1991] 1999).

As with habitus, Bourdieu's use of the term "field" is often seen as overly deterministic, leading to the inevitable reproduction of social inequalities (Sewell 2005). Yet, field, as a heuristic construct, does not necessarily refer to a monolithic sphere of constraints. In my analysis, I emphasize that fields represent the constantly shifting, socially constructed nature of the cultural/institutional spheres of human experience. I examine the extent to which fields present fragmented, hybrid, and contradictory logics and rules as well as how these fault lines shape everyday politics. Most importantly, fields should be understood not as fixed constraints, but rather as spheres of interaction (processes not things).

The remainder of this chapter empirically examines everyday responses or accommodation to neoliberal reform through the lens of Bourdieu's theory of practice. The first type of accommodation—

discursive—relies mainly on the concept of field to better understand the inherent tensions created by the collision of contradictory logics and rule systems: collective versus neoliberal natural resource management. The second type of accommodation—spatial—details how rural producers appropriated elements of both collectivism and neoliberalism in response to a federal land titling program, consolidating internal distributions of economic capital in the process. The third type of accommodation—organizational—analyzes the formation of entrepreneurial sub-groups and an internal timber exchange within certain ejidos. Again, I use Bourdieu's presentation of social and economic capital to examine how these creative responses impacted everyday politics. Finally, I use the concept of habitus to explore the extent to which these discursive, spatial and organizational responses constitute elements of a "culture of accommodation" evident in longstanding practices that pre-date neoliberal reforms.

Forests as Trees: Collectivism Meets Neoliberalism

This section explores the overarching and shifting meta-narratives or logics that have framed agrarian affairs, forest policy and community-based forest management in Mexico since the early 1980s. To illustrate these dynamics, I focus mainly on debates surrounding the bureaucratic field of agrarian policy development. As I noted above, fields present logics—defining discourses and modes of thought—that characterize a domain of social interaction. The logic of a field presents certain taken-for-granted assumptions regarding legitimate action. My main point in this section is that policy compromises negotiated at the national level facilitated local constructions of hybrid, and in many ways contradictory, logics of action for forestry ejidos and their support organizations, producing unresolved tensions in everyday practice. I first contrast the logics of collectivism and neoliberalism and then examine how local actors accommodated elements of the two approaches in their daily activities.

Community Forestry as State-led Populism

Community forest management in Mexico emerged within the context of, but also in response to, regimes built around state-centered development. Although state agencies were still viewed as central to regional development, in the late 1970s administrators began signaling support for policies and programs that would both empower communities and enhance production. The issue of equity emerged as an important message coming from what was widely perceived as an authoritarian state. In remarks to Mexico's national association of foresters in 1976, Undersecretary of Agriculture (and future presidential candidate) Cuauhtémoc Cárdenas (1976:14) emphasized this theme of distributional justice.

[Mexico is] presented with the opportunity and need to join forces among the different types of forest property owners as well as the financial agencies, industries, and investors that currently run the [forestry] industry. [That opportunity and need] refers to the democratic and equitable participation [of all sectors] in the distribution of benefits generated by forest exploitation.

The language of reform used by Cárdenas and others argued that "the true owners and users of the forest" should have greater management power. The passage of the 1986 Forest Law (*Ley Forestal*) marked the culmination of more than a decade of community-level organizing aimed at devolving forest management responsibilities to ejidos following some 25 years where forests on ejido lands were controlled by state-owned enterprises (Klooster 2003). The 1986 law reaffirmed the importance of the ejido as a collective natural resource management entity. It transferred management responsibility from concessionaires to ejidos and their support organizations. Legal responsibility for technical services passed from the state to specially sanctioned for-profit, communal associations (*sociedades civiles*). Thus, a major power shift occurred where forestry ejidos gained formal control over the means of production and state-owned forest industries lost their ability to fully control supply.

In Quintana Roo, the pilot community forestry initiative (known as the Plan Piloto Forestal) that ran from 1983 to 1986 shifted forest management from a state-run enterprise—Maderas Industrializadas de Quintana Roo (MIQRO)—to 10 ejidos following the expiration of a 25-year forestry concession. The Plan Piloto program built on existing community level organizations that had been encouraged during the 1970s by the agrarian reform ministry. It was instrumental in helping the 10 communities to establish local forestry enterprises based on 25-year management plans. The close of the Plan Piloto Forestal in 1986 inspired the creation of a second-tier producer association as a means of ensuring continued economic and political collaboration among the 10 communities. The Sociedad de Productores Forestales Ejidales de Quintana Roo (hereafter the "Sociedad") continues to provide technical support in forest management as well as political representation for nine member communities (the 10th ejido withdrew from the association in 1996). The Sociedad finances its operations mainly through a combination of community payments (fees for technical services) and federal government grants.

As I discuss further below, community forestry in Quintana Roo (and Mexico generally) was set up largely around a logic of collectivism. The organization and governance approach drew on the precedent of the ejido, which emphasized elected leadership, collective ownership and communal decision-making. Moreover, although community forestry enterprises had a clear market orientation, they were governed collectively with daily operations run by an elected executive who responded to the ejido assembly.

The Logic of Neoliberal Reform

In sharp contrast to the advocates of community forestry, neoliberal reformers saw the ejido system as the Achilles' heel of proposed market-centered policies and programs that gained prominence in the late 1980s. Critics argued that the country's agrarian code—which provides the legal basis for the ejido—stifled private sector investment and bred uncompetitive production practices (Cornelius and Myhre 1998; Randall 1996). The writing of Luis Téllez Kuenzler (1994:12), one of the principal architects of neoliberal reform in Mexico's agrarian sector, exemplifies the rationale behind the shift to market-centered development:

> The modernization of the countryside required a redefinition of the role of the state in agrarian activities, both in the policy arena and in its direct intervention via governmental agencies. In the policy arena, the need for a more flexible institutional environment was evident so that producers might fully realize their production potential. In particular, it was necessary to modify constitutional article 27 and its regulations in order to eliminate uncertainties regarding land tenure associated with land distribution, guarantee decision-making and administrative freedom for ejido members [ejidatarios], and permit the transfer of agricultural plots. These measures permit more efficient utilization of natural resources, discourage local land concentration [minifundismo], create new perspectives for ejido members, and enhance family well-being in the countryside.

As I discuss further below, Tellez's view—while influential in the reform process—represented a more "hardline" stance regarding collective land tenure compared with other "softline" factions that supported change without discarding the existing ejido governance structure (Cornelius and Myhre 1998). However, Tellez's version of neoliberalism was strongly represented in key aspects of the constitutional and statutory revisions that appeared in 1992, including an emphasis on private property and private enterprise (the disaggregation of existing collective property and collective enterprises), and the retraction of state "interference" in community affairs (deregulation). I touch briefly on each in turn.

The cornerstone of the agrarian sector reforms of 1992 centered around changes to Article 27 of the federal constitution, formally ending agrarian reform and altering the legal underpinning of the ejido (collective land grant) system to permit the privatization and disaggregation of collective property. Under the Constitution of 1917 and the Agrarian Code of 1936, ejido lands were inalienable. The revisions to agrarian law made it possible for (but did not compel) ejido assemblies to dissolve their communal landholdings and obtain private property titles to the individual plots of land. Changes to the constitution and the agrarian code also allowed foreign and/or domestic corporations to own land in Mexico and enter into commercial partnerships with ejidos, both of which were illegal before 1992 (Cornelius and Myhre 1998; Randall 1996).

Regarding retraction of the state, the changes to agrarian law also signaled the final phase in the dismantling of associated programs and responsibilities carried out by the agrarian reform ministry. Particularly during the height of state-led development from the 1960s until the late 1980s, agrarian reform officials oversaw development activities in individual ejidos, participated formally in assembly meetings, encouraged the formation of ejido unions, and facilitated access to state-run banks. Over time, the ministry shifted from a highly paternalistic role to one that deferred to elected ejido authorities but still facilitated access to loans and programs. Although the withdrawal of the state from community affairs was gradual in many respects, revisions to the Agrarian Code (1992) restructured the ministry in significant ways. While the agency continued to exist, its mission changed dramatically to emphasize certification of ejido land rights and land titling (a national program known as PROCEDE). Some of the legal changes were more subtle, but significant nonetheless. The 1936 Agrarian Code, for example, stipulated that formal acts (decisions) by ejido assemblies had to be ratified by agrarian reform officials and that ejido rightsholders (ejidatarios) had to personally work their own agricultural parcels rather than sublet them to others. Under the 1992 revisions, both of these requirements were removed (Cornelius and Myhre 1998).

Discursive Accommodation: Hybrid Logics, Shifting Fields

Obvious tensions and contradictions surface when contrasting the governing logics of neoliberalism and collectivism. Whereas neoliberal policies emphasized individuals and small producer groups as their main economic subjects, collectivism focused on the role of ejidos and ejido associations. As a result, the collectivist foundations of community forestry reaffirmed communal property and decision-making while the neoliberal policy reforms promoted the disaggregation of these collectivities into private property holdings and individual and/or family choices. The language of neoliberalism evident in the Téllez quote stresses "modernization" and "efficiency" via increased individual freedoms, clearly defined property rights and free exchange. By contrast, the language associated with collectivism, exemplified by the Cárdenas quote, speaks of equity and collective resource control by devolving management responsibility to the "true forest owners". On one level, these quotes illustrate the unresolved tension between entrepreneurial desires and collective responsibility that marks a core conundrum of community-based conservation/development in rural Mexico. At the same time, however, the two logics co-exist and interface within the same locally enacted bureaucratic field. As one of the primary formal institutions that define the bureaucratic field of community forestry in Mexico, changes to national forest policy illustrate this hybridization of neoliberalism and collectivism. Moreover, the example highlights the dynamic and shifting nature of fields.

In 1992, with Luis Téllez Kuenzler serving as Undersecretary for Agriculture, the Mexican government heavily revised the national forest law to fall in line with neoliberal designs. In contrast to the 1986 forest law, which emphasized community forestry enterprises, the 1992 law promoted the development of forest plantations, commercial partnerships between ejidos and the private sector, creation of small private forestry operations, and the deregulation of forestry technical services, harvesting, transport and sale of wood products. As with the changes to the constitution and agrarian code, the revised forest law emphasized clearly defined property rights and deregulation to encourage private sector investment. At the same time, since agrarian law facilitated but did not mandate the dissolution of ejidos, collective property and resource management regimes remained as the foundation of the forestry sector. Additionally, a key compromise provision in the revised agrarian code required ejidos to maintain rather than divide collective holdings such as rangelands and forests.

Interestingly, the 1992 forestry law prompted a significant increase in unregulated logging and a backlash from community forestry advocates, which ultimately led to another revision of the national forest law in 1997 (Bray, Merino and Barry 2005; Klooster 2003). The 1997 law reinstated many regulations that had been dismantled by neoliberal reformers and created subsidy programs that promoted both collective forest management and small enterprise development. Between 1996 and 2001, the nine ejidos included in this study received US$1.13 million (2001 dollars; US$1.38 million adjusted to 2008) in federal and state support for projects focused on forest inventories, management plans, handicraft workshops, timber marketing, and tree nursery maintenance, among others.

In the wake of these policy reforms, local discourse presented similar hybrid constructions that embraced elements of both collectivism and neoliberalism. In general, local leaders in Quintana Roo's community forestry sector embraced an "entrepreneurial" strategy (*una estrategia empresarial*) in dealing with development challenges. During 2000, for example, the Sociedad carried out a series of workshops in most of its nine member ejidos to solicit input regarding the organization's successes and failures. The association's elected leaders and technical staff proposed a significant organizational restructuring to improve efficiency on all fronts: financial, technical and political. The following quote from a senior member of the organization's technical staff exemplifies this line of reasoning:

> If the Sociedad is going to survive, it must become much more entrepreneurial. It can't depend only on ejido payments [for technical services] and government support. The Sociedad should develop its own set of enterprises—such as timber extraction and transport—and charge the ejidos for those services. Also, the technical services part of the organization should be administratively separate from the political part. We should set it up as a firm [*un bufete*] and charge fees based on the services we provide (interview, 12 May 2000).

At the same time, none of the restructuring proposals sought to completely transform the association from a campesino organization with a collectivist governance system into a private capitalist firm. Numerous workshop participants made reference to the association's importance in terms of collective representation. During one public meeting in February 2000, a community leader commented on the urgency of making organizational changes ("We have to do something right away . . . otherwise nothing is going to happen. I've heard this story too many times before") but also pointed out that, "We must take care of our organization [*nuestra Sociedad*]. Our organization keeps us strong" (meeting, 18 February 2000).

The intersection of the logics of collectivism and neoliberalism in these and similar discussions did not produce rival factions or confrontational debates regarding the two approaches' apparent incompatibilities. Rather, participants tended to construct responses that could potentially increase efficiency without compromising collective security. Ultimately, the 2000 restructuring process—which was first initiated in 1994—lost momentum in the face of more immediate fiscal problems. The following email message sent by the Sociedad's president in 2008 suggests that the entrepreneurial approach continues to dominate thinking in what has become a constantly recurring but unresolved discussion on how to restructure the organization:

> I have been seeking support to continue the restructuring project at the Sociedad with the idea of giving it an entrepreneurial turn [*un giro empresarial*]. I am administering financial support to create a holding facility and marketing fund for forest products . . . I would also like to identify contacts for offering environmental services through a private foundation as well as obtain a fleet of extraction, transportation, and secondary processing machinery so we can rent them to member ejidos and others in order to generate more income. For me it is clear that we cannot depend solely on technical services fees; they do not allow us to sustain ourselves as an organization (email, 27 February 2008).

Breaking Up is Hard to Do: Spatial Accommodation and Hybrid Forms of Property

Like the Sociedad, the nine ejidos that comprise the community forestry association's membership draw on both the logics of collectivism and neoliberalism in their everyday affairs. In particular, in the course of managing collective resources, such as land and timber, ejidatarios (rights holders within ejidos) have creatively accommodated specific neoliberal policy changes in ways that complement their desires for both small enterprise development and collective security. In this section, I focus on practices of spatial accommodation where ejidatarios reaffirmed longstanding tendencies of maintaining hybrid property arrangements in response to the federal government's land titling program, PROCEDE. I illustrate how, in the process of rejecting the opportunity to dissolve collective holdings, the nine ejidos adopted a combination of formal and informal responses that in some cases led to the *de facto* subdivision of communal lands within the legal shell of the ejido. For the most part, however, these responses maintained pre-existing property arrangements and thus reinforced established distributions of economic capital among ejidatarios.

Of all the changes encoded in the neoliberal policy reforms of the early 1990s, the possibility of dissolving communally held land grants (ejidos) and creating privately owned holdings under Article 27 of the Mexican constitution has garnered the most scholarly attention (eg Haenn 2006; Nuitjen 2003a; Perramond 2008; Vásquez Castillo 2004). The modifications to Article 27, which removed the legal platform guaranteeing the inalienability of ejido lands, set the stage for a potentially massive redistribution of land ownership as part of the land-titling program, PROCEDE. In response to the legal reforms, ejido assemblies had three broad choices: petition the government to dissolve an ejido and divide its assets among legally registered rights holders (*dominio pleno*); maintain the ejido intact and participate in PROCEDE (including certification of rights and land titling); or do nothing—maintain the ejido as it was and refuse participation in PROCEDE. Those ejidos that chose to participate in PROCEDE were presented with three possible outcomes, including: certification of individual agricultural plots and common property rights; certification of households (*solares*) and delineation of settlements (*áreas urbanas*); and demarcation of ejido boundaries.

As with other regions of the country, most of Quintana Roo's ejidos opted to formally delineate boundaries, define household lots and urban areas, and establish common property rights but decided *not* to formally delineate and certify their agricultural lands. Publicly available data from the federal government's final report on PROCEDE (2006) indicate that 273 out of 277 ejidos participated in Quintana Roo, representing 51,714 "beneficiaries". Interestingly, only 4678 of those individuals (9%) opted to formally certify their rights to agricultural parcels while 31,229 ejidatarios (60%) chose to certify their rights to communal lands (RAN 2006). As of mid 2007, six ejidos (2%) had legally dissolved their collective holdings in favor of private property. In each case, the ejidos were located in urban areas (eg Chetumal), coastal tourism zones (eg Isla Mujeres) or both (eg Playa del Carmen) where active real estate markets provided strong incentives for privatization (RAN 2007). The small number of communities that refused to participate fell into two categories: those facing inter-ejido land disputes and those citing possible political or economic repercussions. Additionally, officials reported that non-participating ejidos either deeply distrusted government or expressed concerns that involvement in PROCEDE would lead to new property taxes even though agrarian policy has no such requirements (Maria DiGiano, personal communication, 1 September 2009; interviews, 20 March 2000 and 18 November 2001).

Not surprisingly, ejidos elected to participate in those components of PROCEDE that enhanced tenure and financial security as well as local autonomy and opted out of those elements that tended to reduce internal decision-making flexibility and challenge local power structures. In Quintana Roo, this selective participation produced limited legally binding spatial realignments but did reduce boundary conflicts among ejidos. Nor did the program significantly clarify individual property rights, promote external investment, or increase production efficiencies as national level reformers had intended. At the same time, however, PROCEDE unintentionally encouraged some ejido assemblies to adopt agreements regarding internal land distribution. As I discuss below, these agreements often conformed to neoliberal ideals in certain ways while reaffirming collective tenure security in other ways. In what follows, I briefly summarize the experiences of the Sociedad's nine member ejidos, including more detailed discussion of two examples.

Table 1: Member Ejidos of the Sociedad de Productores Forestales Ejidales de Quintana Roo

Ejido	Area (ha)		Population (2000)	Ejidatarios (2006)	Authorized volume (m³)	
	Ejido total	Forest commons			Mahogany	Tropical hardwoods
Botes	18,900	8,000	1,771	298	341	1,217
Caoba	68,553	32,500	1,535	311	301	2,745
Chacchoben	18,450	6,000	1,140	294	135	960
Divorciados	12,000	1,500	961	187	22	491
Manuel Avila Camacho	12,000	1,500	829	189	–	–
Nuevo Guadalajara	28,500	6,000	1,390	276	–	–
Petcacab	46,000	32,500	947	206	2,000	3,927
Plan de la Noria Poniente	9,450	5,000	228	52	–	–
Tres Garantías	43,678	28,000	820	105	718	3,488
Total	257,531	121,000	9,621	1,918	3,517	12,828

Sources: Fieldwork 1999–2006; INEGI (2000).

Given important differences among the Sociedad's member ejidos—including total area, extent of forest commons, population, number of rights holders, and authorized annual timber harvest volumes—communities have pursued forestry differently and also have adopted diverse internal property arrangements (Table 1). Those ejidos with the largest forest commons and the highest annual timber harvest volumes have invested more in the development of community forestry enterprises compared with lesser endowed ejidos that either contracted third parties to harvest their timber allotment or decided to abandon forestry altogether (Table 2).

Historically, most ejido assemblies distributed agricultural plots by family under agrarian law decades prior to the neoliberal policy reforms. Common areas, such as forests, were managed by the ejido's executive committee on behalf of all ejidatarios, with any economic benefits divided equally among rights holders. Thus, the selective participation of most ejidos in PROCEDE suggests, in part, that a majority had already accommodated individual and family needs to the constraints of collectivism. This was the case with all nine of the Sociedad's member ejidos. They all demarcated ejido boundaries, delineated settlements and house lots, and certified common property rights. Each assembly

Table 2: Spatial and organizational accommodations of Sociedad member ejidos

Ejido	Production status	Mode of production	Agricultural lands	Forest commons	Work groups	Timber exchange
			Land distribution			
Botes	Active	Sub-contract	Family plots	Intact	No	No
Caoba	Active	CFE[a]	Family plots	Intact	Yes	Informal
Chacchoben	Active	Sub-contract	Family plots	Intact	No	No
Divorciados	Active	Sub-contract	Family plots	Intact	No	No
Manuel Avila Camacho	Inactive	n/a	Family plots	Intact	n/a	n/a
Nuevo Guadalajara	Inactive	n/a	Subdivided	Subdivided	n/a	n/a
Petcacab	Active	CFE	Family plots	Intact[b]	Yes	Formal
Plan de la Noria Poniente	Inactive	n/a	Family plots	Intact	n/a	n/a
Tres Garantías	Active	CFE	Family plots	Intact	Yes	Informal

Source: Fieldwork 1999–2006.
[a]CFE: community forestry enterprise.
[b]Work groups subdivide annual cutting area.

declined to formally delineate and certify rights to agricultural plots (Table 2).

The individual freedoms touted by reformers like Téllez did not entice ejidatarios from the Sociedad's member communities to formally abandon collectivism. Producers saw significant risks and limited benefits in seeking legal titles to private property. As one ejidatario remarked in 2000, "With the price of corn and other products so low, we can barely manage as it is. At least now the government provides some support to ejidatarios. As an individual, I would never be able to get a loan from a bank" (interview, 2 June 2000). In addition to risk aversion, however, other observers suggested that ejidatarios' decisions to avoid formal certification of agricultural plots had a lot to do with internal politics and conflict avoidance. One of the Sociedad's technical staff observed in 2003:

For ejidos to certify agricultural plots, it would force certain individuals to openly address inequities that have occurred in the distribution of farmland over the years. In many communities, the families that have dominated ejido affairs are also the ones with the best lands (interview, 27 June 2003).

By sticking with the status quo, ejidatarios maintained access to both the quasi-private spaces found in family-controlled agricultural plots and the communal spaces containing valuable forest resources. In doing so, they avoided bureaucratic hassles, financial risk and internal conflicts. Moreover, collective lands (*tierras de uso común*) remained inalienable under the 1992 Agrarian Law so those holdings would most likely either return to the public domain or transfer to a private commercial association (*sociedad mercantil*) if a forestry ejido opted for dissolution.

Although the Sociedad's nine member ejidos participated partially in PROCEDE and rejected avenues that would convert collective holdings into private property, two communities—Nuevo Guadalajara and Petcacab—developed internal agreements that liberalized property arrangements but also maintained the legal façade of the ejido. Nuevo Guadalajara went the farthest of the nine ejidos toward disaggregating their collective lands by extending the distribution of family plots to include the forest commons, even though agrarian law dictates that common lands must remain intact. In this case, however, longstanding internal conflicts led ejidatarios to divide all of the community's land. As one of the ejido's leaders noted in 2000, "we could never agree on how to make it work as an ejido. There was too much conflict. So we decided to divide things internally to avoid any further problems" (interview, 28 April 2000). The arrangement allowed Nuevo Guadalajara's members to maintain virtual private property rights but also gain access to public programs. At the same time, the agreement to divide the forest commons contravened agrarian law, and thus in 2000, the Sociedad was unable to work with a group of ejidatarios interested in restarting forestry activities because timber extraction permits could only be granted to an ejido.

In Petcacab, community members maintained the ejido's large forest commons intact but adopted measures permitting semi-autonomous producer groups (discussed below) and, in some cases, individuals to claim temporary control over harvestable trees located on an assigned tract within the annual harvest area (*área de corta*). Under this hybrid arrangement, the ejido's elected leader solicited an annual timber harvesting permit on behalf of all sub-groups but group leaders organized harvesting and milling in conjunction with the Sociedad in order to maximize individual and group autonomy. Group leaders thus referred to work being done on "my plot" or "my trees" and tended to subvert collective responsibilities, such as helping to maintain the communally owned sawmill.

In summary, the diverse spatial accommodations made by the Sociedad's member ejidos suggest an incremental liberalization of both property relationships and control of economic capital. During the 1980s, when community forestry supplanted state-sponsored industrial forestry as the dominant paradigm, conservation and rural development advocates—including many community members—relied heavily on the logic of collectivism as a means of empowering the "owners of the forest" (*los dueños del bosque*). Although most of the Sociedad's nine ejidos established community forestry enterprises within the constraints of the ejido governance structure, they also faltered repeatedly in the face of internal divisions among competing families. The neoliberal institutional shift in the early 1990s helped reaffirm existing configurations of quasi-private and collective lands but also facilitated novel adaptations regarding land and resource tenure that attended to political divisions. While the internal disaggregation of ejido lands in Nuevo Guadalajara proved to be an exception, most of the Sociedad's member communities developed flexible resource management approaches similar to the system adopted in Petcacab. In so doing, they disseminated resource access and control to sub-groups, families and individuals while leaving the remnants of collective property regimes more or less intact.

Organizational Accommodation: Hybrid Governance Regimes and New Modes of Exchange

Compared with the revisions to Article 27 of Mexico's constitution, other changes to agrarian law have received much less attention but have precipitated significant local responses nonetheless. During the agrarian reform period beginning in the 1930s, the ejido governance system—including the ejido assembly and an elected executive committee—administered development activities. The president of the executive committee served as the ejido's legal representative for all official transactions. On paper, ejido governance procedures were democratic, including monthly assembly meetings, regular elections, an internal oversight committee, and external checks and balances by the agrarian reform ministry. In practice, the system presented numerous opportunities for elite control by powerful families, misappropriation of communal resources, and interference by state agencies. As a result, internal conflicts were common (eg Klooster 2000a, 2000b).

In revising the nation's agrarian code, reformers made a subtle but important change that transformed collective governance practices in the Sociedad's largest forestry ejidos. Under Title 4, producer groups are legally permitted to form profit-seeking associations independent of the ejido assembly and executive committee. The change was intended to promote small enterprise development but also facilitated internal reorganization in unintended ways. Prior to the revisions, state agencies, NGOs and other external actors promoted the formation of cooperatives and producer groups but these entities ultimately responded to the ejido assembly. With the changes to the agrarian code, producers can legally form commercial groups that operate independently from the ejido assembly.

In this section I discuss the formation of semi-autonomous commercial groups and the emergence of internal timber exchanges within two of the Sociedad's larger forestry ejidos—Petcacab and Caoba. My main point is that these two organizational accommodations to neoliberal reform constitute hybrid constructions that have liberalized natural resource governance and shifted internal power dynamics. As such, this discussion builds on existing work, including my own, that examines the impacts of commercial groups on ejido governance and community forestry enterprises (eg Taylor 2003; Wilshusen 2005).

Work Group Formation as Liberalized Governance

The emergence of semi-autonomous forestry groups (known locally as "work groups"— *grupos de trabajo*) within the Sociedad's largest ejidos occurred in the mid 1990s following the changes to agrarian law described above (see Table 2). The initial impetus for forming groups stemmed from producers' desires to more directly control the economic capital derived from their share of the ejido's annual timber harvest. In examining the formation and development of work groups, I compared the experiences of Petcacab and Caoba during the period 1999–2006.

In Caoba, three groups emerged in 1997, representing the community's principal factions. One of those groups subsequently subdivided, creating a total of seven work groups ranging in size from 137 to 10 individuals. Three of these groups (68% of ejidatarios) were legally registered with government agencies and could thus operate as independent commercial interests, including administration of loans, grants and timber sales. The remaining four groups also operated independently but were legally required to carry out commercial transactions via the ejido.

Organizational accommodation produced even greater internal division in Petcacab compared with Caoba, including 11 groups formed mainly around family clans. Several of these groups further subdivided into "sections" representing individual families. Section leaders (usually heads of households) managed their own timber profits but otherwise participated as members of a work group. As of 2006, three of the 11 groups (41% of ejidatarios) were legally registered as commercial associations while the rest employed the legal status of the ejido to complete transactions. Different from Caoba, several ejidatarios (8%) chose not to affiliate with a group. In both cases, groups were not entirely autonomous because they continued to collectively manage their forests in line with a 25-year plan and annual harvest permits under the aegis of the ejido (Taylor 2001; Wilshusen 2005).

In comparative terms, the formation of work groups produced differing degrees of change with respect to natural resource governance.

In both cases, most decision making about forest management shifted from the ejido's executive committee to a council of work group leaders. As such, the ejido's elected leadership and assembly (comprised of all rights holders) lost a considerable amount of power. As one older ejidatario from Caoba remarked:

> It used to be that when someone rang the bell to call an assembly meeting, everyone would show up shortly thereafter. Nowadays we're lucky if we meet once or twice all year . . . The comisariado [ejido executive committee] doesn't have the same importance as it once did. Now they just sign off on whatever the [work group] leaders tell them to (interview, 20 July 2000).

Ejidatarios in both communities indicated that the formation of groups distributed resource control and decision-making power in ways that tended to diffuse conflicts. Another ejidatario from Caoba described the change in these terms:

> When we operated as an ejido, each leader took a turn as president and took advantage of the powers that came with the job. Now that we're in groups, political control is more evenly distributed and we don't have to worry so much about competing . . . Folks used to fight to become ejido president. Now we have a tough time rounding up a candidate (interview, 26 January 2000).

Caoba's shift in natural resource governance was less extensive than Petcacab's. In Caoba, two large groups (65% of ejidatarios) dominated forest management activities. They each formally elected executive committees and functioned as ejidos within an ejido. Given a per capita timber volume of just under 1 m^3 (valued at US$200 in 2000), work group leaders in Caoba carried out forestry activities almost exactly as they had when the ejido executive committee oversaw timber harvests, including hiring of a manager (*jefe de monte*) and numerous seasonal employees.

In Petcacab, a more complex set of arrangements emerged. Like Caoba, two groups (23% of ejidatarios) developed formal procedures that mimicked those of the ejido but they competed with an array of individual timber buyers (discussed below) and other, less formally organized groups. With one of the largest annual timber harvests in the region and a per capita allotment of 7.3 m^3 in 2000 (valued at US$1460), Petcacab's ejidatarios tended to minimize the collective aspects of forest management. As a result, all aspects of timber harvesting, transport and sale were divided among groups and individuals. In addition to temporarily subdividing the annual harvest area (discussed above), Petcacab's ejidatarios distributed all labor responsibilities, including delineation of lots, tree spotting, felling and moving logs to a central log yard based on the size and timber volume of each group. Some groups invested in machinery such as trucks and skidders while others sub-contracted with private interests or teams from other ejidos to carry out these activities. Groups also took turns using the ejido's sawmill and were responsible for paying laborers and covering maintenance costs.

This decentralized approach to internal organization and decision making helped diminish domination by local elites but also contributed to production inefficiencies, miscommunication, and tensions among groups. One of the Sociedad's senior technical staff explained the multiple levels of division in Petcacab in the following terms:

> I really can't understand the social processes that are occurring. What is clear is that to the extent that (group) members lose confidence in their leaders, something has to give. People [*la gente*] defend more what they feel is theirs compared to something that they claim rights to as a member of a group or community . . . I think that the ejidatarios are thinking more like small private property owners [*pequeños proprietarios*] than community members [*comuneros*], because with the latter there is a lack of definition regarding property. Under these conditions, it worries me to think about what will happen if we continue to promote the creation of profit-oriented enterprises within ejidos (interview, 2 July 2002).

Internal Timber Markets as Liberalized Exchange

The internal subdivisions within the Sociedad's larger ejidos precipitated new modes of exchange that further exacerbated the social complexities highlighted above. Petcacab's ejidatarios developed a vibrant futures market centered on the sale of timber volume that was ultimately copied to a certain extent in Caoba. Because of its high value, the main form of economic capital in play was mahogany (*Swietenia macrophylla*). In 2000, timber exchanges occurred informally where cash-poor ejidatarios sold all or part of their 7.3 m^3 of volume to local buyers with access to financial capital. One of the few regulations imposed by ejidatarios dictated that only rights holders from Petcacab could buy and sell timber. During 2001, 57% of the ejido's authorized mahogany volume was actively traded. Three buyers controlled 29% of this total (Wilshusen 2005).

Initially, exchanges took place via a simple transfer of cash for a handwritten receipt indicating how much volume had been sold for a determined price. Individuals sometimes sold the combined volume for their family group as much as 3 years in advance (volume for 2008 was sold in 2005). In 2000, timber buyers paid an average of US$125 per m^3 of standing volume. In some cases they paid as little as US$90. By contrast, the average sale price to external buyers for most forestry ejidos in the region was US$200. Once sawn, 1 m^3 of mahogany was worth on average between US$335 and 420, which translated to a potential profit of about US$310. In 2000, two buyers controlled 90 and 77.2 m^3 of standing volume, which carried a base value of US$18,000 and US$15,440 respectively. Both individuals ultimately sold the wood as sawn timber, netting a potential profit of US$37,800 and US$32,425. In contrast, one family of eight, with an initial volume of 58.4 m^3 (valued at US$11,680) sold its entire allotment to buyers, producing about US$913 for each family member.

By 2006, timber exchanges in Petcacab had become more formalized. Buyers and sellers jointly registered exchanges with a newly hired secretary who maintained a detailed database to ensure that sellers had not overdrawn their allotted timber quota. Two of the three buyers mentioned above continued to dominate the local market, however new actors also emerged, including one family group that had not historically participated strongly in community affairs but had accrued financial capital between 2000 and 2005 by reinvesting forestry profits in harvesting and woodworking machinery. Timber volume data from 2006 illustrate the relative change in accumulations of economic capital. In 2001, Petcacab's total permissible mahogany harvest increased, generating a per capita volume of 9.5 m^3. For the 2006 harvest season, one of the two timber buyers mentioned above controlled almost twice the amount of volume compared with 2000 (168.7 m^3 up from 90 m^3) while the other buyer held slightly less than in 2000 (70 m^3 down from 77.2 m^3). Most significantly, a work group that administered 156.6 m^3 in 2000 more than doubled its holding to 364.6 m^3 for 2006, indicating a surge in purchased volume on top of the increase in the total amount of timber harvested.

The experiences of Caoba and Petcacab with respect to organizational accommodation illustrate both differences and dynamism in local responses to neoliberal policy reforms between 1999 and 2006. Ejidatarios in Caoba maintained much of the collective governance system intact even as they decentralized decision making through the formation of work groups. Collective forest management practices remained the same with the two largest groups claiming most of the key positions related to harvesting, milling and transporting wood. Instead of taking turns controlling the ejido executive council, Caoba's two largest groups established a power sharing arrangement. The remaining groups received their portion of timber profits but were otherwise marginal participants in forestry activities. As a result, natural resource governance practices shifted to some extent with the diminished importance of the executive council and ejido assembly but local power dynamics remained relatively stable. Although individuals engaged in some timber sales, transfers of economic capital (cash for timber volume) were limited to a small number of informal exchanges.

In sharp contrast, ejidatarios in Petcacab constructed complex natural resource governance procedures and an active internal exchange in timber futures. Unlike their peers in Caoba, community members in Petcacab maximized individual and group autonomy and largely eschewed collectivism. The emergence of timber buyers, entrepreneurial groups, non-entrepreneurial groups and unaffiliated individuals produced a highly decentralized and dynamic decision-making environment, vibrant internal exchanges of economic capital, and shifting power dynamics. While some local elites—such as the two timber buyers mentioned above—adapted to liberalized governance and maintained positions of dominance, other actors swung from being marginal participants in forestry to developing into key players with significant endowments of economic capital.

Practices of Accommodation: A Longer View

Although neoliberal policy measures facilitated unintended local responses such as the formation of work groups and creation of an internal timber exchange, the underlying practices that enable accommodation (or resistance) come from a deeper cultural repertoire. When viewed from a broader historical perspective, the three types of accommodation that I have presented unfold incrementally over time and thus do not represent a simple set of reactions to one wave of institutional change. Rather, I argue that discursive, spatial and organizational accommodations are path dependent, meaning that practices related to natural resource governance persist over time, resurfacing repeatedly in local responses to multiple waves of state-sponsored reform. This is especially evident when examining modes of local organization and resource governance over time. Thus, even as the overarching logics of reform and the specific contents of responses change—as with the shift from collectivism to neoliberalism or the move from ejido to work group governance—certain practices persist. It was in this sense that Bourdieu used the term habitus to capture the "durable dispositions" of actors as they respond both creatively and routinely, in this case, to shifts in the bureaucratic field of agrarian policy making. I contend that these durable dispositions and practices comprise elements of a "culture of accommodation" to state-led reform efforts. In what follows, I support this assertion by illustrating how work groups emerged in the wake of earlier attempts to organize local producers including cooperatives, credit groups and specialized "production units".

Over the period from approximately 1930 through 1980, state-led development in rural Mexico experienced an incremental "loosening" in which local producers gained progressively more control over natural resource governance. This process of devolution emerged in the wake of multiple waves of state-sponsored policy reforms in which the agrarian reform ministry established different types of commercially oriented groups of ejidatarios. In most cases, these groups were formal or semi-formal entities that combined an economic development mission with a communal governance structure modeled after the ejido. Over some five decades, this hybrid organizational form emerged repeatedly.

Beyond the creation of communal land grants like Petcacab and Caoba in the late 1930s, the federal government's first efforts to organize rural producers (*campesinos*) in Quintana Roo centered on the creation of 70 community-based cooperatives, the majority of which were dedicated to chicle extraction (chewing gum resin) from the region's forests. Coupled with the creation of ejidos, this first wave of agrarian reform reallocated land and resources to rural producers but also established a government-controlled production system. Moreover, chicle cooperatives, such as the ones formed in Caoba and Petcacab, joined most ejidatarios within a communal decision making structure under which local leaders (separate from the ejido executive committee) organized chicle extraction, processing and transportation. While federal officials controlled all financial aspects of chicle production, local producers gained valuable experience and training that would, in part, enable the shift to community forestry in the 1980s.

A second precedent for the formation of work groups and decentralized governance can be found in the creation of "credit groups" (*grupos de crédito*). By the mid 1960s, the federal government encouraged ejidatarios to set up credit groups for activities such as livestock production as a means of diversifying local economies (the state maintained tight control over the far more lucrative timber trade). State-sponsored credit groups joined certain families, provided lines of credit, facilitated purchase of livestock, fencing and other materials, and in some cases, encouraged the creation of new settlements within ejidos to support the new enterprise investments. This was the case in Caoba, where the agrarian reform ministry and the state-owned regional agricultural development bank organized 30 community members within a credit group during the second half of 1967. The bank fronted the necessary funds and resources to clear pasture, purchase heads of cattle, build fencing and relocate families to a new settlement called San José de la Montaña. While the livestock group operated for several years, it contributed to a rift with families that remained in the ejido's main settlement and ultimately became mired in debt. Most importantly, the credit group's activities prompted repeated challenges from the ejido assembly because it sought to operate independently from the rest of the ejido. In one instance, the group claimed profits from mahogany harvested from newly cleared pasture while the ejido assembly contended that the money should be divided equally among all members. A similar fate befell the livestock credit group in Petcacab.

Beginning in the late 1970s, a third wave of state-sponsored organizing within ejidos built on experiences with both cooperatives and credits groups, producing what were known as specialized "production units" (*unidades de producción*). Community members and agency representatives began using the term "work group" at this juncture as the number and types of groups expanded. In Caoba, for example, agrarian reform ministry officials provided credit to support cattle-raising work groups (*grupos de trabajo ganadero*) as well as groups focused on bee-keeping. Specialized production units also emerged within newly established community forestry enterprises in both Caoba and Petcacab in the early 1980s. As with the other types of groups, agrarian reform officials created "forestry production units" including divisions with responsibility for timber extraction, administration and machinery respectively.

In line with a national initiative aimed at stimulating economic diversification and consolidation at the local level, agrarian reform representatives sought to formalize governance procedures within production units to match similar efforts aimed at improving ejido administration. As part of the requirements for receiving lines of credit, the elected leadership for each production unit established formal procedures (*un reglamento interno*) governing their operations. Government officials sought primarily to put in place strict accounting standards to prevent misuse of funds, equipment and other resources. Similar to credit groups, the creation of production units allowed sub-groups of ejidatarios to manage their own small enterprises. At the same time, disagreements regarding collective responsibilities to the ejido assembly under agrarian law continued to produce internal conflicts.

In each of the three examples discussed above—chicle cooperatives, credit groups and production units—the agrarian reform ministry sought to both empower and control rural producers in a manner that progressively devolved more decision-making power to the local level. Both chicle cooperatives and credit groups operated under the tutelage of the agrarian reform ministry while the production units had greater latitude in decision making. These experiences with local organization and modes of natural resource governance directly informed producers' approaches in creating forestry work groups following the neoliberal policy reforms of 1992. In both Caoba and Petcacab, producers adopted collectivist governance practices while simultaneously seeking to liberalize or deregulate capital flows. This deregulation within a collectivist framework allowed producers to overcome longstanding stalemates within ejido assemblies regarding the use of communal resources. Thus, in line with understandings of path dependency mentioned above, ejidatarios tended to replicate many of the formalized organizational forms and governance practices introduced by the state. For example, in 2000, several forestry work groups adopted formal operating procedures modeled after the collectivist tenets of the ejido even as they sought to minimize communal responsibilities.

Responses to state-sponsored efforts to organize and formalize community development activity produced a range of practices that attended to the fault lines of local power dynamics. The pursuit of credit groups and production units, for example, allowed certain factions—such as the cattle raising group in San José—to separate themselves from the larger group and avoid potentially violent conflicts. At the same time, participation in state programs allowed ejidatarios low risk access to economic capital. On multiple occasions, chicle cooperatives, credit groups and production units defaulted on loans that the federal government ultimately forgave or only partially recovered because sub-groups within ejidos had insufficient collateral. Working in sub-groups helped to address inter family conflicts and gain access to state-controlled funds without discarding the collective securities of the ejido.

Fault Lines of Power: Accommodation as an Accretion of Practices

In contrast with much of the literature on neoliberalism and environment generally, and capitalism and conservation specifically, my analysis suggests that attempts to institute neoliberal reforms can result in fragmented policies and partial accommodation by agrarian communities. Just as importantly, in this case, neoliberal reform did not produce a sudden and complete transformation at the local level but rather represented one of many historical waves of state-led institutional changes that evoked local responses within the context of existing political histories and cultures.

The three types of accommodation to state-sponsored reforms that I have discussed regarding community forestry in Quintana Roo, Mexico, point to three important conclusions regarding processes of neoliberalization and environmental governance. These conclusions relate to: locally constructed responses to neoliberal reform (creative accommodation); the merging of logics, property arrangements and governance regimes in everyday practice (hybridity); and the incremental emergence of cultures of accommodation (path dependency). Bourdieu's theory of practice offers a specific vocabulary capable of capturing both the material and symbolic dimensions of the

everyday power dynamics evident across all three of these dimensions—
particularly his presentation of field, capital and habitus.

First, the collective experience of the community forestry association
and nine member ejidos discussed in this chapter suggests that in
spite—and in part because—of complex internal political dynamics,
rural producers responded to institutional reforms in highly creative
ways. In situating accommodation in relation to writings on resistance,
I do not mean to suggest that producers never engaged in acts of
resistance or that they uncritically accepted the contents of neoliberal
policies and programs (acquiescence). Rather, I emphasize practices
of accommodation to capture the creative agency that occurred on
the receiving end of reform. In each of the nine ejidos, local actors
grappled with the complex specifics of neoliberal reforms in ways that
made sense within the context of each community's political history.
Thus, for example, the configuration and practices of work groups in
Petcacab differed in important ways from those in Caoba. Interestingly,
the relative lack of state presence in ejidos during the post-reform
period permitted community members to construct hybrid discourses
and experiment with novel property arrangements, governance regimes,
and modes of exchange (some of which ran counter to agrarian law).

Second, in all cases, creative accommodation produced complex
hybrid arrangements that were not intended by national level
reformers. Moreover, as I discuss throughout the chapter, these
hybrid responses constitute a progressive liberalization of everyday
affairs at the receiving end—both in material and cultural terms—
and have had important impacts on local power relationships. To the
extent that hybrid accommodations have favored entrepreneurialism,
individual tenure claims, semi-autonomous sub-groups, and freer
modes of exchange, they also have contributed to the persistence of
longstanding power relationships and have exacerbated local economic
differentiation. However, neither the Sociedad nor its member ejidos
has entirely discarded the logic of collectivism. As a result, the
focus on entrepreneurial approaches depends on collectivist governance
structures (the association and the ejido), quasi-private spaces within
ejidos complement but do not replace common property (with one
exception), and semi-autonomous groups and internal timber markets
endure by virtue of collective holdings.

Third, an examination of longstanding dispositions and social
practices linked to responses to neoliberalization suggests that, in
addition to "cultures of resistance", one also finds "cultures of
accommodation". I have argued that the everyday practices associated
with cultures of accommodation have emerged over decades in response
to state-led development programs. The types of practices in question
are often informal exchanges situated within local fields of play where

individuals and groups respond both intentionally and routinely to constantly unfolding opportunities and constraints. In the example that I offer, work groups represent a certain liberalization of natural resource governance but their structure and administration replicate those of collectivist entities such as the ejido, cooperatives, credit groups and production units. Thus while neoliberal policy reforms shaped these opportunities and constraints to some extent, local practices such as the formation of entrepreneurial sub-groups emerge repeatedly in the context of struggles, negotiations and alliances among local individuals and families. Ultimately the term "culture of accommodation" seeks to capture the historical trajectory of this creative process of partial assimilation of discourses, practices and organizational forms.

The hybrid discursive, spatial and organizational accommodations that emerged at the receiving end of reform in Quintana Roo, Mexico highlight the importance of carrying out ethnographies of neoliberalization. Conceptually, I have captured everyday responses to neoliberalization in terms of locally situated social processes that intersect with broader fields of play—in this case the shifting bureaucratic field of agrarian policymaking in Mexico. Bourdieu's heuristic constructs—field, capital and habitus—provide a vocabulary for examining the dynamic, microprocesses of everyday politics (actor-centered, reciprocal) while accounting for the historical, cultural and institutional forces that shape social interchange. Metaphorically speaking, Bourdieu's theory of practice uncovers a "geological" view of everyday politics in the sense that everyday interactions—both purposeful and routine—produce friction along innumerable fault lines. These tensions may produce "normal" or regularized responses in accord with longstanding practices (political cultures) where power relationships appear settled over time, building up like layers of sediment. This understanding of power as relatively stable suggests that social differences and inequalities tend to be reproduced over time even as the friction of daily encounters generates periodic shifts in power relationships and incremental social change. Everyday politics in this sense unfolds as accretions of practices rather than sudden transformations. The ongoing accommodations to neoliberalization in Quintana Roo, Mexico represent one set of examples of the myriad ways in which local producers creatively respond to this constant unfolding.

Acknowledgements

Earlier versions of this chapter were presented at the April 2008 meeting of the Association of American Geographers (AAGs) and the symposium on capitalism and conservation held at the University of Manchester (UK) in September 2008. Sincere thanks to the original journal's three referees for challenging and insightful comments. Thanks also to Tony Bebbington, Bruce Braun, Steve Brechin, Dan Brockington, Noel Castree, Catherine Corson, Maria DiGiano, Rosaleen Duffy, Brian King and Nancy Peluso. I also extend my sincere appreciation to the many individuals in Quintana Roo, Mexico who have contributed to this research over the last decade. The project from which this work draws was financed by the Social Science Research Council (SSRC) with funds provided by the Andrew W. Mellon Foundation, a Fulbright-Garcia Robles grant, an Inter-American Foundation fellowship as well as travel grants from Bucknell University.

References

Agrawal A (2005) *Environmentality: Technologies of Government and the Making of Subjects.* Durham, NC: Duke University Press

Alatorre G (2000) *La Construcción de una Cultura Gerencial Democrática en las Empresas Forestales Comunitarias* [The construction of a democratic managerial culture in community forest enterprises]. Mexico City: Casa Juan Pablos, Procuraduría Agraria

Antinori C and Bray D (2005) Community forest enterprises as entrepreneurial firms: Economic and institutional perspectives from Mexico. *World Development* 33(9):1529–1543

Bakker K (2005) Neoliberalising nature? Market environmentalism in water supply in England and Wales. *Annals of the Association of American Geographers* 95:542–565

Bakker K (2007) The "commons" versus the "commodity": Alter-globalization, anti-privatization and the human right to water in the global South. *Antipode* 39(3):430–455

Bourdieu P (1977) *Outline of a Theory of Practice.* Cambridge, UK: Cambridge University Press

Bourdieu P (1984) *Distinction: A Social Critique of the Judgment of Taste.* Cambridge, MA: Harvard University Press

Bourdieu P (1986) The forms of capital. In J G Richardson (ed) *Handbook of Theory and Research for the Sociology of Education* (pp 231–258). New York: Greenwood Press

Bourdieu, P (1990) *The Logic of Practice.* Stanford: Stanford University Press

Bourdieu P [1991] (1999) Rethinking the state: Genesis and structure of the bureaucratic field. In G Steinmetz (ed) *State/Culture: State Formation after the Cultural Turn* (pp 53–75). Ithaca: Cornell University Press

Bourdieu P and Wacquant L (1992) *An Invitation to Reflexive Sociology.* Chicago: University of Chicago Press

Braun B (2002) *The Intemperate Rainforest: Nature, Culture, and Power on Canada's West Coast.* Minneapolis: University of Minnesota Press

Bray D and Merino L (2004) *La Experiencia de las Comunidades Forestales en México* [The experience of forest communities in Mexico]. Mexico City: SEMARNAT, INI, CCMSS

Bray D, Merino L and Barry D (eds) (2005) *The Community Forests of Mexico: Managing for Sustainable Landscapes.* Austin: University of Texas Press

Brenner N and Theodore N (2002) Cities and the geographies of "actually existing neoliberalism". *Antipode* 34:349–379

Bridge G and Jonas A (2002) Governing nature: The reregulation of resource access, production, and consumption. *Environment and Planning A* 34:759–766

Brockington D, Duffy R and Igoe J (2008) *Nature Unbound: Conservation, Capitalism and the Future of Protected Areas*. London: Earthscan

Bury J (2005) Mining mountains: Neoliberalism, land tenure, livelihoods, and the new Peruvian mining industry in Cajamarca. *Environment and Planning A* 37(2):221–239

Cárdenas C (1976) Opiniones del Ing. Cárdenas, Subsecretario del Ramo [The opinions of Ing. Cárdenas, undersecretary of the Forest Service]. *El Mensajero Forestal* 35(367):14

Castree N (2008a) Neoliberalising nature: The logics of deregulation and reregulation. *Environment and Planning A* 40:131–152

Castree N (2008b) Neoliberalising nature: Processes, effects, and evaluations. *Environment and Planning A* 40:153–173

Cornelius W and Myhre D (eds) (1998) *The Transformation of Rural Mexico: Reforming the Ejido Sector*. San Diego: Center for U.S.–Mexican Studies, University of California, San Diego

Escobar A (2008) *Territories of Difference: Place, Movements, Life, Redes*. Durham: Duke University Press

Essex J (2008) The neoliberalization of development: Trade capacity building and security at the US Agency for International Development. *Antipode* 40(2):229–251

Goldman M (2005) *Imperial Nature: The World Bank and Struggles for Social Justice in the Age of Globalization*. New Haven: Yale University Press

Guthman J (2007) The Polanyian way? Voluntary food labels as neoliberal governance. *Antipode* 39(3):456–478

Haenn N (2005) *Fields of Power, Forests of Discontent: Culture, Conservation, and the State in Mexico*. Tucson: University of Arizona Press

Haenn N (2006) The changing and enduring ejido: A state and regional examination of Mexico's land tenure counter reforms. *Land Use Policy* 23:136–146

Harvey D (2003) *The New Imperialism*. Oxford, UK: Oxford University Press

Harvey D (2005) *A Brief History of Neoliberalism*. Oxford, UK: Oxford University Press

Harvey N (1998) *The Chiapas Rebellion: The Struggle for Land and Democracy*. Durham: Duke University Press

Heynen N, McCarthy J, Prudham S and Robbins P (2007) *Neoliberal Environments: False Promises and Unnatural Consequences*. London: Routledge

Instituto Nacional de Estadística, Geografía e Informática (INEGI) (2000) *XII Censo General de Población y Vivienda: Principales Resultados por Localidad* [Twelfth general population and housing census: Principal results by locality]. Aguascalientes: INEGI. Data available at http://www.inegi.gob.mx (last accessed 7 October 2009)

Klooster D (2000a) Community Forestry and Tree Theft in Mexico: Resistance or Complicity in Conservation. *Development and Change* 31:281–305

Klooster D (2000b) Institutional choice, community, and struggle: A case study of forest co-management in Mexico. *World Development* 28:1–20

Klooster D (2003) Campesinos and Mexican forest policy during the 20th century. *Latin American Research Review* 38:94–126

Kosek J (2006) *Understories: The Political Life of Forests in Northern New Mexico*. Durham: Duke University Press

Li T (2008) *The Will to Improve: Governmentality, Development, and the Practice of Politics*. Durham: Duke University Press

McCarthy J (2004) Privatizing conditions of production: Trade agreements as neoliberal environmental governance. *Geoforum* 35(3):327–341

McCarthy J (2005) Devolution in the woods: Community forestry as hybrid neoliberalism. *Environment and Planning A* 37:995–1014

Mansfield B (2004) Rules of privatization: Contradictions in neoliberal regulation of North Pacific fisheries. *Annals of the Association of American Geographers* 94:565–584

Mansfield B (2007) Property, markets, and dispossession: The western Alaska community development quota as neoliberalism, social justice, both, and neither. *Antipode* 39(3):479–499

Martin P (2005) Comparative topographies of neoliberalism in Mexico. *Environment and Planning A* 37:203–220

Mathews A (2005) Power/knowledge, power/ignorance: Forest fires and the state in Mexico. *Human Ecology* 33(6):795–820

Mathews A (2008) State making, knowledge, and ignorance: Translation and concealment in Mexican forestry institutions. *American Anthropologist* 110(4):484–494

Merino L (1997) *El Manejo Forestal Comunitario en México y sus Perspectivas de Sustentabilidad* [Community forest management in Mexico and its prospects for sustainability]. Cuernavaca: UNAM, CRIM, SEMARNAP, WRI

Moore D (2005) *Suffering for Territory: Race, Place, and Power in Zimbabwe*. Durham: Duke University Press

Nuijten M (2003a) Family property and the limits of intervention: The article 27 reforms and the PROCEDE program in Mexico. *Development and Change* 34:475–497

Nuijten M (2003b) *Power, Community and the State: The Political Anthropology of Organisation in Mexico*. London: Pluto

Ortner S (1999) *Life and Death on Mt. Everest: Sherpas and Himalayan Mountaineering*. Princeton: Princeton University Press

Ortner S (2006) *Social Theory and Anthropology: Culture, Power, and the Acting Subject*. Durham: Duke University Press

Peck J (2004) Geography and public policy: Constructions of neoliberalism. *Progress in Human Geography* 28:392–405

Peck J and Tickell A (2002) Neoliberalizing space. *Antipode* 34:380–404

Perramond E (2008) The rise, fall, and reconfiguration of the Mexican ejido. *The Geographical Review* 9(3):356–371

Perreault T (2005) State restructuring and the scale politics of rural water governance in Bolivia. *Environment and Planning A* 37(2):263–284

Perreault T and Martin P (2005) Geographies of neoliberalism in Latin America. *Environment and Planning A* 37(2):191–201

Primack R, Bray D, Galletti H and Ponciano I (eds) (1998) *Timber, Tourists and Temples: Conservation and Development in the Maya Forest of Belize, Guatemala and Mexico*. Washington DC: Island Press

Prudham S (2007) The fictions of autonomous invention: Accumulation by dispossession, commodification and life patents in Canada. *Antipode* 39(3):406–429

RAN (Registro Agrario Nacional) (2006) Avance del PROCEDE Histórico: 1993 al 29 de Diciembre del 2006 [Progress report on PROCEDE: 1993 to December 29, 2006], http://www.ran.gob.mx/ran/programas_sustantivos/ran_procede.html (last accessed 9 January 2009)

RAN (Registro Agrario Nacional) (2007) Núcleos Agrarios que adoptaron el Dominio Pleno de parcelas Ejidales [Agrarian populations that adopted full ownership titles to ejido plots], http://www.ran.gob.mx/ran/pdf/registro/19_Dominio_Pleno_Desagregado.pdf (last accessed 9 September 2009)

Randall L (ed) (1996) *Reforming Mexico's Agrarian Reform*. Armonk, NY: M.E. Sharpe

Roberts A (2008) Privatizing social reproduction: The primitive accumulation of water in an era of neoliberalism. *Antipode* 40(4):535–560

Sawyer S (2004) *Crude Chronicles: Indigenous Politics, Multinational Oil, and Neoliberalism in Ecuador*. Durham: Duke University Press

Sewell W (2005) *Logics of History: Social Theory and Social Transformation*. Chicago: University of Chicago Press

St Martin K (2007) The difference that class makes: Neoliberalization and non-capitalism in the fishing industry of New England. *Antipode* 39(3):527–549

Swartz D (1997) *Power and Culture: The Sociology of Pierre Bourdieu*. Chicago: University of Chicago Press

Swyngedouw E (2005) Dispossessing H20: The contested terrain of water privatization. *Capitalism, Nature, Socialism* 16(1):81–98

Taylor P (2000) Producing more with less? Community forestry in Durango, Mexico in an era of trade liberalization. *Rural Sociology* 65:253–274

Taylor P (2001) Community forestry as embedded process: Two cases from Durango and Quintana Roo, Mexico. *International Journal of Sociology of Agriculture and Food* 9(1):59–81

Taylor P (2003) Reorganization or division? New strategies of community forestry in Durango, Mexico. *Society and Natural Resources* 16(7):643–661

Taylor P and Zabin C (2000) Neoliberal reform and sustainable forest management in Quintana Roo, Mexico: Rethinking the institutional framework of the Forestry Pilot Plan. *Agriculture and Human Values* 17:141–156

Téllez Kuenzler L (1994) *La Modernización del Sector Agropecuario y Forestal* [The modernization of the agricultural, livestock and forestry sectors]. Mexico City: Fondo de Cultura Económica

Tsing A (2005) *Friction: An Ethnography of Global Connection*. Princeton: Princeton University Press

Vásquez Castillo M T (2004) *Land Privatization in Mexico: Urbanization, Formation of Regions, and Globalization in Ejidos*. London: Routledge

Wainwright J and Kim S-J (2008) Battles in Seattle *redux*: Transnational resistance to a neoliberal trade agreement. *Antipode* 40(4):513–534

West P (2006) *Conservation is Our Government Now: The Politics of Ecology in Papua New Guinea*. Durham: Duke University Press

Wilshusen P (2005) Community adaptation or collective breakdown? The emergence of "work groups" in two forestry ejidos in Quintana Roo, Mexico. In D Bray, L Merino and D Barry (eds) *The Community Forests of Mexico: Managing for Sustainable Landscapes* (pp 151–179). Austin: University of Texas Press

Wilshusen P (2009a) Shades of social capital: Elite persistence and the everyday politics of community forestry in southeastern Mexico. *Environment and Planning A* 41:389–406

Wilshusen P (2009b) Social process as everyday practice: The micro politics of conservation and development in southeastern Mexico. *Policy Sciences* 42:137–162

Wolford W (2005) Agrarian moral economies and neoliberalism in Brazil: Competing worldviews and the state in the struggle for land. *Environment and Planning A* 37(2):241–261

Wolford W (2007) Land reform in the time of neoliberalism: A many-splendored thing. *Antipode* 39(3):550–570

Index

Printed in the United States
By Bookmasters